SERICULTURE AND SERI-BIODIVERSITY

by

P.K. Srivastav
Regional Sericulture Research Station
Sahaspur, Dehradun-248 197

and

K. Thangavelu
Director
Central Sericultural Germplasm Resources Centre
Hosur-635 109

Foreword

S.S. Baghel
Vice-Chancellor
Central Agricultural University
Imphal-795 004

2016

Associated Publishing Company

A Division of

Astral International Pvt. Ltd.

New Delhi - 110 002

© AUTHOR
First Published, 2005
Reprinted, 2016

ISBN: 978-93-5130-987-1 (International Edition)

Published by   : **Associated Publishing Company**
                        A Division of
                        Astral International Pvt. Ltd.
                        – ISO 9001:2008 Certified Company –
                        4760-61/23, Ansari Road, Darya Ganj
                        New Delhi-110 002
                        Ph. 011-43549197, 23278134
                        E-mail: info@astralint.com
                        Website: www.astralint.com

Lasertypeset   : **Divya Computers, New Delhi-110005**

Digitally Printed at : **Replika Press Pvt. Ltd.**

Foreword

Chinese sericulture is over 3500 years old, whereas it dates back to the second century B.C. in India. China has been the highest silk producing country in the world, since the discovery of mulberry silkworm by Chinese empress Si Ling Si, the **Goddess of the Silkworms.** Likewise, Chinese oak tasar silkworm was also discovered more than 2000 years ago, while Indian oak tasar silkworm was synthesized recently in 1969 only. Yet, India is a unique country as it is blessed with the occurrence of all the varieties of silkworms i.e., mulberry, tropical tasar, oak tasar, muga and eri. Sericulture has indeed been a tradition in India. Although, the history of sericulture is lost in antiquity, its growth had been phenomenal in India since 1938, when it produced 691 metric tonnes (mt) of raw and ranked sixth. In 1987, the raw silk production went upto 8,455 mt. which was the second highest production in the world, only next to China. The total raw silk production reached 15,214 mt. during 1999-2000 and 15,900 mt. during 2000-01, as against the total consumption of more than 22,000 mt. raw silk by indigenous weaving sector. Although, both demand as well as production, have been steadily increasing within the country, the deficit of more than 6,000 mt. raw silk is met by either legal or illegal import of raw silk from China and other countries, adversely affecting Indian silk industry as well as sericulturists, since Chinese silk besides being better in quality is also cheaper than Indian silk.

No doubt, Indian scientists, sericulturists, technologists and industrialists toiled hard to bring the country at second position, however, we should remember that the Chinese productivity of cocoons and raw silk per hectare of mulberry area is approximately 1.5 and 1.25 times higher than Indian productivity respectively. Hence, quantitative as well as qualitative improvement is still required in on-farm as well as off-farm technologies.

The total export of silk products was Rs. 1250.5 crores during 1998-99 which increased to Rs. 1,501.8 crores during 1999-2000, thereby registering an increase of 20.1%. The direct employment generation in sericulture also increased from 59.5 lakhs persons during 1995-96 to 66.11 lakhs persons during 1999-2000, however, the employment potential of this sector can be increased manifold through integrated research and developmental activities from various stages of host plant cultivation to silk fabrication. Development of bivoltine sericulture technology, evolution of high yielding, stress and disease resistant varieties of silkworms and their host plants suitable for various agro-climatic conditions, improvement in host plant cultivation, seed technology, silkworm rearing, reeling, weaving and wet processing techniques are all set to bring about further improvements in productivity and quality of silk. Integrated approach of scientists, extension workers, farmers, textile technologists, industrialists and policy makers are required to give further boost to Indian silk industry in current GATT era. Non-mulberry sericulture is the strength of only Indian sericulture and hence needs to be strengthened further due to its eco-friendly nature. Due to limited land resources, improvement in vertical production of quality silk deserves special attention. This will also increase income of farmers. Production of more and more international quality 2A-4A bivoltine white silk is indeed the need of the hour to capture the world market *at par* with China. Hence, our human resources should be properly educated and mobilized to make India a global power in silk production and marketing in the current millennium.

I am extremely happy that Dr. P.K. Srivastav and Dr. K. Thangavelu have timely brought out this book entitled **"Sericulture and Seri-biodiversity"**. This is indeed the first book which incorporates the details of mulberry as well as non-mulberry sericulture (tasar, muga and eri); silkworm-host plant interaction; past, present and future trends of global as well as Indian sericulture; genetic resources, utilization and conservation of seribiodiversity; blending of sericulture, horticulture, dairy/animal husbandry, pisciculture, pharmaceutical, leather, cosmetic and oil industries; role of sericulture on stabilization of environment and rearing and seed technologies, etc. Besides, an eagle's eye view of sericulture in India has also been given in the last to enable readers to understand the whole mulberry as well as non-mulberry sericultural scenario in our country quickly.

I am sure that sericulturists, post-graduate students, teachers, researchers, policy makers, administrators and other people engaged in various activities of sericulture will be immensely benefited by this unique book.

I congratulate Dr. Srivastav and Dr. Thangavelu for compiling and presenting hitherto scattered information on various aspects of mulberry sericulture, tasar culture, muga culture and ericulture in a single comprehensive volume in a very precise and simplified manner.

S.S. Baghel
Vice-Chancellor
Central Agricultural University
Imphal-795004, Manipur (India)

Preface

China accounts for approximately 75% of the world silk production. It has been dominating in silk production for more than 3500 years and accumulated to more than US $ 30 billions over the last 40 years through silk export to more than 130 countries. However, after reaching its highest historical records of 87,767 mt. in 1994-95, it experienced a decline of 35-40% during 1997 to 2000 as compared to that of 1994-95. The cocoon and raw silk production in Japan, the second largest producer of silk during 1970 to 1986, has also been consistently decreasing to the tune of 20-30% each year since 1987. The production of silk surpassed the demand within these countries which caused a severe setback to silk industry in both the countries. On the contrary, the consumption of silk in India is more than that of its production and demand as well as production have been steadily increasing. India has indeed become second largest producer of silk since 1987. Yet, it imported approximately 7000 mt. raw silk though it produced 15,214 mt. raw silk during 1998-2000. The production was 15900 mt. while internal demand of raw silk was 23300 mt. during 2001 which may shoot up to 43200 m.t. by 2010. The trends indicate that the gap between demand and supply of silk in India will continue to increase in future due to customary saree culture which may be encashed by proper planning and management. Availability of cheap labour in our country is an additional advantage to bring it at the top of the global sericultural map. It is the current improvement in productivity and quality over several past years that sericulture in India has become highly attractive to the farmers. Advent of bivoltine sericulture with the help of Japan International Cooperation Agency, (JICA) has indeed laid the foundation of another white revolution in the form of silk revolution in our country. Development of high yielding bivoltine "CSR" and "Dun" hybrids suitable to the tropical and sub-temperate climate of India, evolution of high yielding mulberry varieties like V_1, S_{36}, S_{54}, S_{1635} and DD, development of suitable cultivation, seed, rearing and reeling technologies and efficient as well as effective extension machinery have contributed together in revolutionizing sericulture industry in our country. The average yield of farmers of Karnataka, Andhra Pradesh and Tamil Nadu has shot up from 28 kg to 68 kg/100 dfls while renditta reduced from 8.5 to 6.0 which has improved socio-economic status of adopted farmers significantly, as they are earning more than 1.2 lakh/year. The success of bivoltine technologies in temperate-sub tropical/temperate states is likely to be more encouraging due to more suitable climatic conditions. The new "CSR" and "Dun" breeds and their hybrids are *at par* with the races/hybrids developed in Japan, China, Korea and Brazil for uniformity in cocoon shape, cocoon grain, larval marking and grainage performance. These hybrids are capable of increasing cocoon and raw silk productivity of international grade (2A-4A) in coming years.

Non-mulberry sericulture is not only our strength in this GATT era but environment friendly also. There is a need to further strengthen these ecofriendly avocations for upliftment of tribals in their own vicinity. Sustainable utilization of our natural resources should be our mantra in the new millennium. Alliance of sericulture with agriculture, horticulture, pisciculture, silviculture, pharmaceutical, tannin and leather industries will not only complement but shall also strengthen each other. Bye-product utilization, use of silk for non-textile purposes and development of entirely new silk fibres shall indeed be adding new dimensions and vistas to the field of sericulture in our country.

Development of human resources and their proper utilization is the first and foremost requirement for bringing revolution in any field of the country. Hence, consistent improvement in the skill of sericulturists and sericulture scientists is need of the hour. Yet, many of us who work in one sector of mulberry sericulture are aloof with the other sector of non-mulberry sericulture, or *vice versa*. Wrong notions and messages are also reaching us due to one or the other reasons. With the intention of providing up to date and adequate information, this book is being presented to the sericulturists, post graduate students, sericulture scientists, planners and administrators of our country and other countries as well. We have made sincere efforts to accumulate hitherto scattered information regarding global and Indian sericulture in a single volume precisely.

It is earnestly hoped that the unit silk productivity as also quality will immensely improve in our country in next few years. Hybridization of sericulture with other industries; bye-product utilizations; use of silk, silkworms and silk plants for other purposes instead of clothing alone; sustainable use of our seri-biodiversity; environmental stability and socio-economic upliftment of tribals and village folks through sericulture shall become our **"guiding force"** in forthcoming GATT era.

Hence, we hope this book "**Sericulture and Seri-biodiversity**" will certainly be immensely useful to all those concerned with sericulture industry in India and also abroad.

P.K. Srivastav
K. Thangavelu

Acknowledgements

We express our deep sense of gratitude to following persons whose kind cooperation and/or discussion, literature, photographs, encouragement and keen involvement enabled us to prepare and publish this treatise entitled "**Sericulture and Seri-biodiversity**" on global and Indian silk industry:

1. Dr. S.S. Baghel, Vice Chancellor, Central Agriculture University, Imphal.
2. Shri P.Joy Oomen, I.A.S., Member Secretary, Central Silk Board, Bangalore.
3. Prof. S.N. Raina, Deptt. of Botany, Delhi University, Delhi.
4. Dr. R.K. Datta, Ex Director, Central Sericultural Research & Training Institute, Mysore.
5. Dr. S.B. Dandin, Director, Central Sericultural Research & Training Institute, Mysore.
6. Dr. S. Saratchandra, Director (Tech.), Central Silk Board, Bangalore.
7. Dr. B.R.R.P. Sinha, Director, Central Tasar Research & Training Institute, Ranchi.
8. Dr.K.V. Benchamin, Director, National Silkworm Seed Project, Central Silk Board, Bangalore.
9. Dr. Raje Urs, Director, Central Sericultural Research & Training Institute, Berhampore.
10. Dr. N. Suryanarayana, Director I/c, Central Muga Eri Reserarch & Training Institute, Lodoigarh, Jorhat.
11. Shri P.K. Das, Joint Director, Regional Muga Research Station, Boko.
12. Dr. H. Premjit, Dept. of Entomology, Central Agriculture University, Imphal.
13. Dr. N. Ibohal Singh, Dy. Director, Regional Tasar Research Station, Imphal.
14. Dr. K. Chaoba Singh, Dy. Director, Central Muga Eri Reserarch & Training Institute, Lodoigarh, Jorhat.
15. Dr. A.A. Siddiqui, S.R.O., Zonal Silkworm Seed Production Office, Dehradun.
16. M.C. Srivastav, Dy. Registrar, Central Agriculture University, Imphal.
17. Mr. Th. Shyamananda Singh, Office Assistant, Regional Tasar Research Station, Imphal.
18. Mr. C.A. Manjunath, Stenographer, Central Sericultural Germplasm Resource Centre, Hosur.

We are also grateful to the members of our family *viz.* Mrs. Kanchan, Ms. Shruti, Ms. Srishti and Mrs. Pandiyammal, Mr. T. Narayanan and Mrs.T.Dheepa who patiently cooperated though missed our help in family affairs during preparation of this treatise for a very long span of time.

Last but not the least, we are also highly grateful to Associated Publishing Company, New Delhi for publishing and presenting this book to the world of sericulture.

P.K. Srivastav
K. Thangavelu

DEDICATED TO OUR PARENTS

Late Brij Bihari Lal
Late Sheelwati Devi

Late S. Kandhasamy
Late K. Mariammal

WITH SWEET MEMORIES

Contents

Illustrations to Figures

Introduction

Sericulture has indeed been a tradition in China, Japan, India, Korea, Thailand, USSR, Brazil, Italy and France though its history has lost in antiquity. For more than 2000 years sericulture was indeed a monopoly of China and export of breeding material and divulsion of its secrets were prohibited at the cost of death. Yet, the Roman emperor Justinian in order to stop the drain of gold from Rome to China, smuggled the secret of silk culture out of China, through monks who brought silkworm eggs and mulberry seeds in a hollow stick. The king of Khotan (Tibet) and Bucharia also married Chinese princess only to smuggle the eggs of silkworms and mulberry concealed in their head dress across the frontiers of China. Louis XI took drastic steps to curb the tremendous out flow of money from France to Italy while Francis I, father of silk industry in Europe, restricted silk imports and enticed Italian craftsmen with promises of more freedom in their work and introduced sericulture in Rhine valley for saving sinking economy of his country.

All these legends/historical landmarks are indeed testimony of the inherent potentialities sericulture has for transformation of economy of any country. Yet, neither India has realized nor recognized such potentialities of sericulture despite of the fact that some scientists believe that the North-eastern region of sub-Himalayan belt is the original home/abode of various silkworms. We have neither explored, nor tapped and nor fully exploited the potentialities of our natural resources for our sustainable development. In this GATT era, all the countries are striving hard to protect their rights on their natural resources and intellectuals are trying to protect their property rights through TRIPS.

We will have to retrieve our mistakes and re-establish our friendship with nature if we want to survive and inherit a balanced ecosystem to our descendents besides improving our economy by sustainable development through balanced exploitation of our natural resources. We will have to pay due attention and recognition to mulberry as well as non-mulberry sericulture for ecological stability as well as economic upliftment of our tribal and village folks in their vicinity. When Japan has already shrunken and China is also shrinking their sericulture industry, India should fully exploit the golden opportunity and capture the global silk market besides restricting the import of Chinese raw silk in our internal market which is possible only through the enhancement of unit productivity of international grade

2A-4A mulberry raw silk. With the development of Tropical Bivoltine Sericulture Technology our country is indeed poised for and now gradually progressing ahead to achieve these goals. It is in this context the authors of this book had been thinking to share their knowledge and the knowledge of eminent sericulturists and sericulture scientists of India as well as abroad on mulberry as well as non-mulberry **"Sericulture and seri-biodiversity"** for bringing a revolutionary change in research and developmental activities of our country. India should acquire the top position in the sericulture map of the world by the next decade—with this dream we are presenting this book to the sericulturists, post-graduate students, sericulture scientists, planners, administrators and executives.

The book contains nine chapters on various aspects of sericulture and seri-biodiversity.

The first chapter on **"Silk, silkworm and host plants"** comprises of properties, structure and economic importance of silk; adulteration, identification and maintenance of silk; various categories of silkworms and their popular races; primary as well as secondary/tertiary/alternative host plants of mulberry as well as non-mulberry silkworms and their popular varieties/species alongwith their distribution and suitable climate. The second chapter on **"Silkworm and host plant interaction"** consists of phylogenetic relationship of various silkworms with respect to their feeding habit and host plant interation; specialization, generalization, feeding behaviour, food preferences, chemical stimuli and chemosensory basis of food selection and scope for biotechnological approaches for improvement of silk productivity. The third chapter on **"Global sericulture : yesterday, today and tomorrow"** deals with centres of origin of various silkworms and silkplants, historical back ground of sericulture in major and minor silk producing countries, silk road, global silk industry, production scenario in various countries and future of global silk industry whereas fourth chapter on **"Sericulture in India : yesterday, today and tomorrow"** deals with phases of development of mulberry sericulture and impact of mulberry varieties and silkworm breeds on Indian sericulture, comparison of sericulture in China, Japan and India, breeding for high yielding bivoltine and/robust hybrids and mulberry varieties, development of cultivation, seed, rearing and reeling technologies, popularisation of bivoltine tropical technology, challenges ahead and strategies required for Indian mulberry sericulture; tropical tasar, muga, eri and temperate tasar production scenario in the country, overall present status, constraints and strategies required for development of non-mulberry sericulture in India. The fifth chaper on **"Utilisation and conservation of seri-biodiversity"** describes about seri-biodiversity with respect to host plant genetic resources and silkworm genetic resources available in the country in mulberry as well as non-mulberry sericulture; survey, exploration, collection, introduction, characterization, classification, evaluation, documentation, cataloguing, database, quarantine and phytosanitation aspects of germplasm enrichment; global distribution, constraints and measures to promote utilization of genetic resources; registration and authorization of host plant and silkworm germplasm; DNA finger printing and biotechnology; conservation scenario of mulberry, tropical tasar, temperate tasar, muga and eri seri-biodiversity and proposals for conservation of seri-biodiversity in the country. The sixth chapter on **"Sericulture and allied industries"** enlightens about intergration of sericulture with agriculture, horticulture, silviculture, pisciculture, dairy/animal husbandry, pharmaceutical, oil, tannin, leather and other industries for enhancement of income of farmers while seventh chaptor on **"Sericulture and environment"** describes about the role played by mulberry, tropical tasar, temperate tasar, muga and eri sericultures on environmental protection and facilitation of ecological stability in various parts of our country. The eighth chapter on **"Silkworm seed and rearing technology"** deals with various technologies developed for efficient grainage operation for production of quality seeds in high quantity by private as well as government's grainage units and rearing of silkworms for improvement of raw silk productivity and income of farmers/ sericulturists which will facilitate increase in unit productivity of mulberry as well as non-mulberry sectors. The ninth chapter **"Sericulture in India: An over view"** has been given in the last in tabulated form to understand whole sericultural gamut of the country, mulberry as well as non-mulberry (tropical tasar, temperate tasar, muga and ericulture) quickly.

We have tried our best to incorporate maximum related information precisely in a single volume and hence, sincerely and warmly present this treatise to all the concerned post graduate students, sericulturists, scientists, planners and administrators for their benefit and above all for national/global development through upliftment of downtrodden people living below poverty line.

P.K. Srivastav

K. Thangavelu

1

Silk, Silkworms and Host Plants

Silk has indeed been referred to as "Queen of textiles/fabrics". No other fabric can match the beauty of silk because of its luxurious look, the snug feel and glittering lustre which are unquestionably unique to silk. Silk has been recognised as the finest natural fabric used for garments and furnishings. Wherever is silk, satisfaction, luxury, sense of pride, comfort, glamour and attraction are associated with it. Indeed, silk has been a way of life and no ceremony, religious function or ritual is complete without silk due to its sanctity in India. Hence, silk has become an inseparable part of Indian culture. The fine quality, lustrous sheen and traditional colour in brocades of Varanasi; crepes, georgettes and chiffons of Karnataka, Banded and temple silks of Kancheepuram, the delicate silks of Kashmir, the tie and dye craft of Andhra Pradesh, Gujarat and Orissa, make the sarees and other forms of silk apparel from India well known worldwide (Silkmans Companion 1992).

Indian subcontinent has the unique distinction of being the only subcontinent in the world which has been blessed by nature with all natural varieties of silk *viz.*, mulberry, tasar, muga and eri. India has now become the second largest silk producing country in the world only next to China. The sericulture, a labour intensive agro-industry dates back to the second century B.C. in India and involves lot of skill, art and technique of silk production, whereas Chinese silk industry is approximately 3500 years old. According to Dr.Y. Tazima (1981), the then Director of National Institute of Genetics, Mishima, Japan, sericulture is by itself a kind of agriculture but closely connected with a technology such as silk reeling and weaving (see Chowdhury 1984). It is rightly called as the "Industry of the poor" and plays an important role in the rural economy of the country by providing gainful employment opportunity in various sectors *viz.*, food plant cultivation, silkworm rearing, silk reeling twisting, weaving, printing, dyeing, finishing and marketing etc.

A. SILKS

Silk is indeed a protein fibre produced by the silkworms for spinning a cocoon which provides a protective covering to the delicate pupal stage which is the most critical period in the life of silkworms. There are four types of commercially produced natural silk *viz.*, mulberry, tasar, eri and muga. The word silk generally refers to the silk produced by the mulberry silkworm as it is the most versatile and popular silk and contributes as much as 91-95% of the total world silk production. The other three commercially important silks fall under the category of non-mulberry silks or wild silk, which contributes 5-9% of the silk production. Tasar silk is again of two types *viz.*, tropical and temperate (Oak) tasar which are cultivated in the tropical parts of Central India and North West to North Eastern sub-Himalayan belt in India respectively.

The apparently single silk filament or bave which constitutes the cocoon is actually composed of two filaments or brins (fibroin) stuck together and covered by sericin or silk gum while issuing from a pair of silk glands of the silkworms through the spinneret. Sericin is composed of a mixture of colouring pigments and hence looks coloured in some varieties. It varies from 22-30% in cocoons of various varieties or breeds of silkworm and can be separated from the brins by a process called degumming. The two components of silk, *viz.*, sericin and fibroin are proteinous substance made of several amino acids derived from the food plants.

Properties of silks

The commercially important silks possess certain physical and chemical properties, which are briefly described hereunder (F.A.O. 1976, Thangavelu *et al.* 1997, 2000):

A. *Physical properties*

1. *Colour*

The mulberry silk fibre is white or yellow or light green according to the presence of carotenoids or its components and flavones in the sericin enveloping fibroin. On the contrary, muga silk is golden yellow, tropical tasar silk is copperish yellow or fawn coloured temperate (Oak) tasar is yellow-grey and eri silk is white, brick red or light brown in colour.

2. *Bave length*

Eri silk fibre is unreelable as one end of its cocoon is open mouthed and the filament strength cannot withstand the reeling pressure. The length of bave varies with the breed in mulberry silkworms. The univoltine have total length varying from 800-1500 m (reelability 75-80%) while bivoltines have 1000-1500 m (reelability 75-80%), multivoltines have 300-700 m (reelability 50-65%) multi-bi hybrids have 700-900 m and new hybrids have 1000-1200 m (reelability more than that of 70%) bave length out of which generally 80-85% is reelable and the rest is waste. The total filament length of bave in muga, tropical tasar and temperate oak tasar varies from 300-500 m, 650-1300 m and 650-750 m according to race/breed out of which

40-50, 50-65 and 50-60 percent is reelable and the rest is removed as waste which is used in the spinning process, to produce spun silk.

3. *Bave size*

The size of bave, reeled raw silk or silk fibre (thread) is expressed in terms of denier, which refers to the weight of a length of 450 m in 0.05 gm units or 9000 m in one gm. unit. Hence, each sample is said to be one denier in size if its 450 m weighs 0.05 g or 9000 m weighs one gm. The bave from floss and outer covering of shell is thicker than that of middle and inner portions of the shell. The denier of univoltine, bivoltine and multivoltine mulberry silk range from 2.0-2.75, 2.0-3.0 and 1.5-2.5 respectively. On the contrary, muga, tropical tasar, temperate tasar and eri silk thread have 4.0-9.6, 8.0-12.0, 4.5-5.5 and 2.2-2.5 denier respectively depending upon races/breeds. The denier of the thread is proportional to the mean diameter of the bave/fibre, hence uni and bivoltines have generally larger denier than multivoltine varieties. Accordingly, univoltine and bivoltine races have 15-20 microns while multivoltine races have 6-14 microns diameter of the bave in mulberry silk.

4. *Impurities*

Mulberry and eri silks are without any type of impurities while tropical tasar, temperate tasar and muga silks have impurities in their threads.

5. *Lustre*

Mulberry silk is highly lustrous while muga and tasar silks have less lustre, and eri silk has no lustre. The ultimate lustre of silk fabric is influenced by silk reeling and processing techniques.

6. *Hygroscopic nature*

Silk is highly absorbent fibre and can absorb as much as 30% of its weight of moisture under saturated conditions of atmosphere. However, bone dry silk can absorb only 11 or 10 percent of its weight of moisture when kept under standard atmosphere having 65% humidity at 25°C; if the silk is raw or in degummed condition respectively. The mercantile weight of silk is calculated according to this moisture regaining capacity of silk. Hence, hygroscopic properties of silk are commercially very important. When wet 20% strength of silk fibre decreases, which is regained upon drying. When soaked in warm water, the silk fibre swells but does not dissolve. The dissolved substances in water are also absorbed by silk fibre along with water, hence quality of water is given high importance at the time of reeling, washing, dyeing and finishing and soft water is preferred during these processes.

7. Density

Density of silk is not absolutely constant since all varieties of silk vary within small limits with silkworm breeds. However, the accepted standards of density of raw silk is 1.33 and that of degummed silk is 1.25. Only 65% volume of silk is solid while the rest is composed of vacuoles.

8. Tensile strength

Before breaking, silk can elongate approximately 20 percent of its original length when moistened. Its tensile strength is enormous and can bear breaking load of approximately 3-4 gms/filament in univoltine and 3.0-3.5 gms/filament in multivoltine mulberry silk. While breaking strength decreases, the weight and elasticity of mulberry silk increase with increase in moisture content. When dry, mulberry silk can be stretched 1-2 percent only.

9. Scroop

Rustle of silk fabrics is called scroop. The crackling sound emitted by the silk thread when it is squeezed or pressed is referred as its scroop. It is not an inherent property of silk but acquired by it due to its treatment in a bath of dilute tartaric or acetic acid and subsequent drying without washing during manufacturing process.

10. Electrical properties

Silk is a poor conductor of electricity and hence extensively used for covering wire in electrical apparatus. However, it accumulates a static charge by friction which can be dissipated by high humidity to the tune of 65% RH at 25°C.

11. Effect of heat

White silk turns yellow on heating at 110.5°C in oven for 15 minutes. It disintegrates when heated at above 170°C while on burning it emits a characteristic peculiar odour of burnt hair.

12. Effect of light

Sunlight damages silk fabrics and 50% strength of silk is lost after six hours exposure to ultraviolet rays.

B. Chemical properties

1. Fibroin and sericin

The percentage of fibroin and sericin varies from 75-80 and 20-25 in mulberry silk fibre respectively. In multivoltine breeds, sericin which is soluble portion of the silk thread, bave varies from 20-30 percent depending upon the season.

2. *Reaction with acids*

Strong (concentrated) hydrochloric acid dissolves silk in 1-2 minutes while treatment of silk with strong sulphuric acid for only a few minutes followed by rinsing and neutralization causes contraction in fibre length from 30-50% and lustre of silk is lost. Treatment of silk with dilute hydrochloric acid for few minutes also causes shrinkage without any loss of strength. Similar effects have been observed with nitric acid and orthophosphoric acids also. This property has advantageously utilised for production of creping in silk fabrics/materials. The treatment of silk with nitric acid of 1.33 specific gravity for one minute at 45°C temperature produces bright yellow colour in silk which is permanent and fast and can further be brightened by treatment with an alkali.

3. *Reaction with perspiration*

When silk is worn, perspiration greatly tenders the silk. The perspiration is alkaline when deodorants containing aluminium chloride are used. This causes tendering of worn silk fabrics.

4. *Reaction with alkali*

Dilute alkalies reduce lustre of silk while strong hot alkalies *viz.*, caustic soda or caustic potash dissolve silk fibres. Ammonia and alkaline soaps do not affect fibroin but dissolve the sericin layer of silk. These soaps, however, affect fibroin also if silk is continuously boiled in such soaps for longer period. There is no action of borax either on sericin or fibroins. Treatment with limewater for few minutes softens sericin and causes swelling in fibre while its continued treatment causes brittleness in silk.

5. *Reaction with metallic salts*

There is strong affinity between silk and metallic salts. This property has been advantageously utilised for increasing the weight of silk and also for improving the draping properties of silk. Silk can absorb 8-10% of tin salts from cold solution of stanus chloride without injuring the material. Iron salt and logwood are used to dye and increase weight of black silk. Such weighting may add as high as 25% of the weight lost during degumming and also increases the breaking strength.

6. *Reaction with dyestuffs*

Among all the textile fibres, silk has greatest affinity to dyes and can be dyed with basic as well as acidic dyes even at low temperatures because of being a protein and possessing both acidic and basic properties.

Structure of silk

The cocoon formed by silkworms is composed of filament of silk, which is reeled by softening the gum of sericin surrounding the filament. During reeling, the silk filament is drawn out and unwound. The filaments of several cocoons are taken together to form a thread of the required denier. The thread sometimes twisted slightly. Several such strands are twisted together to form a thicker and stronger yarn by a process called "throwing" in the twisting unit. The sericin is retained till yarn or fabric stage and affords protection during processing. Finally, it is washed out by boiling in soap water and such removal of sericin softens the silk and imparts lustre.

The waste, damaged or pierced cocoons etc. are used to make 'spun silk' which is made from discontinuous thread twisted together to make yarn by different devices. The thickness of the spun silk is expressed in terms of 'count' as against 'denier', which is used to express the thickness of the reeled silk, which is derived from compact cocoons.

The microscopic examination of silk fibre revealed that longitudinally inner brins are composed of two smooth and transparent cylinders of fairly uniform thickness clearly separated from each other for considerable lengths. Sericin layer is irregular and lumpy in places and surrounds both brins and forms outer covering.

Generally the shape of the bave of mulberry and eri silk are circular. However, detailed cross sectional studies from various portions of mulberry cocoons revealed that shape of the bave/fibre is not uniform. The silk fibre from floss layer is elliptical, from middle compact shell is circular, from pelade is flat and ribbon like. Both the brins however, look like equilateral triangles bearing rounded off angles and their flat sides face each other.

The microscopic longitudinal view of degummed silk fibre exhibits two lustrous, homogeneous and compact nods (brins) of nearly uniform thickness, which are indeed fibrillons in structure because they break up into numerous minute fibrils when soaked in 25% solution of sodium hydroxide for 24 hrs and subsequently crushed. Each constituent fibril is about one twentieth of the normal filament or less than one micron in thickness and run parallel to the axis. The fibre splits into constituent fibrils, which project out from the main fibre as minute hairs in certain silkworm varieties, which produce thick fibre (bave). This undesirable characteristics known as lousiness is more frequent in the middle layer and less frequent in the inner most layer of the cocoon shell. Shape and size of brins affect dyeing of fabrics because yarn made from brins of large diameter or flatter brins dye darker while those from smaller brins with almost circular shape dye lighter. When fabrics woven with lousie silk are dyed, they appear as if they are covered by dust or show a paler shade than the main fabric. Hence, lousiness usually becomes a problem.

The crystalline structure of fabroin has been revealed through x-ray photography. The crystals are few thousandths of a micron thickness and less than one tenth of a micron length. Large number of crystals form a fibre in such a way that sum total of all the surface area of crystals is very much high as compared to the outer visible surface of the fibre.

Economic importance of various silks

Natural silk comprises a mere 0.2% of the worlds total production of textile fibres and is used to make not only dress materials, readymade garments, sarees, hosiery, knitwear, scarves, stoles, ties and curtains but also used for insulating material for electric wire and cables, type ribbon, industrial beltings, cartridge bag and cloth for naval guns. Besides, it is also used by anglers as fly-fishing lines for attaching their hooks, for sutures during surgery and for preparing parachutes due to high tensile strength.

Silk based Chinese cold cream and American powder have also come in the market. Silk has also found its way into sports as tennis, racket strings, bicycle tires in French racing. In China, even a damaged artery can be replaced by a prosthesis of silk and in Italy the teeth are realigned with silk braces.

Silk has also played very important role in scientific discoveries due to its strength comparable to filament of steel despite of being light in weight. Benjamin Franklin conducted his famous experiments with electricity using a silk kite. The technology developed for silk weaving helped the scientists to soar to even greater heights as they could make the fabrics for the nose of the Concorde, or the balloon that lifted the atomic bomb into the atmosphere before they began underground testing. Long before any one knew of microbes, Louis Pasteur discovered two silkworm diseases that he brought under control in 1857 and also made correlations that led to his monumental findings on infectious diseases. A dish of silkworm pupae in China is regarded very good for health because of being full of proteins and capacity to control high blood pressure (Hyde 1986).

Recently, research and development work conducted by P.A. Technology Company, Royton (London) reveals that the synthetic silk much similar to the silk produced by spiders may be produced through genetic engineering. They reported that bacteria and fermentation technology can be used to make high strength and impact resistant silk for commercial exploitation and such synthetic silk can be used not only for bullet-proof clothing but also for aerospace and automobile industries since a strand of such synthetic silk would be at least five times as strong as one made of steel. Hence, if it is silk, it need not be soft always (Mathur and Singh 1989). Use of silkworms as a very good biological tool in genetic engineering is increasing day by day due to its short and polyvoltine life cycle. The softness is associated with the exquisite quality of mulberry silk and to some extent with the eri silk. Apart from lustre, sheen, brilliance and elegance that associate with silk, its soft feel makes it uniquely a different fabric. Crepe-de-chine, Georgette, Satin, Chiffon and Chinnon silk fabrics do not possess exquisite softness. Soft silk sari with attractive gold lace borders (Jari) not only enhances glamour to the wearer but also bestows dignity to the fair sex. The amount of work and care bestowed in producing this quality are indeed incredible but uneven

Indian silk has been fully exploited to look at its best through weaving of this quality (Kanan 1989).

Tasar and muga silk fabrics have the following common draw backs not encountered in mulberry silk:

i.   Due to heavy texture they are not popular as lady's dress materials.

ii.   They have rough texture and poor finishing due to coarse denier, unevenness, lousiness and variation of colour.

iii.   They lack anti-crease or crease resistant property.

iv.   They are not convenient for all time use.

v.   They are very costly for a common man.

vi.   The cocoons have poor reelability, silk yield and quality of silk.

vii.   The elasticity is low.

viii.   The poor cohesion causes entanglement and more breaks in preparatory processes resulting in more wastage of hairy thread.

ix.   The degumming neither makes the cloth pure white nor yield brilliant colours after dyeing.

x.   They have defects in abrasion resistance.

xi.   Denier variation is so high that frequent thick and thin length occur in reeled yarn resulting in poor performance in weaving.

xii.   Improper joining of filament during reeling makes the yarn more fibrous and increase the breakages in weaving.

xiii.   Tenacity is poor.

However, tasar silk is warm and hence this property can be utilised for the production of winter garments by developing a suitable blend with woolenised jute, which will be cheaper than pure wool and give a silky appearance. Tasar polyester blend will improve the wash and wear and crease resistance, but systematic trials are needed to achieve these characteristics (Majhi & Thangavelu 1991).

Muga silkworms exclusively reared on Mejankari (*Litsaea citrata* Blume) yield an unique variety of Mejankari silk which is creamy white, very tough but soft in comparison to conventional golden yellow Muga silk produced by silkworm reared on primary food plants *viz.*, Persea (*Machilus*) *bombycina* and *Litsaea polyantha*. Due to its exquisite quality, Mejankari silk was the exclusive wear of the Ahom kings and the nobles in Assam.

Eri silk is otherwise called as Ahimsa silk and it can only be spun and by and large spun on Takli or Amber charkha or spinning wheel by poor persons which produces very coarse yarn of 30 s and production is also considerably low (20-25 gm/day). By using amber charkha eri silk yarn of 120-140 gm/manday (8 hrs.) of 60 s can be produced for which local people need training. At the time of processing eri cocoons for silk extraction, the cocoon is cut and pupa is removed so that there is no killing of pupa for the sake of silk yarn and hence the name is given as Ahimsa silk.

The yarn of all non-mulberry silks is excellent material for shirtings, suitings, neckties, bedspreads, curtains and other furnishings. The eri chadhar/ shawls (Fig. 1) are preferred by common people for protection against cold in the winter, particularly the monks.

Adulteration in silk

The manufacturers adulterate the textile and sell them as pure "Mysore Silk Sarees" to earn heavy profit. The cheapest filament rayon blends very efficiently with silk in warp as well as weft with 50/50 bleached and dyed silk, mixed with dyed rayon and when twisted can easily be passed off as 100% pure silk. Silk is a costly raw material hence, price of the saree is fixed on the basis of its weight and rayon filament adds to the weight of the fabric when woven with jari border besides looking as perfect as silk. Such fabrics deteriorate in quality and get damaged with pinholes after a few months of wear.

Another kind of adulteration is weighing of the silk fabrics and increasing their weights by adding metallic salts through tin-phosphate-silicate process at the time of dyeing by the traders to recover the loss of weight due to degumming. The advantages claimed include improvement in draping properties of weighed cloth and some properties of the fabric intended for special uses *e.g.*, tie, but such fabric is sensitive to light and other tendering agents and often results in a direct loss of strength of the weighed yarn or cloth if not cared well.

The Government should initiate effective preventive steps to check all kinds of adulterations by certifying the purity status of the fabrics so as to protect the consumer's rights and eliminate the fraudulent activities (Kanan 1986).

Identification of silk

The polyester and fine viscose items are mistaken for silk. The most decisive test of pure silk is that it burns slowly and yields smell of a burnt hair and forms black crushable beads when ignited; whereas viscose rayon burns quickly, emits a smell of burnt paper and forms black crushable ashes. The polyester at first sinks melts and smells differently and forms uncrushable beads at the burnt end. Distinct vertical warp and horizontal weft lines are seen throughout the body of fine silk fabric when it is held before a bright light. Silk fabrics are usually costly and pleasantly bright as compared to viscose and polyester fabrics which are extra shiny that strain eyes. Silks have paper like finish and yield a characteristic sound called "Scroop". On rubbing with glass rod silk acquires a positive charge while glass rod is negatively charged. When the fibre is treated with cold concentrated hydrochloric acid or boiled in 1% sodium hydroxide solution, silk dissolves quickly. Silk does not change colour when boiled with 5% sodium hydroxide solution followed by addition of a drop of lead acetate solution, whereas wool fibre and the solution turn black after such treatment (Manna 1999a).

Maintenance of silk

Special care and delicate handling are required for silk fabrics. Besides, human perspiration that degrades silk, cockroaches, rats, moulds and fungi are usual enemies of silk fabrics. Few simple and practical tips for proper maintenance of silk fabrics (Manna 1999b) are detailed below:

A. *Keeping*

1. Silk fabrics should be stocked in dry atmosphere free from dampness, moisture and aforesaid natural enemies.

2. Naphthalene balls should be kept inside the cupboards and fresheners should be used in surroundings.

3. After using silk fabrics should not be thrown away in rolled mass of ball to avoid formation of temporary crease. Hanging them under shade after proper folding for only short period will drive out perspiration and prevent undue crease formation. Subsequently, they should be transferred to cupboards.

4. Contact of silk with naked plaster or lime-wash on the walls should be avoided.

5. During hot summer, new garments should be worn only for 2-3 days. Subsequently they should be given first wash and stored properly.

6. Perfumes, sprays or heavy sweat may lead to formation of faint stains on the fabrics, so heavily sweated silk garments should not be stored unwashed.

B. *Washing*

1. Silk fabric should be soaked in soft water for 30 min. and subsequently they should be transferred to mild detergent solution prepared in soft warm water (35°c). Hard water or iron content in water cause yellowing of silk fabrics so they should be avoided. Actual recipe should be prepared according to the instructions given on detergent cover.

2. Instead of washing machine, washing by hands should be preferred for silk fabrics otherwise stiches may open up due to harsh treatment of the former. Gentle squeezing and releasing of silk fabrics under mild detergent solution/ recipe will release perspiration, sprays, dirt and dust easily.

3. The wet fabric should be hung under the shade without squeezing unnecessarily to avoid crease formation. Sundrying deteriorates the strength and colour of the silk due to presence of cosmic rays (UV) hence, should be avoided.

C. *Pressing*

1. Only thermostatically controlled iron box should be used and its knob should be set at the point marked for silk.

2. Little soft water should be sprinkled in the fabric by hand followed by rolling and keeping of the fabric for a while.

3. A flat bed should be prepared with a thick blanket and a white cotton cloth should be spread over it. Subsequently, silk fabric should be spread over it and a fine cotton sheet like "dhoti" should be evenly spread on silk fabric for safe and crease free pressing.

4. The iron box should be moved non-stop, gently and uniformly over whole fabric which should be subsequently folded and repressed and shifted to cupboard after inserting news papers between the folds for after press stability.

5. Silk fabrics may also be washed dry.

D. *Removal of stains*

Knowledge of removal of stains is useful in industry as well as home because they reduce the price of fabrics as well as garments and consumers avoid using such fabrics/garments. Methods of removal of different stains advocated by Central Silk board (Silk in India: Statistical Biennial, 1986) are detailed below:

Sl. No.	Source of Stain	Treatment for Removal
1	Blood	Sponge with a few drop of NH_3 in 10 vol hydrogen peroxide.
2	Butter	Use carbon tetrachloride
3	Cocoa chocolate	Wash in hot water and soap
4	Coffee	Let the fabric dry, then sponge with carbon tetrachloride.
5	Cream	Sponge with carbon tetrachloride
6	Egg	Sponge with cold water
7	Fruit	Sponge with alkali and alcohol mixed in equal ratio
8	Grass	Apply ammonia and cold water then wash in warm water with soap or H_2O_2 with NH_3
9	Lipstick	Sponge with carbon tetrachloride
10	Machine oil	Cover the stain with an absorbent, keep for some time, subsequently brush off and remove traces with carbon tetrachloride
11	Metal copper	Treat with 1/2 % hydrogen cyanide solution
12	Milk	Sponge with lukewarm water, dry and then sponge with carbon tetrachloride
13	Mud	Allow to dry, brush and then sponge with carbon tetrachloride
14	Nail polish	Sponge with acetone
15	Varnish oil	Sponge with carbon tetrachloride
16	Paint or varnish	Dip in a mixture of turpentine and kerosene
17	Perspiration	Use dilute hydrochloric acid
18	Shoe polish	Sponge with carbon tetrachloride

Thus, silk is a very delicate fabric and it needs to be cared like the queen of textiles indeed.

B. SILKWORMS

The living machines, which synthesize silk are known as silkworms. Yet, they are actually not worms but caterpillars and belong to two families: Bombycidae (the commercial mulberry silkworm) and Saturniidae (the wild silkworms of the order Lepidoptera which consists of butterflies and moths). A list of all commercially exploited silkworms along with their food plants is given in Table 1. Although all Lepidoptera produce silk but they are not comparable to the long lustrous fibre of silk created by the silkworms. There are more than 500 species of wild silkworms feeding mostly on various trees or shrub species and occasionally on herb species. They are more robust than mulberry silkworms and produce tougher and rougher silk, which is not as easily bleached and dyed as the mulberry silk. Silkworms have been designated as extraordinary eating machines. Mulberry silkworms, *Bombyx mori* (Fig. 2) increase their body weight 10,000 times in their 25-28 days lives by feeding mulberry leaves voraciously (Figs. 3-4). Subsequently, they spin cocoons and undergo hibernation (Fig. 5).

Table 1: The important parameters of common *Antheraea* and *Samia cynthia* species

Parameters	A. frithi	A. roylei	A. proylei	A. pernyi	A. assamensis	A. mylitta	S. cynthia ricini	S.cynthia
Fecundity (No.)	154	232	129	132	140	257	300-500	300-400
Hatching (%)	70.45	89.66	57.00	58.00	75.50	67.65	85-90	90
Effective rate of rearing (%)	27.90	31.73	42.46	40.0	35.10	19.84	80	75
Cocoon yield/dfl	20	66	31	30	37	35	250	200
Single cocoon weight (g)	3.92	8.5	5.91	6.80	5.0	11.48	2.8-3.2	3.0
Single shell weight (g)	0.51	1.05	0.69	0.76	0.35	1.90	0.38-0.42	0.4
Shell ratio (%)	13.01	12.35	11.67	11.27	7.0	16.55	12-13	13

Source: FAO 1987, Sarkar 1988, Ibohal Singh *et al.* 1988.

The second category of silkworms, known as eri silkworms consists of both domesticated as well as wild silkworms. The domesticated eri silkworm/silkmoth, *Attacus ricini* (Figs. 6-11) is now known as *Samia cynthia ricini* Hutton. Jr. syn *Philasomia ricini* Boisd; whereas wild eri silkworm *Attacu ailanthus* is now known as *Samia cynthia* (Drury) syn. *Philosamia cynthia* Drury. Both eri silkworms originated in South-East Asia. While *Samia cynthia ricini* syn. *P. ricini* is multivoltine (many crops in a year) and non-hibernating type, *Samia cynthia* syn. *P.cynthia* is uni and bivoltine (one or two crops in a year) and hibernating type. They mainly feed on Castor (*Ricinus communis*, Figs. 12, 13), Kesseru (*Heteropanax fragrans*, Fig. 14)) and Tapioca (*Manihot utilitissima*, Fig. 15).

The third category of silkworms belong to genus *Antheraea* which are very important from economic point of view since they produce tropical tasar, oak tasar and muga silk primarily by feeding on *Terminalia/Shorea* species, *Quercus/ Lithocarpus* species and) *Persea (Machilus)/Litsaea* species respectively. Sixteen species of *Antheraea viz.*, *A. assamensis* (Helfer), *A. compta* Roth schild, *A. frithii* (Moore), *A. godmani* Druce, *A. harti* Moore, *A. knyvettii* Hampson, *A. montezuma* (Salte), *A. mylitta* (Guerim Meneville), *A. paphia* Linnaeus, *A. pernyi* (GM), *A. polyphemus* (Cramer), *A. proylei* Jolly, *A. roylei* Moore, *A. soemperi* Felder, *A. yamamai* (G.M) and *A. oculea* (Neumoegen) have been reported to feed upon various oak species (Ibohal Singh 2000, Ibohal Singh *et al.* 1998). Of the sixteen oak fed *Antheraea* species, *A. yamamai* is reared in Japan, *A. oculea* is reared in Mexico and *A. polyphemus* is reared in America while the remaining 13 species are found in India and may be reared on other tree species, some times even better than on oaks.

The important characteristics of commonly found *Antheraea* and *Samia* species (non-mulberry silkworm) found in India are described in Table 1 following Ibohal Singh *et al.* (1998, 2000), Sarkar (1988) and FAO (1976).

The characteristics of wild silkworms in India change according to ecological conditions and food plants. Accordingly, *Antheraea mylitta* (Figs. 16-17) which is primarily a tropical tasar silkworm and feeds on variety of food plants has been found to consist of many ecoraces confined to different ecological niches. The important ecoraces of tropical tasar silkworm are Andhra Local, Nowgaon, Daba, Munga, Sarihan, Bhandara, Bogoi, Sukinda, Sukli and Moonga feed on *Terminalia* species (Figs. 18-20); Barharwa, Laria-P, Moddia, Mugia, Raily-N, Modal and nalia feed on *Shorea robusta* (Fig. 21); Giribum feed on *Zizyphus mauritiana* and Tira feed on *Lagestroemia* species (Figs. 22-24). The distribution and important parameters of these ecoraces are enumerated in Table 2 following Sinha (1998) and Thangavelu *et al.* (2000a). Evaluation studies of seven ecoraces conducted at Central Tasar Research and Training Institute, Ranchi revealed that Modal is best in respect of fecundity and hatching percentage, while Raily is best for larval weight and absolute silk yield and Laria is best for effective rate of rearing, cocoon weight, shell weight and silk ratio. On the basis of pooled score allotted to them their comparative superiority was found in following descending order: Raily > Daba > Modal > Bhandara > Sarihan > Sukinda > Laria. This order may help the breeders in selecting a race for different breeding programmes (Sinha 1999). Tropical tasar silkworms are distributed in Central India *viz.*, Bihar, Jharkhand, Chattisgarh, Madhya Pradesh, Orissa, Maharashtra, West Bengal, Uttar Pradesh and Andhra Pradesh. Occasionally, they are also reported from Assam and Manipur. Tropical tasar silkworms are either bivoltine (two crops) or trivoltine (three crops), of which the first crop is seed crop while the second and the third crops are regarded as commercial crops. On the contrary, muga silkworm *Antheraea assama* Ww. found in North-eastern region, mainly in Assam, is a multivoltine species with 5-6 generations a year. Of these two are commercial crops while the remaining are either pre-seed or seed crops. The rearers of the region generally take three crops

Table 2: Ecoraces of Indian wild silkmoths

Ecoraces	State	Cocoon Wt. (g)	Shell Wt. (g)	SR (%)	Peduncle Length (cm)	Cocoon Size Length (cm)	Width (cm)	Filament Length (m)	Reelability (%)	Denier	Cocoon colour	Primary host plant
A. TROPICAL TASAR (*Antheraea mylitta*)												
Andhra Local	Andhra Pradesh	8.14	1.37	16.93	3.6	3.0	2.5	660	69	7.0	Whitish grey	*Terminalia* spp.
Nowgaon	Assam	8.15	0.94	11.82	3.5	3.8	2.9	—	—	—	Grey	*Terminalia* spp.
Barharwa	Bihar	10.35	1.90	18.82	5.8	4.8	3.4	1234	64.3	NA	Grey	*Shorea robusta*
Daba	Bihar	11.95	1.79	16.06	6.8	5.7	3.7	850	56.7	10	Grey	*Terminalia* spp.
Laria-P	Bihar	7.78	1.63	20.97	5.2	5.0	3.1	590	NA	8	Blackish grey	*Shorea robusta*
Moddia	Bihar	12.91	2.84	22.25	3.2	5.7	3.5	1234	60.3	NA	Grey	*Shorea robusta*
Mugia	Bihar	8.81	1.93	22.05	6.5	5.2	3.8	1239	60.2	10	Grey	*Shorea robusta*
Munga	Bihar	5.98	0.76	12.79	5.8	4.8	3.3	—	—	—	Grey	*Terminalia* spp.
Sarihan	Bihar	7.30	1.03	14.08	4.1	3.6	2.6	570	50.4	8	Grey	*Terminalia* spp.
Raily-N	Madhya Pradesh	11.86	2.15	18.44	3.5	5.3	3.4	1400	62.7	10	Blakish Grey	*Shorea robusta*
Bhandara	Maharashtra	7.30	1.51	20.76	4.5	4.1	2.4	670	68.4	7	Grey	*Terminalia* spp.
Giribum	Manipur	9.82	1.62	16.85	3.4	2.9	2.2	—	—	—	Whitish grey	*Zizyphus jujuba*
Bogoi	Orissa	7.61	1.33	18.04	6.4	4.3	3.8	—	—	—	Whitish grey	*Terminalia* spp.
Modal	Orissa	13.61	2.98	22.12	6.4	5.4	3.5	1483	65.4	11.5	Blackish grey	*Shorea robusta*
Nalia	Orissa	11.36	2.14	18.10	7.2	4.1	2.4	647	56.0	8	Grey	*Shorea robusta*
Sukinda	Orissa	11.22	1.71	14.49	5.3	4.9	3.1	845	65.0	10	Yellow	*Terminalia* spp.
Sukly	Orissa	7.38	1.16	15.83	3.4	3.7	2.8	—	—	—	Whitish grey	*Terminalia* spp.
Moonga	Uttar Pradesh	6.49	1.10	17.07	5.8	4.5	3.7	834	58.6	9	Grey	*Terminalia* spp.
Tira	West Bengal	5.82	0.92	15.86	4.6	2.5	2.3	674	55.0	10	Whitish grey	*Lagerstroemia* spp.
Shiwalika	J&K, H.P	5.65	1.13	20.37	—	—	—	—	—	—	—	*Zizyphus jujuba*
NG 94	Nagaland	9.17	1.56	17.26	—	—	—	—	—	—	—	*Zizyphus jujuba*
NEI 95	Assam	8.15	1.01	13.32	—	—	—	—	—	—	—	*Terminalia* spp.
Japla	Bihar	9.47	1.08	11.78	—	—	—	—	—	—	—	*Zizyphus jujuba*
Palma	Jharkhand	10.44	1.94	18.72	—	—	—	—	—	—	—	*Shorea robusta*

Ecoraces	State	Cocoon Wt. (g)	Shell Wt. (g)	SR (%)	Peduncle Length (cm)	Cocoon Size		Filament Length (m)	Reelability (%)	Denier	Cocoon colour	Primary host plant
						Length (cm)	Width (cm)					
Korbi	Chattisgarh	9.69	1.83	18.58	—	—	—	—	—	—	—	*Terminalia* spp.
Dadar & Nagar Haveli	Dadar & Nagar Haveli	6.89	1.15	16.67	—	—	—	—	—	—	—	*Terminalia* spp.
B. MUGA *Antheraea assamensis*	Assam	6.30	0.50	9.50	Rudi-mentary	5.2	2.4	204-500	40-50	4.0-9.6	Light brown	*Machilus* & *Litsaea* spp.
C. ERI *Samia cynthia ricinii*	Bihar Bengal Assam	3.00	0.40	13.00	Absent	4.0	2.5	Unreelable	—	2.2-2.5	White, brick red	*Ricinus communis*
D. OAK TASAR *Antheraea proylei*	N.E. states Uttaranchal hills J&K, H.P.	6.22	0.76	12.90	Weak & short	6.5	2.5	650-750	50-60	4.0-5.5	Grey	*Quercus* spp.
Antheraea roylei	N.E states Uttaranchal hills J&K, H.P.	5.30	0.50	9.40	Thin & short	4.0	2.0	—	—	—	Grey	*Quercus* spp.

After Sinha 1998, Thangavelu *et al.* 2000

due to adverse climatic conditions, out of which two are commercial and one is seed crop.

Systematic studies on the existence of ecoraces of muga, *A. assama* (Figs. 25-27)) and oak tasar, *A. proylei* J. (Figs. 28-31) are lacking. However, colour polymorphism (green, blue, yellow and cream) has been reported in both the species also like that found in *A. mylitta* D. Mejankari silk obtained from rearing of muga silkworms on Mejankari (*Litsaea citrata* Blume) and Dighloti (*L. salicifolia* Roxb.) are unique and different from the one reared on Som (*Persea bombycina* King, Fig. 32) and Soalu (*L. polyantha* Juss, Fig. 33) (Sahu and Das 1999, Saikia and Goswami 1997).

Out of 16 oak fed species, 4 species *viz.*, *A. proyeli*, *A. pernyi*, *A. frithii* and *A. roylei* (Figs. 34-36) and six evolved breeds of *A. proylei viz.* PRP2, PRP3, PRP5, PRP12, RPP4 and Blue along with two breeds of *A. pernyi viz.*, Blue-6 and Blue Yellow-1 received from China are maintained in germplasm bank of Regional Tasar Research Station, Imphal. Among the evolved breeds PRP5 and PRP12 with average cocoon yield of 46 and 49/dfl respectively against 31 cocoons/dfl of *A. proylei* are at the stage of multilocational field trial. *A. proylei*, *A. pernyi* and other evolved breeds have only one successful crop on *Q. serrata* (Fig. 37) and *Q. grifthii* (Fig. 38) during spring crop (March-April) in the sub-Himalayan belt of North-eastern region. In the high altitude regions of North-western sub-Himalayan region spring crop rearing is conducted on *Q. incana* (Fig. 39) and late spring rearing is conducted during May-June on *Q. semecarpifolia*—a late sprouting and late maturing oak species. Two additional crops of oak tasar may be raised during summer and autumn seasons through 16-17 hrs. of photoperiodic light treatment of the diapausing pupae at 22± 2°C for 18-25 days. *A. roylei* and *A. frithii* are wild and bivoltine in nature while *A.pernyi* and *A.proylei* are weak bivoltine commercially cultivated species (Ibohal Singh 2000). *A frithii* feeds on *Lithocarpus dealbata* wildly in the nature (Fig. 40). Attempts are being made by the scientists to cultivate and semi-domesticate them in raised plantations for commercial exploitation.

Eight ecoraces of eri silkworm *Philosamia ricini* Boisd. syn. *Samia cynthia ricini* Jr. *viz.*, Titabar, Mendi, Dhanubhanga, Sille, Nongpoh, Borduar, Khanapara and Khaldang were accorded accession numbers from ESW 01 to ESW-08 and characterised by observing the colour and markings on the cuticle of fifth instar larvae. The ecorace, ESW-08, khaldang has been found to be most superior with respect to fecundity (465.78), hatching percentage (94.92), shell weight (0.512 gm) and SR percentage (14.64). Morphological characteristics of larvae, cocoon, fecundity, hatching percentage of these eri silkworm ecoraces are given in Table 3 following Sarmah (2000). Six pure line strains *viz.*, yellow-plain (YP), yellow spotted (YS), Yellow-Zebra (YZ), Greenish Blue Plain (GBP), Greenish Blue Spotted (GBS) and Greenish Blue Zebra (GBZ) were isolated from Titabar (ESW-01) and Borduar (ESW-06) ecoraces on the basis of larval body colour and markings at Regional Eri Research Station, Mendipather (Sarmah 2000). According to Benchamin (2000a) eri regional stocks of "Borduar" and "Titabar" are more productive amongst six evaluated from different zones.

Table 3: Morphology and evaluation of ecoraces of eri silkworm *Samia cynthia ricinii* Boisd.

Acc. No.	Race Name	Origin	Larval body colour	Larval body markings	Cocoon colour	Fecundity (No.)	Hatching (%)	Larval weight (g)	Cocoon Wt. (g)	Shell weight (g)	SR (%)
ESW-01	Titabar	Assam (Titabar)	Yellow and greenish blue	Spotted and plain	White	463.68	91.35	6.280	3.790	0.493	13.01
ESW-02	Mendi	Meghalaya (Mendipathar)	Greenish blue	Plain	White	453.55	95.51	6.033	3.643	0.470	12.76
ESW-03	Dhanubhanga	Assam (Dhanubhanga)	Yellow and greenish blue	Plain	White	443.05	91.93	5.983	3.603	0.477	13.24
ESW-04	Sille	Arunachal Pradesh (Sille)	Yellow	Plain	White	437.53	94.05	5.877	3.687	0.467	12.67
ESW-05	Nongpoh	Meghalaya (Nongpoh)	Yellow	Plain	White	447.99	91.91	5.147	3.583	0.463	12.92
ESW-06	Borduar	Assam (Borduar)	Yellow and greenish blue	Zebra and Plain	White	464.90	93.06	6.560	3.987	0.537	13.47
ESW-07	Khanapara	Assam (Khanapara)	Yellow and greenish blue	Plain	White	461.27	92.03	6.213	3.890	0.500	12.85
ESW-08	Khaldang	Meghalaya (Khaldang)	Yellow and greenish blue	Zebra	Brick red	465.78	94.92	5.900	3.498	0.512	14.64

After Sarmah 2000

The mulberry silkworm (*Bombyx mori*) races/breeds are univoltine, bivoltine and multivoltine. In multivoltines, eggs complete their embryonic period within 10 days and embryo hatches out as larva. Such eggs are called non-hibernating eggs and such breeds complete more than 5 life cycles in a year and are called multivoltines or polyvoltines, and reared in tropical areas. On the contrary, in univoltine and bivoltine breeds, the embryo continues to grow halfway, subsequently enters a period of diapause and resumes growth only in early part of the spring and hatches out as larva. Due to hibernating character of eggs, univoltine breeds complete only one life cycle in a year in colder regions. Bivoltine breeds consist of alternating life cycle of hibernating and non-hibernating eggs and complete two life cycle in a year. Temperature, photoperiods and hereditary factors control the voltinism in silkworm breeds.

The important multivoltine breeds of mulberry silkworm are Pure Mysore (PM), Hosa Mysore, Mysore Princess, Kolar Gold, Kollegal Jawan, MY-1, P2D1, OS616, G, Nistari, Nistari (M), Nistari (P), CB5, RD1, Moria white, Sarupat, Tamil Nadu (W), C.Nichi while important bivoltine breeds of mulberry silkworm are NB4D2, SH6, YS3, SF19, PAM-101, PAM-111, P5, NB18, CC1, CA2, KA, JD6, NB7, KPG-B, CSR-2, CSR-5 etc. which are being maintained at Central Sericultural Germplasm Resources Centre (SMGS/CSGRC Newsletter 1(1):1999). At present 64 multivoltine and 293 bivoltine races are maintained in the germplasm collection at CSGRC, Hosur.

Multivoltine "Pure Mysore" breed which is flossy greenish in colour is reared in Karnataka, Tamil Nadu and Andhra Pradesh. On the contrary "Nistari" multivoltine race which is yellow coloured and flossy is reared in West Bengal. Though silk contents in multivoltines is poor as compared to bivoltines. Yet, they are hardy and well acclimatised to tropical conditions. In Karnataka, bivoltine races *viz.*, Kalimpong A, NB7, NB4D2 and NB18 are also used whose cocoons are oval shaped white in two former races and peanut shaped white in latter two races. However, except in Kashmir, in all other regions, where sericulture is practised, the hybrids of multivoltine and bivoltine *viz.*, PM x NB18, PM x NB7, KA x NB4D2, KA x NN6D and NB7 x NB18 or bivoltine x bivoltine hybrids *viz.*, KA x NB4D2, NB7 x NB4D2, CSR2 x CSR4 and CSR2 x CSR5 are more popular during favourable seasons (Aug-Feb) they yield more silk as compared to multivoltine breeds. CSR18 x CSR19 a robust bi x bi hybrid is relatively tolerant to summer and rainy seasons (Datta 2000 a, b,c). In temperate areas like Jammu & Kashmir, hybrids of bivoltine races SF19, YS3, B40 (Japanese) and Yawki, Haulak, Changnunj (Chinese) are reared (Ullal and Narasimhanna 1994). Recently, the F1 bivoltine of YS3 x SF19, CA2 x NB4D2, CC1 x NB4D2, SH6 x NB4D2, PAM-111 x SF19, and PAM-101 x NB4D2 produce cocoon yield of 65.16, 60.24, 58.56, 53.52, 52.72 and 52.2 Kg./ 100 dfls during autumn crop 1994 in Jammu & Kashmir which shows that leaf quality and climatic conditions during autumn season are quite conducive for bivoltine crop in the region apart from spring crop (Shamim Baksh *et al.* 1995).

C. HOST PLANTS OF SILKWORM

Food (host) plants of silkworm may be categorised under mulberry and non-mulberry groups and non-mulberry food plants are further categorised under tropical tasar, temperate (oak) tasar, muga and eri food plants on the basis of the silkworms reared upon them (Table 4).

Table 4: Food plants of silkworms in Indian sub-continent.

SILKWORMS		FOOD PLANTS (TREES)		
Vernacular name	Zoological name	Popular names	Botanical names	Family
1. Mulberry Silkworm	*Bombyx mori L.* (Family: Bombycidae)	**A. PRIMARY FOOD PLANTS**		
		1. Eng : Mulberry	*Morus alba* L.	Moraceae
		Hindi : toot, Tutri	*M. indica* L.	Moraceae
		Urdu : Shahetut	*M.laevigata* Wall.	Moraceae
			M.nigra L.	Moraceae
			M.serrata Roxb. ex Brandis	Moraceae
		B. ALTERNATIVE FOOD PLANTS		
		1. Eng : Peepal Hindi : Peepal	*Ficus religiosa* L	Moraceae
		2.	*Cudrania triloba.*	Moraceae
		3. Eng: Lettuce Hindi : Salad	*Lactuca saliva* L.	Asteraceae
		4. Eng : Osage orange		
2. Tasar Silkworm	*Antheraea mylitta D.* (Family: Saturniidae)	**A: PRIMARY FOOD PLANTS**		
		1. Eng : Indian Dammer Hindi: Sal, Sakhu	*Shorea robusta* L.	Dipterocarpaceae
		2. Eng: White Murdah Hindi: Arjun, Kahua	*Terminalia arjuna* Bedd.	Combretaceae
		3. Eng: Laurel Hindi: Asan, Saja	*Terminalia tomentosa* W & A	Combretaceae
		B: PRIMARY/SECONDARY FOOD PLANTS		
		1. Eng: Alex wood Hindi: Dhaura	*Anogeisaus latifolia* Wall.ex Bedd	Combretaceae
		2. Eng: Anjan Hindi: Anjan	*Hardwickia binata* Roxb.	Caesalpinaceae
		3. Eng: Lendia Hindi: Sidha, Sidi	*Lagerstroemia parviflora* Roxb.	Lythraceae
		4. Eng: Indian jujube Hindi: Ber	*Zizyphus mauritiana* Lam. Syn. *Z.jujuba* Lam. non Mill.	Rhamnaceae
		C: SECONDARY FOOD PLANTS		
		1. Eng: Mountain Ebony Hindi: Kachnar	*Bauhinia variegata* Linn.	Caesalpinaceae
		2. Eng: Silk cotton tree Hindi: Semul,Shembal	*Bombax ceiba* Linn. Syn. *B.malabaricum* DC., *Salmalia malabarica* (DC.) Schott & Endl.	Bombacaceae

SILKWORMS		FOOD PLANTS (TREES)		
Vernacular name	Zoological name	Popular names	Botanical names	Family
		3. Eng: Kumbhi Hindi: Kumbi	*Careya arborea* roxb.	Lecythidaceae (Juglandaceae)
		4. Eng: Sissoo Hindi: Sisam	*Dalbergia sissoo* Roxb.	Papilionaceae
		5. Eng: Castor Hindi: Erandi	*Ricinus communis* Linn.	Euphorbiaceae
		6. Eng: Lac tree of South India	*Shorea talura* Roxb. Syn *S.roxburghii* G.Don.	Dipterocarpaceae
		7. Eng: Black plum, Java plum Hindi: Jamun, Jam	*Syzygium cumini* (L.) Skeels Syn.*Eugenia* *jambolana* Lam., *E.cumini* Druce.	Myrtaceae
		8. Eng: Teak Hindi:Sagwan	*Tectona grandis* Linn.f.	Verbenaceae
		9. Eng: Belleric myrobalan Hindi: Bahera	*Terminalia bellerica* Roxb.	Combretaceae
		10. Eng: Chebulic myrobalan Hindi: Harre, Hirda	*Terminalia chebula* Retz.	Combretaceae
		11. Eng: Flowering murdah Hindi: Kinjal	*T. paniculata* Roth	Combretaceae
		12. Hindi: Banber	*Z. jujuba* Mill. syn. *Z.* *sativa* Gaertn., *Z.vulgaris* Lam.	Rhamnaceae

D. TERTIARY FOOD PLANTS

SILKWORMS		FOOD PLANTS (TREES)		
		1. Hindi: Char, Achar, Chironji	*Buchanania latifolia* Roxb. syn. *B.lanzan* Spreng	Anacardiaceae
		2. Hindi: Rangruri	*Canthium dicoccum* (Gaertn.) Merril syn. *C.didynum* Roxb.	Rubiaceae
		3.	*Careya sphaerica* Roxb.	Lecythidaceae (Juglandaceae)
		4. Eng: Bengal currants Hindi: Karaunda	*Carrica carandas* Linn.	Apocyanaceae
		5. Eng: Black oil plant Hindi: Malkangni	*Celustrus* *paniculatus* Linn. syn. *C.paniculata* Willd.	Celestraceae
		6. Hindi: Nalbali	*Cipadessa fruticosa* (Blume)	Meliaceae
		7. Eng: Indian Satinwood Hindi: Bhirra, Girya	*Chloroxylon* *swietenia* DC. syn. *Swietenia* *chloroxylon* Roxb.	Rutaceae

SILKWORMS		FOOD PLANTS (TREES)		
Vernacular name	Zoological name	Popular names	Botanical names	Family
		8. Eng: The Ebony of Northern India Hindi: Tendu, Abnus	*Diospyros melamoxylon* Roxb. syn.*D.wightiana* Wall., *D. dubia* Wall., *D.exsulpta* Bedd.	Ebenaceae
		9. Hindi: Aliar, Sinatha	*Dodonaea viscosa* Linn.	Sapindaceae
		10. Eng: Indian Gooseberry Hindi: Amla	*Embelica officinalis* Gaertn.	Euphorbiaceae
		11.	*Ficus benjamina* Linn.	Moraceae
		12. Hindi: Kat Gular	*Ficus hispida* Linn.f.	Moraceae
		13. Eng: Peepal, Peepul Hindi: Pipal, Pipali	*Ficus religiosa* Linn.	Moraceae
		14. Hindi: Kamrup, Chilkhan	*Ficus retusa* Linn.	Moraceae
		15. Hindi: Piper, Piperi, Jari	*Ficus tsiela* Roxb.	Moraceae
		16. Vern: Kel, Karal	*Ficus tsjakela* Burm.f.	Moraceae
		17. Hindi: Dikamali	*Gardenia lucida* Roxb. syn. *G.resinifera*	Rubiaceae
		18. Hindi: Kharpat, Kaikar	*Garuga pinnata* Roxb.	Burseraceae
		19. Eng: Common Crepe Myrtle Hindi: Saoni	*Lagerstraemia indica* Linn. Linn.	Lythraceae
		20. Eng: Queens Crepe Myrtle, Queens flower, Pride of India Hindi: Jarul	*Lagerstroemia flosreginae* Retz.syn. *L.speciosa*	Lythraceae
		21. Eng: Indian Butter Tree	*Madhuka indica* J.F.Gmel. syn.*M.latifolia* Macbr. *Bassia latifolia* Roxb.	Sapotaceae
		22. Hindi: Phutki	*Melastoma malabathricum* Linn.	Melastomaceae
		23. Hindi: Bija Sal	*Pteracarpus marsupium* Roxb.	Papilionaceae
		24. Vern:Ora, Orcha	*Rhizophora caseolaris* Linn.Sp.pl syn.*Sonneratia acida* Linn.f., *Sonneratia caseolaris*	Lythraceae
		25. Hindi: Bhand, Churna, Suran	*Zizyphus rugosa* Lam.	Rhamnaceae
		26. Hindi: Kat Ber, Kathber	*Zizyphus xylopyrus* Wild	Rhamnaceae Rhamnaceae

SILKWORMS		FOOD PLANTS (TREES)		
Vernacular name	Zoological name	Popular names	Botanical names	Family
3. Oak Tasar Silkworm	1. *Antharaea frithii* (Moore) (Indian) Family: Saturniidae	**A: PRIMARY FOOD PLANTS**		
		1.Vern: Sahi, Ding-sai, Keuko, To-ik	*Lithocarpus dealbata* (Hk.f. & Th.ex Miq.) Rheder. syn.*Quercus dealbata* Hk.f. & Th.ex Miq, *Pasania dealbata* (Hk.f. & Th.ex Miq.) Kanjilal *et al.*	Fagaceae
	2. *A.proylei* Jolly(Indian) 3. *A.pernyi* GM (Chinese) 4. *A.roylei* Moore (Indian)	2. Vern: Dingrittiang, Uyung	*Quercus acutissima* Carr. syn. *Q.serrata* auct. non Thunb.	Fagaceae
		3. Vern:Dieng-Wah Dingim	*Q.grifithii* Hook & Thom ex Miq.	Fagaceae
		4. Eng: Green Oak, Holly Oak Vern: Moru, Chora	*Q.himalayana* Bahadur syn.*Q.floribunda* Lindle ex A. Camus Chenes, *Q.dilatata Wall ex A.DC.*	Fagaceae
		5. Eng: White Oak Hindi: Banj, Ban	*Q.leucotrichophora* A. Camus Syn. *Q.incana* Roxb.	Fagaceae
		6. Eng: Brown Oak Vern:Kharsu, Banjar	*Q.semecarpifolia* Sm.	Fagaceae
		B: SECONDARY FOOD PLANTS		
		1. Eng: Indian Chestnut Vern: Bank-katus, Serang, Sareng, Hinguri	*Castanopsis indica* A.D.C.	Fagaceae
		2. Vern:Chaukhu, Bon-belphoi Arkhala, Tangji	*Castanea purpurella* (Miq.).Balak syn. *C.hystrix* DC, *Castanea purpurella* Miq.	Fagaceae
		3. Vern: Siri, Shirang, Shi-rang	*Castanopsis lancaefolia* Hickl. & Camus syn. *Q.lancaefolia* Roxb., *C.roxburghiana* Biswas	Fagaceae
		4. Vern: Sahi, Diang-sai Keuko, To-ik	*Lithocarpus dealbata* (Hk.f. & Th.ex Miq.) Rheder. syn. *Q.dealbata* Hk.f. &	Fagaceae

SILKWORMS		FOOD PLANTS (TREES)		
Vernacular name	Zoological name	Popular names	Botanical names	Family
			Th.ex Miq., *Pasania dealbata* (Hk.f. & Th.ex mig.) Kanjilal *et al.*,	
		5. Vern: Kuhi	*L.fenestrata* (Roxb.) Redder.	Fagaceae
		6.	*L.xylocarpa* (Kurz.) Markgr.	
		7. Eng: Evergreen/Holly/ Holm Oak Vern:Brechur, Irri	*Q.baloot Griff.* sym. *Q./ilex* non Linn.	Fagaceae
		8. Eng: Blue Japanese Oak Vern:Siri, Baran, Banni, Inai	*Q.glauca Thunb.*	Fagaceae
		9. Vern: Phalut, Phalant, Srikung	*Q.karmroopi* D. Don. syn. *Q.lineata* Bl.var. *thomsoniana* (A.DC). Wengzig, Q.lineata (non bl.), *Q.thomsoniana* A.DC.	Fagaceae
		10. Vern: Bajrant, Buk, Shalsi	*Q.lamellosa* Sm.syn. *Q.imbricata* Buch-Ham. ex D.Don, *Q.lamellata* Roxb.	Fagaceae
		11. Eng: Wooly Oak Vern: Ranj, Sanj, Riani, Kiani, Banja	*Q.lanata* Sm.syn. *Q.lanuginosa* D.Don, non Thuill, *Q.nepaulensis* Desf.	Fagaceae
		12. Eng: English Oak	*Q.robur* Linn. (Exotic)	Fagaceae
		13. Eng: American Red Oak	*Q.rubra* Linn. (Exotic)	Fagaceae
		14. Eng: Cork Oak	*Q.suber* Linn. (Exotic)	Fagaceae
		15. Eng: English Willow Vern:Bibsu, Kumanta	*Salix vimnalis* Linn.	Salicaceae
4. Muga Silkworm	*Antheraea assama West.* syn. *A.assamensis* Helf, *A. mejankari Moore Moore* Family: *Saturniidae*	**A. PRIMARY FOOD PLANTS**		
			Machilus bombycina king ex Hook.f. syn. *Persea bombycina* Kostern.	Lauraceae
		1. Vern: Som, Artucheknan		
		2. Vern: Soalu, Meda, Patoia, Kakuri	*Litsaea polyantha* Blume syn. *L.monopetala* (Roxb) Pers., Tetranthera monopetala Pers.	Lauraceae

SILKWORMS		FOOD PLANTS (TREES)		
Vernacular name	Zoological name	Popular names	Botanical names	Family

B. SECONDARY FOOD PLANTS

		Popular names	Botanical names	Family
		1. Vern: Mejankari, Sillimbar	*Litsaea citrata* Blume syn. *L.cubeba* Pers.	Lauraceae
		2. Vern: Kothalua	*L.nitida* Roxb.	Lauraceae
		3. Vern: Digloti, Dighleti	*L.salicifolia* Hook.f.	Lauraceae

C. TERTIARY FOOD PLANTS

		Popular names	Botanical names	Family
		1. Eng: Pisa Vern:Petarichawa, Bangnala	*Actinodaphne angustifolia* Nees	Lauraceae
		2. Vern: Pattikuta, Patihanda	*A.obovata* Blume	Lauraceae
		3. Vern: Malligiri	*Cinnamomum cecicodaphne* Meissn.	Lauraceae
		4. Vern: Gansarai	*Cinnamomum glanduliferm* Meissn. syn. *C.glaucescerns* syn. *Camphora glanduliferum* Nees.	Lauraceae
		5. Vern: Tejpata, Chammejam, Pattichanda	*Cinnamomum obtusifolium* Nees.	Lauraceae
		6. Vern: Bhemloti	*Celastrus monosperma* Roxb.	Celastraceae
		7. Eng: Gumhar Vern:Gambari	*Gmelina arborea* Roxb.	Verbenaceae
		8. Vern:Panchapa, Chapa	*Magnolia pterocarpa* Roxb. syn. *M.sphenocarpa* Wall.	Magnoliaceae
		9. Vern: Bor Soppa	*Michelia oblonga* Wall. ex Hook.f. & Thomas	Magnoliaceae
		10. Eng: Champak Vern:Champa	*M.champaka* Linn. syn. *M.aurantiaca* Wall.	Magnoliaceae
		11.	*Symplocas grandiflora* Wall.	Symplocaceae
		12. Eng: Sapphire Berry, Sweet leaf	*Symplocas paniculata* Miq. syn. S.*crataegoidas* Buch-Ham. ex D.Don	Symplocaceae
		13. Eng: Machilus Vern: Kawala	*Machilus odoritissima* Nees	Lauraceae
		14. Vern:Tejphal, Tejbal, Tumru, Darmar, Nepali Dhaniy	*Zanthoxylum armatum* DC syn. *Z.alatum* Roxb., *Z.planispinum* sieb. & Zucc.	Rutaceae

SILKWORMS		FOOD PLANTS (TREES)		
Vernacular name	Zoological name	Popular names	Botanical names	Family
		15. Vern: Bajramani	*Zanthoxlum limonella* (Dennst.) Alston syn. *Z. budrunga* Wall. ex DC, *Z.rhesta* DC.	Rutaceae
		16. Vern: Banber, Pitni, Beri, Bagori	*Zizyphus jujuba* Mill. syn. *Z.sativa* Gaertn., *Z.vulgaris* Lam	Rhamnaceae
		17. Eng: Indian Jujube, Common jujube Vern:Ber	*Zizyphus mauritiana* Lam. syn.*Z.jujuba* Lam. non Mill.	Rhamnaceae
5. Eri Silkworm	*Samia cynthia ricini* Jr. syn.*Philosamia ricini* Bois syn., *Attacus ricini (cultivated)* *P.cynthia* Drury syn. *Samia cynthia* (D.) (wild) Family: *Saturniidae*	**A. PRIMARY FOOD PLANT**		
		1. Eng: Castor Vern: Erandi	*Ricinus communis* Linn. syn. *R.inermis* Jacq., *R.lividus* Jacq., *R.speciosus* Bl., *R.viridis* Willd., *Croton spinosus* Linn.	Euphorbiaceae
		B. SECONDARY FOOD PLANTS		
		1. Vern: Payam	*Evodia fraxinifolia* Hook.f.	Rutaceae
		2. Vern: Kesseru, Tarla	*Heteropanax fragrans* Seem	Araliaceae
		3. Eng: Cassava, Manioc, Tapioca Vern:Simul-alu, Sakarkanda	*Manihot esculenta* Crantz. syn. *M.utilissima* Pohl., *M.aipi* Pohl., *M.dulcis* Pax.	Euphorbiaceae
		C. TERTIARY FOOD PLANTS		
		1. Eng: Allanto, Tree of heaven Vern:Barkesseru, Barpat	*Ailanthus altissima* (Mill) Swingle syn. *A.glandulosa* Desf.	Simaroubaceae
		2. Vern: Maharukh, Barkesseru	*Ailanthus excelsa* Roxb.	Simaroubaceae
		3. Vern. Gogul	*Ailanthus grandis* Prain	Simaroubaceae
		4. Vern: Guggal Dhup	*Ailanthus tryphysa* Alston syn. *A.malabarica* DC.	Simaroubaceae
		5. Eng: Papaya Vern: Papeeta	*Carica papaya* Linn.	Caricaceae
		6. Eng: Gumhar Vern: Gambhari, Gumhar, Khambari	*Gmelina arborea* Roxb.	Verbenaceae

SILKWORMS		FOOD PLANTS (TREES)		
Vernacular name	Zoological name	Popular names	Botanical names	Family
		7. Vern: Thebow	*Hodgosonia heteroclita* Hook.f. & Thomas	Cucurbitaceae
		8. Eng: Physic nut, Purging nut Vern:Bagbherenda, Jangali arandi, Safed arand	*Jatropha curcas* Linn. syn.*J.mollucana* Herb. Russ.	Euphorbiaceae
		9. Eng: Coral plant Vern: Bhotera	*Jatropha multifida* Linn.	Euphorbiaceae
		10. Eng: Temple Tree, Pagoda Tree Vern:Gulanhi Phool, Champa chameli	*Plumeria rubra* L. forma *acutifolia* Poir. syn. *P.acutifolia* Poir, *P.acuminata* Roxb., *P.acuminata* Ait.	Apocyanaceae
		11. Vern: Korha	*Sapium eugenifolium* Buch-Ham.	Emphorbiaceae
		12. Eng: Chinese Tallow tree Vern:Pippalyang, Vilayati shisham, Pahari shisham	*Sapium sebiferum* Roxb.	Euphorbiaceae
		13. Vern: Darmar, Tumru, Nepali Dhaniya	*Zanthoxylum armatum* DC syn. *Z. alatum* Roxb.	Rutaceae
		14. Vern: Bajramani	*Zanthoxylum limonella* (Dennst.) Alston syn. *Z.budrunga* Wall. ex DC.syn.*Z.rhesta* DC.	Rutaceae
		15. Eng: Indian Jujube, Common Jujube Vern: Baer, Ber	*Zizyphus mauritiana* Lam. syn. *Z.jujuba* Lam. non Mill.	Rhamnaceae

Source : Anonymous 1949-79, FAO 1976, Sengupta and Dandin 1989, Jolly *et al.*1974, Seth 1995 a,b,c,d, Thangavelu *et al.* 1988, Sarkar 1988 and personal observations.

Mulberry silkworms are reared mainly on five species of mulberry *viz.*, *Morus alba* L., *M. indica* L., *M. laevigata* Wall. (ex Brandis), *M. nigra* L. and *M. serrata* in Indian subcontinent. However, under stress conditions they may also be reared on *Lactuca sativa* L. (Lettuce), *Ficus religiosa* L. (Peepal), *Cudrania triloba* and osage orange but the rearing performance is very poor and mortality is very high (Ullal and Narasimhanna, 1994). While lettuce belong to family Asteraceae, Mulberry, Peepal and *Cudrania triloba* belong to family Moraceae. Geographically the following mulberry varieties are found suitable to various regions (Prasad 1989, Ullal & Narasimhanna 1994, Krishnaswami 1994, Chowdhury 1984, Ahsan 2000, Datta 2000b, Dhingra *et al.*, 2002, Patil *et al.* 2002, Juyal *et al.* 2003, Singh *et al.* 2003).

1. South India (Karnataka,Tamil Nadu, Andhra Pradesh): S36, K2, S13, S30, S54, MR2, Mysore Local, V-1,

2. West Bengal (Malda, Murshidabad, Birbhum): S1, S799, Berhampore local, C776, S1635.

3. Darjeeling : Kosen (Japanese), Tr-10.

4. North Western Himalayan region: Goshoerami, Kokuso-27, Kairyoroso, K2, Chak majra, KNG, Tr-10, S146.

5. North Eastern region: S1635, S1, BC259, Tr-8, Tr-10.

6. Chotanagpur (Jharkhand): S1, C763, K2, S54.

Following varieties of mulberry are found suitable and recommended for various purposes/climatic conditions:

i. Irrigated regions—K2, S36, V-1, S1635.

ii. Rainfed regions—Ber. S799, S1, K2 (M5), Sujanpur-5, S13, BC259, S1635, Tr-8, Tr-10, S14, S34, RFS135, RFS175.

iii. Hilly regions—Kosen, MR2, K2, S54, BC259, Tr-10.

iv. Sub-tropical regions—S146, Chinese white, Himachal local, Tr10 and Chak majra.

v. Temperate regions—Chinese white, Goshoerami, Ichinose, KNG and Kekuso.

vi. Intercropping with tea—S1301, S36, S523, TR-4, C776, S1, K2, S41, S30, TR-8, S799, RFS135.

vii. Fungal resistant: MR2, Kosen.

The food plants of non-mulberry silkworms (tasar, muga and eri) may indeed be categorised as primary, secondary or tertiary food plants on the basis of preferential feeding and/or adaptability of these silkworms to their host plants (Table 4). The polyphagus nature of all the non-mulberry silkworms is not only interesting but also a baffling problem of sericulture. While Daba, Sarihan, Bhandara, Andhra local and Sukinda ecoraces of tropical tasar silkworms (*Antheraea mylitta* D. have adapted to *Terminalia arjuna* Bedd. and *T. tomentosa* W. & A., Laria, Barharwa, Raily, Modal and Nalia ecoraces have adapted to *Shorea robusta* L/*Anogeissus latifolia* wall. ex Bedd and Tira, Jiribam and Belguam ecoraces adapted to *Lagerstroemia parviflora* Roxb, *Zizyphus jujuba* Lam. non Mill. and *Hardwickia binata* Roxb. (Fig. 41) host plants, hence they serve as primary host plants of these ecoraces respectively, besides serving as secondary host plants to other ecoraces (Table 2). Principally, however, *T. arjuna* Bedd., *T. tomentosa* W. & A. and *Shorea robusta* L. only are regarded as primary host plants of *A.mylitta* D. On the contrary, Oak tasar silkworm (*Antheraea proylei* J.), muga silkworms (*Antheraea assama* W.W.) and eri silkworms (*Samia cynthia ricinii* Hutt./*S. cynthia* (Drury) do not exhibit much ecoracial adaptation with respect to host plant, while *Quercus serrata* auct non Thumb. syn. *Q. acutissima* Carr. and *Q. grifithiii* Hook. and Thom. ex. Miq serve as primary host plant of *A.proylei* J. in North-Eastern region, *Q. incana* Roxb.

syn. *Q. leucotrichophora* A. camus and *Q. semecarpifolia* Sm. serve as their primary host plants in North-Western region. *Lithocarpus dealbata* (Hk.f. & Th.ex Miq.) Rhedder Syn. *Q. dealbata* Hk. f. & Th.ex Miq. serves as primary host plants of another commercially unexploited wild oak tasar silkworm, *A. frithii* in North-Eastern region. *Machillus bombycina* King ex. Hook.f. Syn. *Persea bombycina* Kostern. and *Litsaea polyantha* Blume syn. *L. monopetala* (Roxb.) Pers. serve as primary host plants of *A. assama* W.W. (muga silkworm) while the food plant status of *L. citrata* blume, *L. salicifolia* Hook.f., *L.nitida* Roxb. and *Cinnamomum glanduliferum* Meissn. either as primary/secondary host plants is still doubtful. Though *Ricinus communis* L. is most favoured yet *Heteropanax fragrans* Seem. has also been regarded as primary host plants of eri silkworms, *Samia cynthia ricinii* Jr. in Upper Assam. *Manihot utilitissina* Pohl. and *Evodia fraxinifolia* Hook.f. are important secondary host plants of eri silkworms next only to above primary host plants and they are used mainly in Meghalaya and Nagaland respectively. Apart from aforesaid primary/secondary host plants, there are several secondary and tertiary host plants of these silkworms but they are less significant (Table 4, Figs. 42-48)).

D. DISTRIBUTION AND CLIMATE

I. Mulberry

China, India, Japan, South Korea and erstwhile Union of Soviet Socialist Republic are sericulturally advanced countries. Besides India in Indian subcontinent, Bangladesh, Sri Lanka, Pakistan and Afghanistan are also engaged in sericulture. Mulberry is supposed to be native of Indo-Chinese region and found widely distributed in both the hemispheres. Of the four species *viz.*, *M. indica*, *M. alba*, *M. serrata*, *M. lavigata* found in India, *M. indica* and *M.alba* are almost not available under natural condition and *M. serrata* is threatened and restricted to only North-Western parts of the country. The original home of the genus is a lower Himalayan belt of Indo-China region where about 6 species occur in wild condition bearing primitive and natural cultivars, hybrids, weedy forms and related genera upto an elevation of 7000 feet. In addition, a few mulberry species occur in South Korea, Japan and Russia. Due to cross fertilization wide variation occurs in natural populations of mulberry species (except *Morus rubra* L.) in most deciduous forests in foot hills of Himalaya, China, South East Asian region and Japan (Jolly and Dandin 1986, Sengupta and Dandin 1989).

In India, Karnataka, West Bengal, Andhra Pradesh, Tamil Nadu, Assam and Jammu & Kashmir are traditional sericulture states while Madhya Pradesh, Uttar Pradesh, Orissa, Maharashtra, Uttaranchal, Chattisgarh, Bihar and Jharkhand etc., are non-traditional states for mulberry cultivation. Rainfall varies from 60-250 cm, R.H. varies from 65-90, temperature varies from 0-48°C (Optimum 24-27°C) and soil varies from clayey-sandy loam, loam, black cotton and alluvial type with 6.0-7.5 soil pH.

II. Tropical tasar food plants

Tropical tasar culture is practised mainly in India in the States of Bihar, Madhya Pradesh, Chattisgarh, Orissa, West Bengal, Maharashtra, Andhra Pradesh, Karnataka and Uttar Pradesh upto an altitude of 2000' ASL. The host plants of tropical tasar silkworm are found in Red loamy, lateritic, black clayey, sandy red, forest and hill to alluvial soil in tropical wet evergreen, tropical moist deciduous, tropical dry deciduous to mountain sub-tropical (Shiwalik) and even thorn forests (Sahabad Rajasthan). The food plants of tropical tasar silkworm grow well in soils having pH ranging from 5.0-7.0. Rainfall of 100-200 cm is suitable for them. Distribution of silkworm races in relation to forest types indicates that the races are restricted mainly in tropical moist deciduous forest in Central India within 16-24°N and 80-88°E where the average rainfall varies between 101-130 cm and deciduous zone of tropical forest where average rainfall is 98-100 cm. The occurrence of tropical tasar ecoraces has been observed within 12-13°N and 72-96°E. The tropical tasar silkworms have also been found in Himachal Pradesh, Sikkim, Assam, Manipur, Meghalaya, Tamil Nadu, Pondicherry, Jammu & Kashmir, Dadar and Nagar Haveli and Nagaland but their commercial exploitation is extremely less in these States/territories (Singh and Srivastav 1997, Thangavelu *et al.* 2000). The available records reveal that they are distributed in Pakistan, Bangladesh, South China, Nepal, Burma and Sri Lanka also (Arora and Gupta 1979 a, b). The optimum temperature ranges from 24-35°C and RH 70-80% for silkworm rearing. However, the tropical tasar food plants have been found to be distributed under extremes of temperature ranging from 10-48°C and RH 45-85%.

III. Temperate tasar food plants

China ranks first in oak tasar production. Out of 35 oak species reported from various provinces 07 species *viz.*, Quercus liaotungenesis Koil., *Q. mongolica* Fisch., *Q. dentata* Max., *Q. acutidentata*, *Q. acutissima* Carr., *Q. variabilis* and *Q. oxyloba* Fisch. are being exploited for rearing of their oak tasar silkworm A. *pernyi* in Liaoning, Shandong, Henan, Guzhon, Heilonjiang, Jilin and inner Mangolia chiefly. In Taiwan, *Q. acutissima* Carr. is the main food plant under exploitation for rearing of A. *yamamai* whereas both in North as well as South Korea *Q. mongolica* Fisch. is main food plant of A. *yamamai*. However, in Japan *Q. fabri*, *Q. himalayana* (Hiragi), *Q. acutissima* Carr. (Kunugi), *Q. serrata* Thunb. (Konara), *Q. liaotungenesis* Koil., *Q. mongolica* Fisch., *Q. dentata* (Kashiwa), *Q. myraefolia* (Shisha kashi) are major oak species exploited for cultivation of A. *yamamai* (Mishra *et al.* 2000, Mathur and Vishwakarma 1997).

Oaks have been found to occur not only in Indian subcontinent but also in Europe, North and South Korea, North-East corner of South America, West Malaysia, China, North America and Japan. The major centres of diversity are Mexico, China, Central America and Malaysia. In India more than 35 species of oaks have been reported (Negi and Naithami 1995). Many of them occur gregariously and make regular altitudinal zones mainly in Western Himalayan regions. The vast natural oak flora is distributed between 2000-9000 ft. ASL in the sub-Himalayan regions from J & K in North West, Garhwal and Kumaon hills of Uttaranchal and Himachal Pradesh in central zone to Manipur and Arunachal Pradesh in North East India. In India oaks (Family: Fagaceae) are represented by four genera *viz., Quercus, Lithocarpus, Castanea* and Castanopsis. Seven species *viz., Q. serrata* auct. non Thumb. syn. *Q. acutissima* Carr., *Q. semecarpifolia, Q. dilatata, Q. leucotrichophora* syn. *Q. incana, Q. grifithii, Q. semiserrata* and *Lithocarpus dealbata* syn. *Q. dealbata* have been found suitable for oak tasar rearing in India. While *Q. incana, Q. dilatata* and *Q. semecarpifolia* occur in North-Western region, the remaining four species occur in North-Eastern region. Out of 18,41,500 ha. of forest land covered by oaks, only 46,500 hect. land is accessible for tasar rearing in India.

The climatic conditions occurring in oak forest tract vary from sub-tropical to sub-arctic depending on altitude, latitude and location aspects and physiography. Three marked seasons *viz.,* summer, monsoon and winter are experienced in these areas.

Lower oak forests have fairly warm temperature upto 35°C during summer in ban (*Q. leucotrichophora*) oak forest as compared to upper oak forests where usually average maximum temperature is not more than 22°C. During winter, snow is commonly experienced even in lower oak forests and heavy snowfall occurs in the upper oak forests of the western Himalaya and temperature remains below 0°C for fairly long periods. On the contrary, no snowfall occurs in lower oak forests of central and eastern Himalaya while upper oak forests experience some snow in winter. All oak forests receive fairly high rainfall which is more than 200 cms every year. Different types of soils which constitute base for oak forests are red and black soils, ferruginous red soils which include red earths and red loam, brown forest soils which show acidic to neutral pH and mountain and hill soils (Negi and Naithani, 1995).

IV. Muga food plants

Som (*Machilus bombycina* King ex hook.f. *Persea bombycina* Koestern.) and *Litsea polyantha* Juss., *L. salicifolia, L.citrata* and *L.nitida* plants used for muga silkworm rearing in North eastern region are distributed in Nepal, Myanmar, Malaysia, Indonesia, Bhutan and Sri Lanka. Besides Assam in India, they have also been reported from Arunachal Pradesh, Manipur, Mizoram, Nagaland, Tripura, Sikkim, Uttaranchal, West Bengal, Himachal Pradesh, Gujarat and Pondicherry. In Uttaranchal the food plants are distributed over two major hills upto 6000 ft. ASL. i.e., Garhwal and Kumaon. The climates of the North Western regions vary according to their elevation, from hot in the foot hills to the freezing point in the Himalayas (Sengupta *et al.* 1993). A natural hybrid of *L. salicifolia* and *L. polyantha* which can ensure higher returns has also been reported by Rajaram and Samson (1999) from Assam. All muga food plants grow well in wet and warm climatic conditions with high rainfall and slightly acidic to acidic alluvial sandy loam and lateritic red loam soil (Das *et al.* 2000). Normally, they thrive well under soil pH

Fig. 1
Eri silk shawl

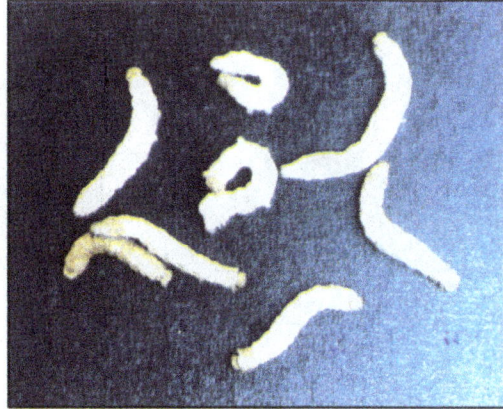

Fig. 2
Larvae of *Bombyx mori*
(Mulberry silkworms)

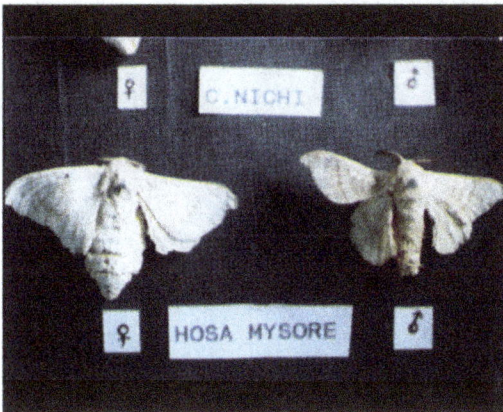

Fig. 3
Bombyx mori - moths
(Mulberry silkmoths)

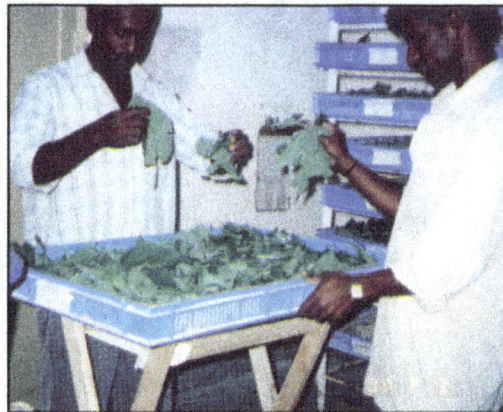

Fig. 4
Rearing of *Bombyx mori*
(Mulberry silkworm)

Fig. 5
Cocoons of mulberry silkworm

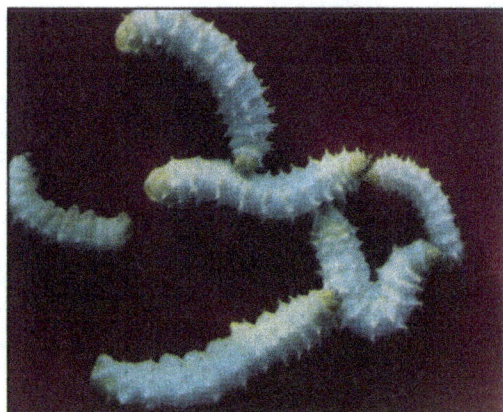

Fig. 6
Eri silkworm of local strain

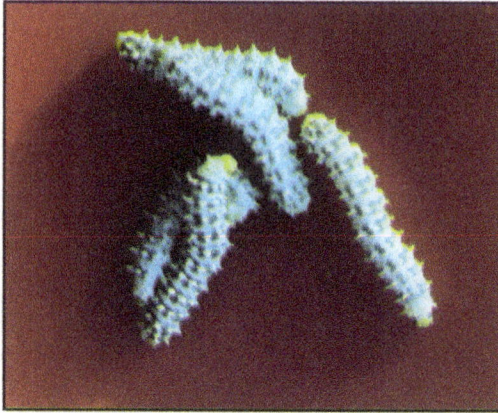
Fig. 7
Eri silkworm of Borduar strain

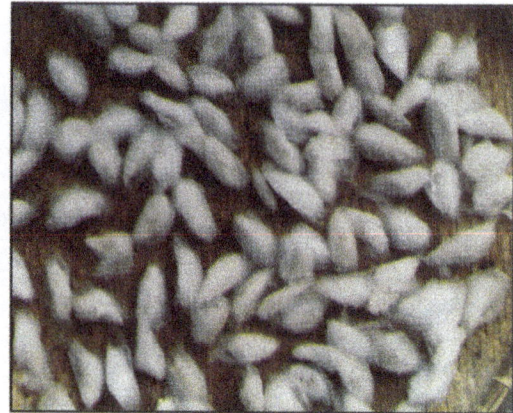
Fig. 8
Eri cocoons of local yellow strain
from Imphal

Fig. 9
Eri cocoons of Borduar strain

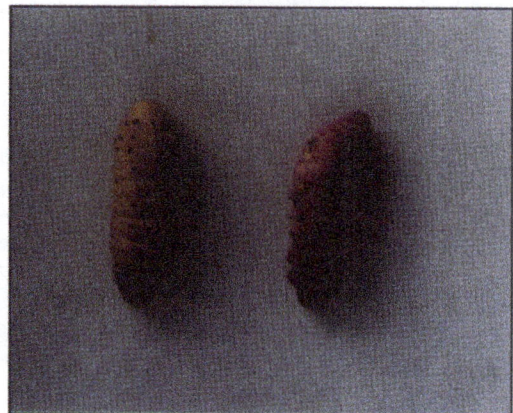
Fig. 10
Eri pupae

Fig. 11
Coupling of Eri silk moths

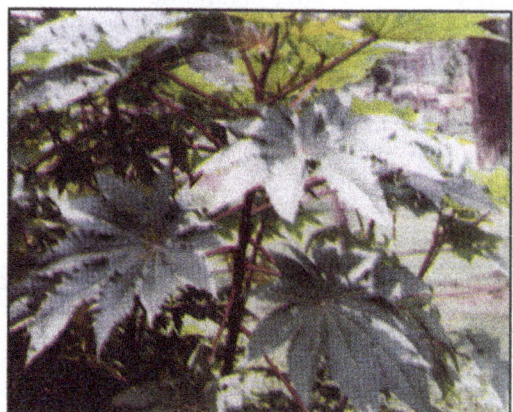
Fig. 12
Ricinus communis (Green variety)

Fig. 13
Ricinus communis (Red variety)

Fig. 14
Kesseru (*Heteropanax fragrans*)

Fig. 15
Manihot utilitissima

Fig. 16
Antheraea mylitta larva (Blue)

Fig. 17
Antheraea mylitta cocoons and moths

Fig. 18
Terminalia crenulata

Fig. 19
Terminalia alata

Fig. 20
Terminalia glabra

Fig. 21
Shorea robusta

Fig. 22
Largerstroemia sarviflora

Fig. 23
Largerstroemia speciosa

Fig. 24
Largerstroemia indica

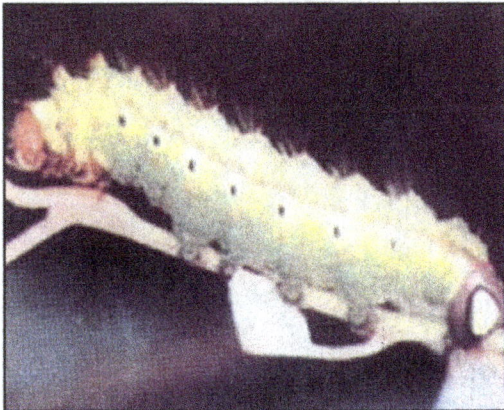

Fig. 25
Muga silkworm

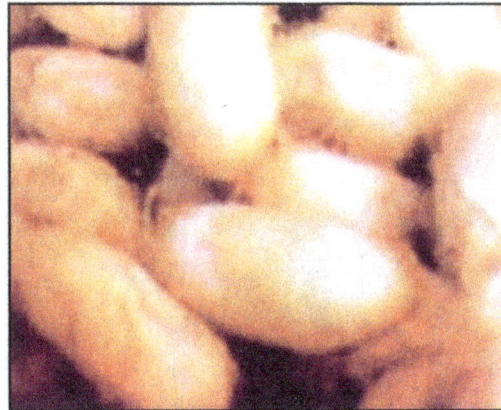

Fig. 26
Muga cocoons

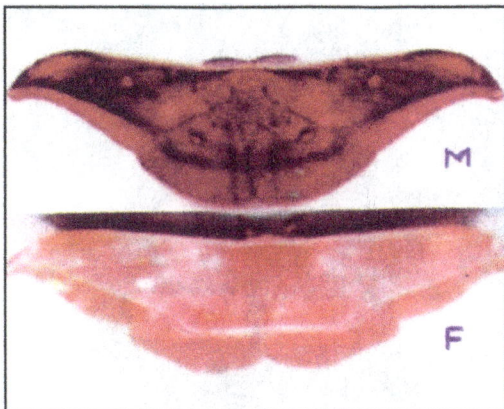

Fig. 27
Muga silkmoths

Fig. 28
Antheraea proylei larva (Green)

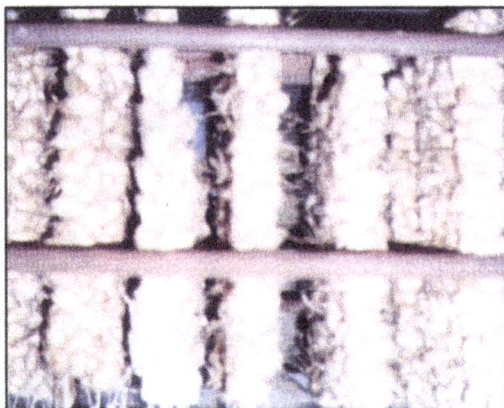

Fig. 29
Antheraea proylei cocoons

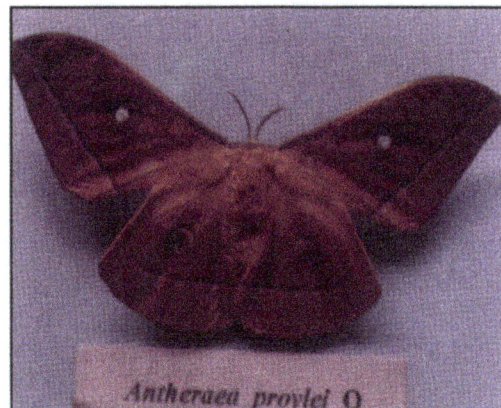

Fig. 30
Tropical tasar silk moth
(*Antheraea mylitta* D)

Fig. 31
Tropical tasar silk moth
(*Antheraea mylitta* D)

Fig. 32
Som (*Machilus bombycina*)

Fig. 33
Litsae polyantha

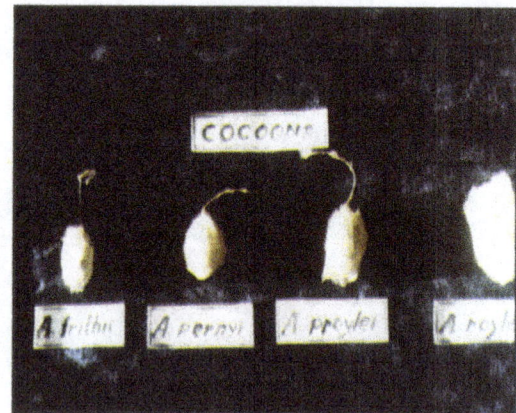

Fig. 34
Cocoons of wild silkworms

Fig. 35
Male and female moth of *Antheraea roylei*

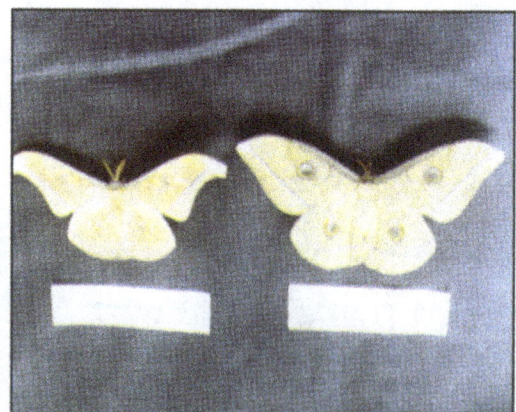

Fig. 36
Male and female moth of *Antheraea frithii*

Fig. 37
Quercus acutissima (Syn. *Q. Serrata*)

Fig. 38
Quercus grifithii

Fig. 39
Quercus incana

Fig. 40
Lithocarpus dealbata (Syn. *Q.Serrata*)

Fig. 41
Hardwicia binata

Fig. 42
Dalbergia sissoo

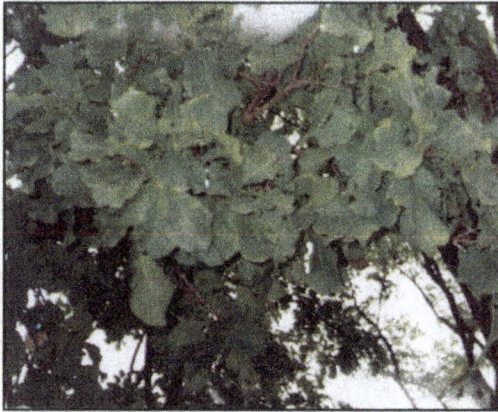

Fig. 43
Kumbhi (*Careya arborea*)

Fig. 44
Terminalia chebula

Fig. 45
Bauhinia variegata

Fig. 46
Tectona grandis

Fig. 47
Michelia champaca

Fig. 48
Ficus religiosa

ranging from 4.0 to 6.8 and rainfall ranging from 160-210 cm. The average annual minimum temperature, maximum temperature and relative humidity has been found to be 10.3-10.4°C, 31-32°C and 81-93% in most suitable genetic resource centres at Boko and Borahibari in lower Assam (Kamrup) and upper Assam (Sibasagar) between 91.24°-94.55° E longitude and 25.95°-26.95° N latitude (Sengupta *et al.* 1993, Rajaram *et al.* 1993).

V. Eri Food Plants

The history of ericulture is lost in antiquity yet it is believed to have originated in India and Assam has been regarded as the original home of eri silk from time immemorial. The major countries now engaged in ericulture are India, China, Brazil, Russia, Thailand, Philippines, Indonesia, Malaysia, Ethiopia and many other countries due to availability of food plants. While Castor (*R. communis* L.) is used throughout India, Kesseru (*H. fragrans*) is used mainly in upper Assam, Tapioca (*M. utilitissima*) is used mainly in Meghalaya and Payam (*E. fraxinifolia*) is mainly used in Nagaland as eri food plants. Borkesseru (*A. excelsa*), Gulancha (*P. acutifolia*) and Gamari (*Gmelina arborea*) are rarely used in Assam and/or Meghalaya as eri food plants (Sarmah 2000). Eri food plants are extensively distributed/cultivated in Assam, Manipur, Bihar, West Bengal, Orissa, Andhra Pradesh and Karnataka upto an altitude of 5000' ASL. They are mainly found in clayey loam to black cotton soil having 5.6-7.0 pH. The annual rainfall ranges from 150-200 cm in suitable places and the climatic conditions have been found to vary very widely from tropical-subtemperate types depending upon altitude, latitude and location.

2

Silkworm and Host Plant Interaction

Insect-host plant interaction has been the most baffling yet interesting topic of study. Why a particular insect attacks/feeds on a specific host plant or many food plants, what factors govern these and does it depend on the physiological requirement of the insects, capability of the host plants to fulfil such requirements and feeding behaviour of the insects etc., still raise curiosity among the entomologists and common man as well.

In mulberry and non-mulberry sericulture, the silkworm and host plant interaction is somewhat more interesting because all of them belong to the order Lepidoptera though their families may be different. While mulberry silkworm belongs to Bombyciidae family, non-mulberry silkworms belong to Saturniidae family and probably this difference also differentiates their feeding behaviour and determines their relationship with their host plants.

Food plants, silkworms and phylogenetic relationship

In mulberry sericulture, there exists gene to gene relationship between the host plant, mulberry (*Morus alba, Morus indica*) and the silkworm *Bombyx mori* because the insect is monophagus. In other words "Lock and Key" model has been proposed to enunciate the relationship between both the organisms. On the other hand, non-mulberry silkworms viz. *Antheraea mylitta* D., *Antheraea assama* and *Philosamia ricini* syn. *Samia cynthia ricini* known as tropical tasar, muga and eri silkworms respectively feed on many food plants (Table 4) and hence they have been designated as polyphagus insects. In such cases, gene to gene relationship or lock and key model does not exist between the food plants and the insects presumably because they have less evolved in comparison to mulberry silkworms. In the evolution hierarchy, the mulberry silkworm has achieved the top position followed by eri silkworms, muga silkworms and tasar silkworms in descending order. This is also authenticated by the fact that while mulberry silkworm is more than 3500 years old, tasar is probably only 2000 years old as per the Chinese records. During the course of evolution mulberry silkworm not only developed gene to gene relationship with its food plant but also acquired indoor sedentary habits and became domesticated by gradual loss of appendages. Eri silkworm on the other hand only developed sedentary habits and became indoor but remained polyphagus by feeding on Castor (*Ricinus communis*) as well as Kesseru (*Heteropanax fragrans*), tapioca (*Manihot utilitissima*) and other food plants. Muga silkworm acquired semi-domesticated habit by feeding most of the time out door on natural or cultivated Som (*Machilus bombycina*) or Soalu (*Litsaea polyantha*) trees and other secondary food plants (polyphagus) and form their cocoon indoor, while tasar silkworms have remained exclusively wild since they not only feed on various primary food plants (polyphagus) *viz.*, Arjun (*T.arjuna*), Asan (*T.tomentosa*), Sal (*Shorea robusta*), Ber (*Ziziphus jujuba*) *Anogeissus latifolia* and Saoni (*Lagerstroemia purviflora*) etc. throughout their lives but also form their cocoons outside *in-situ*. Temperate tasar silkworm *Antheraea pernyi/A.proylei*, on the other hand feeds on many *Quercus* species but can be reared outdoor as well as indoor completely and acquired both wild and domesticated habits.

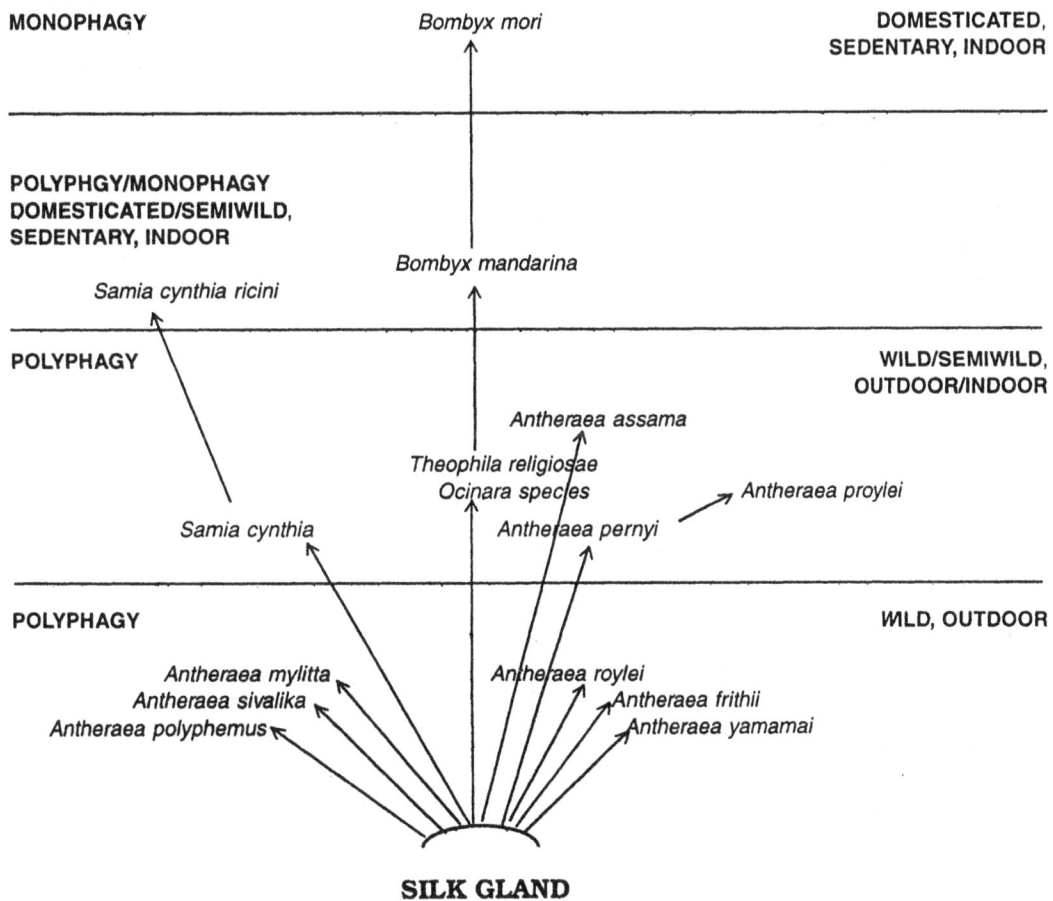

MONOPHAGY *Bombyx mori* **DOMESTICATED, SEDENTARY, INDOOR**

POLYPHGY/MONOPHAGY DOMESTICATED/SEMIWILD, SEDENTARY, INDOOR

Samia cynthia ricini *Bombyx mandarina*

POLYPHAGY **WILD/SEMIWILD, OUTDOOR/INDOOR**

Antheraea assama
Theophila religiosae
Ocinara species *Antheraea proylei*
Samia cynthia *Antheraea pernyi*

POLYPHAGY **WILD, OUTDOOR**

Antheraea mylitta *Antheraea roylei*
Antheraea sivalika *Antheraea frithii*
Antheraea polyphemus *Antheraea yamamai*

SILK GLAND

Diag. 1: Hypothetical phylogenetic status of various silkworms with respect to silkgland and host plant interaction.

On the basis of above assumptions, tentative phylogenetic relationship between different silkworms and their origin has been proposed in diagram-1, which is at hypothetical stage. However, whether origin of various silkworms is monophyletic or polyphyletic, is indeed a subject of long debate and requires many evidences from other fields also.

Specialisation and generalisation in silkworms

Kogan (1977), Slansky (1976) and Futuyma (1983) have defined species of herbivorous insects as specialists (stenophagic) if they consume plants within one species, one genus or one family respectively and mulberry and oak tasar silkworms may therefore, be regarded as specialists while other silkworms may be categorised as generalists (euryphagic). Australian scientists have found that many species of phytophagus insects on *Eucalyptus* could be considered as generalists on the smaller scale of their geographical ranges as they feed on numerous eucalypt species. But when a local community is considered, many of these species could be regarded as specialists because they feed only on a subset of the eucalyptus available at that site. Therefore, both geographical scale and taxonomic level are important for identification of ecological properties of all the silkworm ecoraces/populations and species as is evident from Table 2 in case of tasar ecoraces.

As the individuals from specialised species are assumed to be more efficient metabolically and behaviourally than individuals from generalised species throughout their entire geographical ranges, the specialists are more effective competitors than generalists, since a specialist exploits its particular environment better than closely related generalists. These assumptions may be applied to herbivorus insects like silkworms also because of the high degree of chemical distinctiveness of many plant species used as food. The non-mulberry silkworms feeding on a variety of plants must be able to tolerate, detoxify or possibly metabolize an array of qualitatively different chemicals that have potentially deleterious effects. Maintenance of metabolic machinery for such an array may involve biochemical and physiological costs thereby reducing the efficiency with which a generalised herbiovore like tasar and muga can process their food.

Primary versus secondary metabolites and feeding behaviour

Intensive work carried out in India as well as abroad on various species of silkworms and their food plants has established that quality of leaves based on estimation of moisture and primary metabolites *viz.*, nitrogen, proteins, minerals, fibre and carbohydrates etc., plays a significant role in promoting healthy development of silkworms which has also got definite bearing on qualitative and quantitative silk production. Further, the information available reveal that food plant selection is solely determined by secondary metabolites such as essential oils, glycosides, alkaloids etc. According to Hamamura (1970), the feeding behaviour of mulberry silkworms is governed by olfactory attractants, biting factors and swallowing factors. The act of feeding involves a consecutive sequence of behavioural components that are triggered and regulated by specific stimuli operating through neural pathways. The behavioural steps are modified somewhat according to the structure and function of the insects mouthparts. For insects with chewing mouthparts the sequence begins with biting and is followed by ingestion (swallowing) and continuous feeding, which eventually leads to cessation of feeding and the locomotion (Kogan 1977, Beck 1965). Each behavioural step can be terminated by the presence of negative stimuli and/or absence of the appropriate positive stimulus. The initial recognition of the food item in insects through contact is by the tactile, thermal and chemical receptors located in the antennae, legs, maxillary and labial palpi of the mouthparts. The combined information received by these receptors helps to initiate the feeding behaviour (Kennedy 1978). Ingestion or engorgement may continue or stop depending upon the continued stimulation of the epipharyngeal receptors by positive stimuli and the absence of negative stimuli. The adequacy of a diet for maintenance of feeding depends on both its chemical and physical characteristics.

Induction of food preference

Induction of food preference is defined as the process whereby ingestion of a particular food during ontogeny creates a preference for that food over others previously (Dethier 1982). Induction of food preference has been documented in lepidoptera insects including *Antheraea polyphemus* as polyphagous species show a greater induced preference than oligophagous species (Hanson 1976). Recent chemical studies showed that a complex of feeding stimulants of both polar and non-polar compounds may be responsible for the induction of preference (Staedler & Hanson 1978). Induced preferences are composed of increased orientational responses and/or increased feeding responses (Saxena and Schoonhoven 1982) supporting the involvement of olfactory stimuli. Electrophysiological studies showed that both olfactory and contact chemoreceptors were involved in host discrimination (Hanson and Dethier 1973). It appears that the induction involves both central and peripheral modifications of the nervous system.

Restriction of potential food range may be achieved by induced preferences since it is physiologically advantageous especially for polyphagous insects like tasar, eri and muga. Since detoxifying toxic secondary plant chemicals is metabolically costly (Schoonhoven and Meerman 1978), by restricting feeding on few plants, energy expenditure can be economised and the potential hazard of ingesting toxins is avoided.

Types of chemical stimuli

Dethier *et al.* (1960) proposed five basic terms *viz.*, attractant, repellent, arrestant, stimulant and deterrent for differentiating the behavioural effects of chemicals on locomotion, feeding and oviposition. Beck (1965) adopted and modified that terminology and added two more terms *viz.*, incitant and suppressant and stated that terms should be applicable to physical stimuli as well.

During the food finding phase, the insects respond by flying or walking to volatile chemicals emitting from their food. The chemicals that elicit initial orientational responses toward or away from its source are termed "attractants" and "repellents" respectively. The terms "arrestant" and locomotor "stimulant" refer to a second set of orientational stimuli that cause a decrease or increase in the rate of movement and eventually lead the insects to accumulate at or disperse away from the food source.

The chemicals that evoke or prevent biting or piercing actions are called "incitant" and "suppressant" respectively. The terms "feeding stimulant (phagostimulant)" and "feeding deterrent" refer to compounds that initiate or inhibit feeding (swallowing).

The duration of feeding is regulated by the presence of "feeding stimulants" as well as other factors. The physical characteristics of food may affect the rate of food intake. Feeding may be enhanced by the presence of feeding "co-factors" or reduced by the action of "feeding deterrents", toxicants and metabolic inhibitors. "Feeding co-factors" are chemicals that do not elicit feeding responses by themselves but produce synergistic action when combined with feeding stimulants.

As the same compound may have multiple effects on feeding behaviour, the designation of various classes of chemical stimuli is not strict, for example sucrose may function as an incitant and a feeding stimulant for the same insect. Likewise, a chemical may serve as an attractant as well as an arrestant. On the other hand, a compound may be classified either as a feeding stimulant or deterrent depending upon its relative concentrations in a diet.

Among phytophagous insects, monophagous and oligophagous species are most rigid in their chemosensory requirements. They are extremely sensitive to the presence of negative stimuli and require the full component of positive stimuli to elicit feeding. Hence, their diets are restricted to few plants that meet these conditions. Polyphagous species, on the other hand are less specific in their requirement of positive stimuli and are able to tolerate many negative stimuli. Hence, they are capable of accepting a greater variety of plants as food.

Bioassay

The ultimate test for olfactory stimulus is to use traps in an open field situation. This method has been used extensively in screening or verifying chemicals as "attractants" for a variety of insects (Beroza and Green 1963, Beroza 1970). Attractant traps are currently used to monitor or direct control of pest species.

Bioassay for gustatory stimuli are based on the biting and feeding responses of test organisms to a substrate that is incorporated with chemical stimulants. Plant parts such as root stem or leaf discs can also be used as substrates (Hsiao & Fraenkel 1968). In case of silkworms, the leaves of food plants may be used as substrates infiltrated with test chemicals under vaccum. Assessment of the activity of chemical stimuli are carried out by a wide range of qualitative and quantitative techniques. The distribution of test insects on control and test substrates, as noted at intervals are used to evaluate stimulatory effects of chemicals. Feeding responses may also be measured by the area of the substrates consumed, by the weight change of the substrate or the test insect, or by the number or weight of faecal pellets produced.

For bioassays of feeding deterrents, leaf-disc test and the sandwich test have been widely adopted (Jermy 1966). For a quantitative measurement of deterrancy, the chemicals are incorporated into artificial diets. The amount of faecal pellets produced or weight losses of insects as compared to control are used to determine the degree of deterrancy (Hsiao & Fraenkel 1968).

Electrophysiological techniques have been extensively used for bioassay of chemicals affecting feeding behaviour. Behavioural observations with intact animals are usually required to assist the interpretation of electrophysiological evidence. Mouthparts and tarsal receptors are amenable to electrophysiological recordings and exhibit a high degree of chemosensory specificity. A good correlation exists between electrical impulses generated from receptor responses to chemicals and the overt behaviour produced by such chemicals (Ma 1972). Hence, reliable bioassays may be developed by the electrophysiological technique.

Chemical stimuli in Bombyx mori

Much work has not been carried out on the chemical stimuli of silkworms pertaining to feeding behaviour except on *Bombyx mori*. A brief account of work done on *Bombyx mori* along these lines is enumerated here:

1. The "green leaf volatile" is a mixture of a large number of saturated and unsaturated aliphatic alcohols, aldehydes, ketones and esters. These chemicals have been demonstrated as olfactory attractants for a variety of phytophagous insects. *B.mori* is attracted by several of these leaf volatiles eg., 3-Hexene-1 ol. and Linanool (alcohols) and Linalyl acetate and Terpinyl acetate (esters) serve as their olfactory attractanfs (Hamamura 1976).

2. Nutrient chemicals such as, sucrose, fructose, raffinose, ascorbic acid and B-sitosterol serve as phagostimulants for *B.mori* (Ito 1960, 61, Hamamura 1970, Nayar and Fraenkel 1962).

3. Secondary plant chemicals such as chlorogenic acid, morin and isoquercitrin (phenol), n-Hexacosanol and n-octacosanol serve as phagostimulants for *B. mori* (Kato and Yamada 1966, Hamamura 1970, Mori 1982).

Chemosensory basis of food selection

It is still not very clear whether food selection by phytophagous insects is determined by secondary plant chemicals serving as token stimuli, by a combination of nutrients and secondary plant chemicals, or by the mere presence of deterrent chemicals. Today we know that these views are not mutually exclusive. The ability of an insect to recognise its host amid a multitude of non-hosts depends on detecting a coding capabilities of chemosensory system and on the integration of the central nervous system. Behavioural and electrophysiological study has revealed that insect chemoreceptors are highly discriminatory and no two species have identical receptor system. That is why each insect species has evolved a unique chemosensory system that enables it to recognise its host plant in accordance with its physiological adaptability.

Generally, insects are adapted to feed on a particular plant species and acquired the ability to tolerate or detoxify secondary plant chemicals which are deterrents or toxic and, in few cases the presence of these chemicals has been used by insects as token stimuli for recognizing their host but such adaptation is a special case rather than a widespread phenomenon. Presence of tannins in all the food plants of *A.mylitta* D. (polyphagous) may serve a very good example of such type of adaptation. Insect species (beetles) feeding on plants that are rich in flavanoids (quercitrin, quercetin, rutin, myricitrin, myricetin and morin) adopted these chemicals as token stimuli (sensory capability) which is the consequence of a long period of association. Due to limited behavioral (sensory) and physiological (tolerating) machinery most oligophagus insects cope with only certain secondary plant chemicals occurring within their normal food range. They are thus, protected from ingesting alien and potentially toxic substances by the possession of "deterrent" receptors that react to a wide spectrum of chemicals (Schoonhoven 1981). Likewise, chemoreceptors that are tuned specifically to common nutrient chemicals *viz.*, sugar, amino acids and sugar alcohols acting as phagostimulants have also been demonstrated in a variety of insects (Schoonhoven 1981) which enables insects to initiate and maintain continuous feeding on the plants or specific tissues of the plants bearing nutritive phagostimulants and cofactors.

Apart from playing a major role in the feeding behaviour, the chemical stimuli also regulate oviposition behaviour of many oligophagus and monophagous species. In such cases, the selection of oviposition site is often the foremost and key step in host selection. "Hopkins host-selection principle" hypothesizes that females preferentially lay eggs on plant species on which they had fed upon as larvae (Jermy *et al.* 1968). However, majority of the experimental evidences do not support this hypothesis except parasitic wasp and oligolectic bees (Smith and Cornell 1979, Feinsinger 1983). In most of the insects, larval feeding and adult oviposition involve separate behavioural and genetic control mechanisms hence oviposition behaviour of females is regulated by chemical stimuli that are often distinct from those involved in feeding. Generally, adults respond primarily to host-specific chemicals and hence are more restricted in their host-plant range than the larva. Therefore, the oviposition behaviour may also influence a degree of host specificity in phytophagous insects. However, not much work seems to have been done on oviposition behaviour of silkmoths *in-situ* hence this area of research needs to be explored in detail.

The olfactory and contact chemoreceptors of insects contain both specialist and generalist receptor cells. Specific chemical such as token stimulants or deterrents at the peripheral level are detected by specialised cells and sensory information is transmitted to CNS directly via labelled lines. On the other hand, generalist cells respond to complex mixtures of chemical stimuli and provide an overall pattern of sensory inputs to the CNS.

The expression of feeding behaviour is under genetic and physiological control. The physiological aspects of silkworms that need to be investigated more thoroughly are the feeding rhythm and the endocrine regulation of the feeding behaviour. On the other hand, genetic aspects must include study of behavioural variations at intra and interspecific levels and mode of inheritance of such traits. Such genetic studies will also enable us to understand the mechanisms of host race formation and speciation, particularly in non-mulberry silkworms, which are polyphagous.

Scope for biotechnological approaches

Keeping in view the aforesaid discussions with respect to insect-herbivory interaction pertaining to silkworm and their food plants numerous fields of investigation are open leading to further biotechnological applications in sericulture industry, some of them are as hereunder:

i.   While lot of work has been done on various types of chemical stimuli in *B.mori,* this field is still barren in case of tasar, muga and eri with respect to feeding behaviour which require attention in future.

ii. Host species populations may exhibit polymorphism with respect to anti-nutritional factors/deterrents. Hence, it can be used as a selectable marker for the selection of genotypes devoid of phenolic feeding deterrents.

iii. Well identified feeding attractants, incitants and phagostimulants of tasar and muga silkworms may open avenues for their domestication in future by spraying them in artificial diets.

iv. Whether feeding stimulants/incitants do also influence fecundity of the silkworms in positive direction by acting as precursors of fecundity enhancing substances (FES) or just contribute towards palatability of silkworms to its feed vis-a-vis silk conversion ratio should be investigated in detail.

v.   Precursors of fecundity enhancing substances, if any, may be isolated from food plants varieties of silkworms and such selected varieties containing high quality of fecundity enhancing substances may be multiplied for their future plantation.

vi. Isolation and characterisation of such specific phytochemicals associated with fertility and fecundity of silkworms (FES) may facilitate increase in oviposition rates and fecundity in grainages to enhance silk productivity.

vii. B-Carotenoids precursor of vitamin A acts as photoreceptor pigments and has been reported to play a significant role in the incidence of diapause in conjunction with the photoperiodism, temperature and humidity. Effect of carotenes on induction/termination of diapause is required to be explored in the wake of the controversy as to whether it is a physiologically or genetically controlled phenomenon.

viii.   B-sitoserol, a phytosteroid is metabolised by the insect into cholesterol, which in turn is converted into ecdysteroids, which are moulting hormones of insects. The study of variation patterns in B-sitoserol among gene pools may lead to the selection of host species genotypes with high B-sitosterol content, which in turn may lead to shortening of the silkworm life cycle.

Recently, a project entitled: "Enhancement of tropical tasar silk production through the manipulation of insect"—herbivory interaction funded by Department of Biotechnology, Govt. of India has been undertaken in 1991 by scientists of Delhi University and CTR&TI, which is looking after chemical stimuli in *Antheraea mylitta* also.

3

Global Sericulture: Yesterday, Today and Tomorrow

Silk accounts for about 0.2% of the total textile fibre production in the world while cotton and synthetics together account for approximately 90% followed by cellulosics and wool which account for 6.8% and 3.6% respectively (Ronald Curry 2000, Table 5). The silk industry consists of five independent parts *viz.*, silkworm rearing, reeling, twisting/throwing, weaving and manufacturing. Silk is one of the world's most ancient and prestigious fibres. The word "Sericulture" has been derived from the word "Su" (Si) which means silk. Hence, sericulture is indeed an art and practice of cultivation of silk which involves a number of activities beginning from production of mulberry leaves, development of breeding silkworm stocks, egg production and rearing of silkworms. It is exclusively a labour intensive agro-based cottage industry taken up either as a major or subsidiary avocation depending upon the intensity, along with agriculture and horticulture.

The centres of origin of various silkworms and their food plants is controversial and the history of sericulture in most of the dominating countries is lost in antiquity. However, it is undisputed that *Bombyx mori*, the mulberry silkworm originated in China and the centre of origin of mulberry is also China-Japan region of Asia. However, only that country will rule the silk trade in the world, which will improve the quality of silk and quantitatively enhance silk production and supply the silk at competitive price in forthcoming GATT and WTO era.

I. CENTRES OF ORIGIN

The mulberry silkworm is sensitive to cold and it cannot survive in colder region without the help of man, yet it is believed that it originated somewhere in the high altitude of the mountais of Eastern Himalayas in the outskirts of mount Everest from where it spread to warmer regions. By the beginning of Christian era, sericulture established itself firmly between Gangetic and Brahmaputra valleys from where it spread to middle east.

Sericulture and Seri-biodiversity

Table 5: Global production of textile fibres (Thousand tonnes)

Year	Natural fibres			Sub total	Man made fibres		Total
	Cotton	Wool	Silk		Cellulosic fibre	Synthetic fibre	
1995	11809	1502	49	13360	2959	7346	23665
1985	17540	1673	59	19272	2999	12515	34786
1995	19200	1600	100	20900	300C	20200	44100
	*(43.6)	(3.6)	(0.2)	(47.4)	(6.8)	(45.8)	(100)

Source: Silk Review 1997 - International Trade Centre (ITC) UNCTAD / WTO
* Figures in brackets refer to percentage to grand total.
Adopted from CSB 1999.

The domesticated variety of silkworm *Bombyx mori* (n=28) evolved from wild silkworm, *Bombyx (Theophila) mandarina* (n=27) which is still extant and endemic to China, Korea and Japan and present in Eastern India. The domesticated silkworm *Bombyx mori* evolved from *B. mandarina* by breakage of one chromosome into two chromosomes (Kawaguchi 1928 quoted by Sarker 1998). Other related wild species *viz.*, *T. religiosae* are available in areas around Eastern Himalayas. The recent surveys conducted by CSGRC, Hosur (Ann. Rep 1995-96) and collection of *Theophila* sp. revealed that hills of Kalimpong (W.B.), Sikkim and Manipur may be the natural abode of various sericigenous insects. The light trap collection of adult bombycid *Ocinara* sp. in the mount Harriet range of South Andaman island has also confirmed the availability of wild relatives of *B. mori* there (Sarker 1998). Prevalence of all the types of mulberry as well as non-mulberry silkworm in Brahmaputra valley also confirms the view that North Eastern India may be the subsequent centre of origin of saturniid silkworms. The studies on cocoon colour pigments and geographical distribution of silkworms indicate that European races contain a high proportion of carotenoids while Indian, Indo-Chinese and Korean races contain high proportions of flavonoids. As the proportion of carotenoids and flavonoids vary widely in Chinese and Japanese races, it was inferred that primary source of material of Europe and India was from different regions of China, while primary source of Japan might have been from different regions of China (Fujimoto and Hayashita 1961 quoted by Sarker 1998). However, studies on racial diversification with respect to amylase activity revealed that the Indian races are 100% homozygous dominant (199%), while Chinese and Japanese races are mixture of homozygous dominant (33%), heterozygous (62.5 to 66%) and very small proportion of homozygous recessive (1 to 4.5%). On the contrary, the European races are mostly heterozygous (63%), homozygous recessive (33%) and very small proportion of homozygous dominant (4%) types. These studies reveal that Japanese and Chinese races are similar while Indian and European races are distantly related (Matsumara 1929-1951, Sarker 1998). According to Vavilovs theory, the most dominant genes of a species occur in the locus where it had originated and the majority of the recessive genes are found near the periphery of the species' Centre of diversity. Hence, further studies may prove origin of silkworms in India/

China on the basis of prevalence of maximum dominant genes in these countries in future.

In Asia, 24 species of mulberry are found in China, 19 in Japan, 6 in Korea, 4 each in Taiwan and India, 3 each in Burma and Indonesia, 2 each in Thailand, Vietnam and Afghanistan and one each in Arabia, Oman and Muscat, while 14 species are found in North America and the other 7 species are found in Central and South America which shows that great diversity of the species occurs in Sino-Japanese region of the old world and the Rocky mountains southwards to Andes of continental Americas in the New World (Sanjappa 1989). Hence, Chinese region is thought to be primary gene centre of *Morus* and the Indian region may be the secondary centre of diversity for mulberry. However, the name Morus is derived from Morea, a place in Peloponnesian Peninsula of Greece, where mulberry trees were abundant.

Whatever may be the centre of origin of mulberry and mulberry silkworms, but there is no doubt that tropical tasar silkworm (*A. mylitta* D.), muga silkworms (*A. assama* W.w.) and eri silkworms (*Philosamia ricini* Bois.) originated in India because of great diversity observed in tasar silkworms, endemism of muga silkworms in Assam and oldest history of ericulture (4000 B.C) by the Indo-Mongoloid and Tibeto-Burmese tribals in North East India.

The Indo-Malayan region has indeed been regarded as the ancestral home of genus *Terminalia* (food plant of tasar silkworms) as a whole, from where it spread to all other geographical regions of the world. All the 18 Australian species of *Terminalia* appear to be endemic. Hence, genus *Terminalia* appears to have differentiated independently in two or more regions and probably has bitopic or polytopic centres of origin rather than monotopic centre of origin. However, Central India has been regarded as the centre of origin of section "*Pentaptera*" of genus *Terminalia* (*T. arjuna* and *T. tomentosa*) due to enormous genetic diversity prevalent there (Srivastav *et al.* 1997a).

Chinese oak tasar (*A.pernyi* G.M.), Japanese oak tasar (*A. yamamai* G.M.) and Indian oak tasar silkworms (*A. roylei, A. frithii, A. proylei J.* etc.) have different centres of origin and different food plants. While Chinese oak tasar silkworm is reared on *Q. acutissima, Q. mongolica, Q. liatungenesis* and *Q. acutidentata* Japanese oak tasar silkworm is reared on *Q. acutissima, Q. serrata, Q. dentata* and *Q. myrainaefolia*. On the contrary, Indian oak tasar is reared on *Q. serrata, Q. grifithii,, Q. semecarpifolia, Q. himalayana* and *Lithocarpus dealbata*.

History of Sericulture/Silk flora

Silk was discovered by the Chinese more than 3500 years ago. The chronological events which led to discovery of silk and expansion of silk industry through out the world may be summarised following Agarwal (1995), Hyde (1986) and Phillip (1988) *et al.* as follows:

A. CHINA

1640 B.C.:	The Chinese empress Si Ling Si, known as the "Goddess of the Silkworm" introduced sericulture in China. According to one legend, her husband Chinese emperor Chin Shih Huang Ti (Di), ordered her to find out the cause of destruction of his mulberry trees. She discovered white worms eating mulberry leaves and spinning shiny cocoons. She further discovered slender thread of silk on unwinding the delicate cobweb leaving behind the small brown chrysalid from the cocoons dropped in a hot cup of tea/water. For more than 2000 years, sericulture was monopoly of China and export of breeding material and divulsion of its secrets were prohibited at the cost of death.
1600 B.C:	Breeding of silkworms was started.
1200 B.C:	Chinese immigrants first started sericulture in Korea.
1200-1050 B.C:	Silkworm was domesticated.
1027 B.C :	Breeding programmes were well organised.
206 B.C:	Weaving and printing attained high degree of perfection (Golden age of Chinese Sericulture).
4th Century A.D:	Koreans deputed from Japan returned to Japan with four Chinese girls who taught the art of silk manufacture to emperor and people of Japan.
618-909 A.D:	Silk was exported to Rome through Silk Road and smuggled out to Constantinople (recent Istanbul).
645 A.D:	Hiuen-T-Sang, the great Chinese pilgrim wrote that the king of Khotan in Tibet requested the Chinese emperor to provide silkworms but his request was turned down. Later, the king married a Chinese princess who brought with her some mulberry seeds and silkworm eggs by concealing them in the lining of her dress in and around 140 B.C. and spread sericulture in Khotan in Sinkiang province. Another legend says that a Chinese princess married to Bucharia smuggled eggs of silkworms in her headdress across the frontiers of China.

From Tibet sericulture spread to the valley between Brahmaputra and Ganges river from where it slowly moved to west into Persia and Central Asia.

During 555 A.D, the Roman Emperor Justinian in order to stop the drain of gold from Rome to China, smuggled the secret of silk culture out of China through two Persian/European/Indian monks who brought silkworm eggs and mulberry seeds in a hollow stick.

B. INDIA

Hindu epics more than 2000 years old refer to silk. The history of sericulture in India is lost in antiquity. However, it may be chronologically summarised as hereunder:

4000 B.C :	Indo-Mongoloid and Tibeto-Burmese group of tribals of North East India practised eri culture even before migration of Aryans to India.
1400 B.C:	Traces of tasar silk was excavated from Nevassa in Maharashtra which indicate that the practice of tasar culture in India is as old as that of China.
321-296 B.C:	"Arthashastra" of Chandraguptas time distinguished cultivated and wild varieties of silk and mentions about production of silk in Magadh, Pundrya and Subarakundya. Lokucha (*Artocarpus lakoocha*), Vakula (*Mimusops elengi*) and Vata (*Ficus bengalensis* and *F. religiosa*) and Naga tree when fed to silkworms yielded silk of wheat, white, butter and yellow colour.

An Indian emperor sent gifts of silk draperies to a Persian King hundreds of years before Christ.

150 B.C.:	Mulberry trees flourished well in Kashmir and woven silk and silk shawls were popular among elite classes of Kashmir.
58 B.C.:	Raw silk was exported from India to Rome during the reign of Kanishka. Ramayana indicate that Lord Rama's nuptial gift to Sita contained 'tasar silk'.
629-645 AD:	Bhaskara, the great monarch of Pragjyotishpura sent silk clothes of all varieties including chinapatta (figured textile), Kauseya (Kosa textiles), Khsauma (fine silk) and Jattipataka (woven silk) produced in Kamlupa (Kamrupa, Assam) to King Harsha of Kanyakubja. The indigenous people of Assam of this period knew the art of producing silk from wild silkworms *viz.*, Doyang muga, Deomuga, Kotkari muga by feeding muga silkworms on digloti (*Litsaea salicifolia*), mejankari (*L. citrata*) etc.
12th Century A.D:	The rearer and reeler communities known as Jogi or Katoni migrated from Bengal to Assam reared silkworms and dyed silk from local vegetable products in collaboration with Singpho, Khamti and Fakial tribes of Assam.
16-17th Century A.D:	Sericulture industry gathered momentum under Moghul regime.
1603-1641 A.D:	The golden muga silk was deposited to the royal treasury for household of Pratap Singh, a king of North East India.

Sericulture flourished in Bengal under the patronage of Nawabs and even earlier.

1780-1790 A.D:	Tippu Sultan introduced sericulture in Mysore from Bengal.
1869 A.D:	Maharaja Ranbir Singh built 127 rearing houses in Kashmir.
1878 A.D:	Pebrine disease badly affected sericulture and the industry also suffered set back due to competition from imported silk from Japan.
1892-1910 A.D:	Sericulture was organised in Kashmir as an industry.
1900 A.D:	Six filatures were set up by East India Company in Bengal.
1914-1918 A.D:	Industry suffered due to World War-I
1930 A.D:	Trade centres at Surat and Mouslipatnam in West and East Coast, silk centres at Malda and Murshidabad and a filature at Patna were established.
1939-1945 A.D:	Industry suffered due to World War-II.
1945 A.D:	Sericulture was encouraged again.
1947 A.D:	India became Independent.
1949 A.D:	Central Silk Board Act was passed by the parliament.
1950 A.D:	Central Silk Board was established under the Chairmanship of Late Pandit Jawahar Lal Nehru, the first Prime Minister of India and country produced 900 MT of mulberry silk during early fifties.
2000 A.D:	India produced approx. 15000 MT of silk and acquired second position in the world next only to China and earned foreign exchange of Rs.1250 and 1502 crores during 1998-1999 and 1999-2000 respectively.

C. JAPAN

History of sericulture in Japan is also lost in antiquity, however few chronological events may be summarised as hereunder:

199 A.D:	One of the descendents of Chinese emperor Kama-o brought some silkworms to Japan and introduced sericulture there.
300 A.D:	Silk fabrics were manufactured.
4th-5th Century A.D:	According to another view, Japanese sent some Koreans to China who brought four Chinese girls to teach the art of silk manufacture to the emperor and Japanese people. The other view mentions that four Korean girls were brought to Krushu islands along with some eggs and sericulture was practised in large scale by queen on a Mikado named Yu-Lyak.
586-587 A.D:	Art of sericulture was taught by Sai-Sai, the prince of O-Mei to Japanese people.

900-922 A.D:	Sericulture became very important throughout Japan.
1500 A.D:	Silk manufacture attained highest peak.
1700 A.D:	Rice, silk and copper became most important commodities in Japan.
1869 A.D:	Pebrine disease spoiled sericulture in Europe and Suez canal was opened.
1870-1890 A.D:	Foreign exchange was earned by exporting raw silk to West.
1893-1914 A.D:	Filature dominated over hand reels.
1930 A.D:	Multiple reeling machines and power looms enhanced the silk production.
1942-45:	Japanese silk market was cut off due to World War-II and all the available silk was diverted to war use *viz.*, silk parachutes and silk powder bags.
1957-1986 A.D:	Japan became second largest silk producing country in the world.
1987-1996 A.D:	Japan became third and 1997 onwards silk production further declined and it has become fourth largest silk producing country in the world.

Silk Road

Silk Road had been the oldest and longest commercial route in the world, extending more than 10,000 km. The silk road was opened actually during 113-118 B.C. by the Han Emperor Wu Ti (Di) and Rome and China, the two powerful civilizations reigned at either end of the silk road which served as a bridge of culture and commerce that peaked in the Tang Dynasty during 618-907 A.D. The silk road was actually a perilous network of routes starting from Xian in Shaonxi Province, passing through Lanchau to Dun Huang and subsequently it was divided into Northern and Southern Roads.

The Northern Road passed through Turfan after crossing Pamir mountains and reached Ferghana and subsequently to Kazhakh steppe, while Southern Road traversed a barren crust of earth through tracherous Pamir mountain and Taklamakan desert across Central Asia through Yarkhand continuing up to Parthia, India to Anticch (Antakya) of modern Turkey. The last lap to Europe and Egypt was by water through Mediterranean Sea ports.

The Silk Road was hazardous to traders, monks and pilgrims carrying Buddhist teachings between India and China. Only few made the entire trip. Caravan loads were passed from trader to trader who intended to exchange gold, wool, horses, Jade and glass for silk at each oasis and stronghold and prices went up with each exchange.

The Silk Road was virtually abandoned during Ming Dynasty (1368-1644 A.D) and Alexandria became the centre of silk distribution when oriental trade routes were changed by sea routes.

D. EUROPE

The smuggling of silkworm eggs and mulberry seeds through the Silk Road from China to Constantinople (Istanbul) in 6th Century by two monks marked the beginning of silk production in Europe. In the 8th and 9th Centuries, Thebes, Greek and Syrian countries enhanced their silk production. Until 1204 A.D, Greeks maintained their supremacy and in 740 A.D sericulture was introduced into Spain. By the 13th Century, Italy emerged as the silk centre of the West and the Italian Silk Industry helped the Italian Economic Renaissance. By the 14th Century Italian sericulture spread to Lucca, Venice, Florence and Genoa cities.

In France, silk weaving was started at Tours in 1480 under Louis XI who took drastic steps to curb the tremendous outflow of money from France to Italy. In 1520 Francis-I, Father of silk industry in Europe, brought silkworm eggs from Milan for rearing in Rhine valley. He restricted silk imports and enticed Italian craftsmen with promises of more freedom in their work. The industry was firmly established by Louis XI who offered premiums for planting of mulberry. In 1854 a deadly plague (Pebrine) began killing silkworms. Pasteur investigated the problem for three years and discovered two diseases that he brought under control and also made correlations that led to his monumental findings on infectious diseases. In the late 18th Century, 18000 looms were under operation in Lyon but after French Revolution and World War-II, cotton superseded silk in demand and the silk industry was almost destroyed.

In 1598, sericulture was introduced into Germany during the reign of Frederick. In England, during 14th Century silk manufacturing began with planting of mulberry and rearing of silkworms was stimulated by James-I and Elizabeth-I during 16th Century. In 1825 attempts were made to introduce silk production into Ireland also but sericulture never became important in British Industry. Sericulture was encouraged in Russia by Michael-I in 1880 A.D. (Phillip 1988, Agarwal 1995).

E. AMERICAN CONTINENT

Sericulture in Mexico was introduced in 1522 A.D by Spanish conquistador Cortes. James-I introduced it in American colonies in 1609-1619 in Virginia in U.S.A. but it could not compete with cotton or tobacco and therefore silk production was abandoned. During 1833-35, a new mulberry variety was introduced due to frenzied speculation, which collapsed due to tedious labour demanding work. Until 1919 A.D. attempts were continued in California but sericulture proved impracticable in U.S.A. As Americans lacked in patient labour which is required in silk production, they diverted their drive to silk manufacture in 1800s in Paterson and New Jersy which also declined in 1913 due to strike in silk factories. In Brazil, sericulture was introduced during World War-II in Parana and Sao Paolo where Italian and Chinese silkworm races were crossed. In Argentina it was introduced in Cordova, Santa Fe, Medoz, Tacunan, Buenos Aires etc. Since 1997, Brazil has become third in silk production due to declining production in Japan.

F. SOUTH EAST ASIA

In South East Asia, sericulture was introduced during 11th Century and aboriginals of Sumatra, Java, Thailand and Myanmar adopted sericulture keenly.

G. MIDDLE-EAST

Sericulture was started in Iran during sixth century and great progress was achieved during 1500-1700 A.D. The soil and climate of Turkey and Cypress are well suited for sericulture and Cypress provided healthy silkworm seeds when Pebrine devastated silk industry in Europe (Agarwal 1995).

II. GLOBAL SILK INDUSTRY

Silk accounts for merely 0.2% of the total textile fibre production in the world (CSB, 1999) as compared to 43.6%, 45.8%, 6.8% and 3.6% shares of cotton, synthetic, cellulosic and wool fibres respectively (Table 5). Though silk is a tiny fibre in quantitative terms, yet its unit value is extremely higher than that of cotton due to its qualities. Silk is comfortable to wear in all the seasons as it is warm in winter and cool in summer, carries prestige, allows for brilliant colours and excellent in draping qualities. The only disadvantage of silk is that it can only be dry-cleaned, cannot be worn every day and difficult to iron and very expensive. The silk is perceived as being exotic, feminine, light, natural, seductive, sensuous and soft. Thus, the silk has very few disadvantages and hence silk has still a bright future. It has survived crises of competition, supply and price for 5000 years. The general guidelines on the positive aspects of silk have been enumerated by Ronald Currie, Secretary General, International Silk Association, Lyon, France (2000) as hereunder:

- Silk is a natural ecofriendly fibre, using few harmful fertilizers and no insecticides.

- Silk is the only mythical fibre with roots anchored in a 5000 year old history.

- The incomparable qualities of silk (softness, brilliance, drape etc.) are more important than its relative difficulties of care.

(a) Production

The world raw silk production has been very much fluctuating since 1938 when it was approximately 56,457 mt. which declined to 32,827 mt. during 1966, subsequently it rose up to 41,315 mt. during 1970. The raw silk production during 1938, 1983, and 1984 was 56,100-56,600 mt., hence silk industry remained stabilised during 1938-1984. When we study the production trends of silk over the last sixteen years since 1983 to 1998, we find that a consistent increase was noticed until 1995 except in the years of 1984 and 1988 which experienced a slight decrease from their previous years. The world silk production was only 56,600 and 62,381 mt. during 1984 and 1987 respectively. Subsequently, the world raw silk production consistently increased year by year and touched its peak during 1995 at 105,138 tonnes which showed an increase of 85.76% over the production of 1983. However, the global silk production after reaching this peak during 1995, started to decline sharply until 1998 during which 71,727 mt. was produced experiencing a decrease of 32% as compared to the production of 1995 (Table 6 & 7). Data of global silk production from 35 countries available for 1999 and 2000, reveal that the production of raw silk was 81,465 m.t. during 2000 which is 5.4 % more than that of the previous year (76,312

Sericulture and Seri-biodiversity

m.t.) during 1999 (Indian Silk Nov. 2001). This increase is mainly due to rise in production of silk in China and India which is heartening to the global silk trade community, worried over the shrinking silk supply in the recent days.

Table 6: Global raw silk production, export and import from 1938-1970 (unit: Metric Tonnes)

Countries	Raw silk production			Export of raw silk		Import of raw silk	
	1938	1966	1970	1966	1970	1966	1970
High Income Countries							
Bulgaria	180	248	245*	146	290	—	—
France	105	NA	NA	54	49	771	619
Greece	255	55	45**	—	—	—	—
Hungary	20	17	4	—	—	—	—
Italy	2738	593	310	384	589	1861	1851
Japan	43152	18694	20515	527	75	1138	3876
Poland	1	7*	7*	—	—	—	—
Romania	15	90*	70*	50	8	—	—
Spain	15	42	31*	—	—	34	11
USSR/Uzbekistan	1900	2644	2940*	—	—	—	—
Yugoslavia	45	50	18	32	—	—	2
F.R.G.	—	NA	NA	2	5	202	183
Switzerland	—	NA	NA	93	49	435	224
United Kingdom	—	NA	NA	21	23	210	185
U.S.A.	—	NA	NA	2	57	1687	457
Australia	—	—	—	—	—	1	1
Austria	—	—	—	—	—	1	1
Netherlands	—	—	—	—	—	—	2
Low Income Countries							
Brazil	33	135	259*	21	170	—	—
China Peoples Rep.	4853	7180***	11124***	3700	4800	—	—
Egypt	—	10	9	—	—	11	25
India	691	1502	2250	—	—	45	28
Iran	210	124*	210	—	—	—	—
Khmer Republic	—	20*	20*	—	—	49	10
Korea, Republic	1824	1154	2846	808	2036	—	—
Lebanon	21	9*	15*	—	—	—	—
Madagascar	—	9*	15*	—	—	—	—
Pakistan	—	40*	40*	—	—	—	—
Syrian Arab Republic	8	8*	11*	5	16	—	—
Thailand	—	50*	200	—	—	—	200
Turkey	213	114*	120*	85	46	—	—
Hong Kong	—	—	—	—	24	106	44
Total (A)	56279	32795	41304	6065	8237	6551	7719
Total (B)***	56457	32827	41315	6069	8237	6551	8107

* Estimated by FAO, ** Estimated on the basis of information from Greek Min.of Agriculture

*** Estimated from a Japanese source

**** Total includes estimate for other countries for which data are not available

Source: FAO Monthly Bulletin of Agricultural Economics and Statistics, Vol-21, No.12 Dec.1972.

Source-FAO 1976

Table 7: World mulberry raw silk production in tonnes

Country	1983	1984	1985	1986	1987	1988	1989	1990	1991	1992	1993	1994	1995	1996	1997	1998
China	28140	28140	32000	35700	35800	34380	40800	43800	48480	54480	69300	72000	77900	59000	55117	49430
India	5681	6895	7029	7905	8455	9300	10020	10200	10800	12600	13200	13200	12884	12927	14048	14048
Japan	12456	10800	9592	8341	7864	6840	6060	5700	5520	5100	4200	3900	3240	2580	1920	1080
Brazil	1362	1458	1558	1780	1780	1740	1680	1680	2100	2280	2340	2520	2468	2270	2120	1821
USSR/Uzbk	3899	3999	4000	4000	4000	4020	4020	4020	4020	2160	1800	1800	1320	2500	2000	1500
Vietnam	NA	NA	NA	NA	NA	NA	NA	NA	NA	NA	NA	NA	2100	1500	834	862
Thailand	NA	NA	NA	NA	NA	NA	NA	1503	1612	1589	1229	1377	1313	1144	1039	900
Iran	NA	NA	NA	NA	NA	NA	NA	381	385	423	427	396	750	600	500	400
S.Korea	2292	2088	1850	1650	1608	1320	1200	780	900	910	840	491	346	146	146	146
N.Korea	NA	NA	NA	NA	NA	1000	1000	1200	1300	1200	1200	1200	600	360	200	150
Others	2770	2720	2671	2874	2874	2660	2120	1719	1615	1677	2801	3504	2217	2165	1666	1438
Total	56600	56100	58700	62250	62381	61260	66900	70983	76732	82419	97331	100388	105138	85192	79590	71727

Source: Silkmans Companion 1992, Kumar and Das 2000, CSB 1999

Many reasons have been assigned for the cut back in global silk production during the last several years (Ronald Currie 1997, 2000, Kumar and Das 2000) which may be summarised as hereunder:

1. Silk production has been dominated by one major player China, which has voluntarily curtailed its 36% silk production and many Chinese farmers have switched over to more lucrative crops. The new policy to decontrol the agriculture sector has compelled the sericulture industry to face stiff competition from other crops and Chinese Government has also been finding difficulties in subsidizing the silk industry. Further, the Chinese Government is removing inferior qualities of raw silk from the market to offer better qualities at higher prices and currently paying more importance to vertical improvement than to horizontal improvement/expansion.

2. India, the world's second largest producer has not noticeably increased its production though Indian Silk industry looks stable and healthy.

3. Brazil and Japan, the other two largest silk producing countries have reduced their production levels and sericulture is slowly diminishing there due to high labour cost and rapid industrialisation.

4. New countries are not taking to silk production, to cover up the short falls.

5. A new synthetic product "sand washed" silk found a ready market during late eighties and early nineties among young people and men at affordable prices with a casual style, which suited them perfectly, but it disappeared quickly because it could not meet the qualities of natural silk.

Sericulture is practised as a small family avocation in India among the marginal and small farmers in the rural sector. Hence, it is less affected by the massive industrial development during 1990s. The efforts are being made to widen the production bases in African countries like Ivory Coast, Kenya, Nigeria, Madagascar and Uganda etc., South East Asian countries like Bangladesh, Philippines and Thailand etc. and Latin American countries like Brazil, Bolivia, Argentina etc., and many other developing countries.

The Chinese government has voluntarily reduced its production to remove inferior qualities of raw silk from the market for offering better qualities at higher prices. Due to membership of the World Trade Organisation (WTO), the raw silk imports of Japan will gradually have to be freed which will make Japanese sericulture less competitive and less profitable. As Indian manufacturers are modernising their plant and prefer to buy imported better quality yarn, Indian sericultural production is also not increasing and this trend will continue unless bivoltine silk production is promoted. The production in Brazil and Vietnam has also remained stable because world prices were too low to offer these countries a reasonable profit on their production. Hence, there is no doubt that the silk consumers in the world will be facing a shortage of raw silk supplies due to reduction in global production. This shortage of supply will in turn increase the price of silk in future and raw silk production will once again become an attractive venture.

The overall demand for silk is expected to increase as consumers of China and India are becoming more affluent. European economy is gradually gathering strength, United States economy is thriving well and consumers ought to renew their interest in silk. The Japanese young people have also acquired a taste for western style silk goods and hence probably they will continue to buy imported silk goods or silk goods manufactured offshore by Japanese companies. According to Ronald Currie, Secretary General of International Silk Association (1997) following steps should be undertaken by the silk producing and processing countries to facilitate silk consumption reach a high level and maintain its position:

1. New strains of silkworm need to be developed and reeling quality and inspection have to be improved to justify the higher prices and regain users confidence in silk.

2. New products of good quality in dyeing, printing and making up should be developed for higher level than the cheap super market and departmental store as their items have not been renewed and have become too familiar.

3. Research must be carried out to make silk more easily washable in water or detergents specifically designed for the care of silk and other delicate fibres should be developed.

4. The international silk mark should be strengthened by transforming it into a label of quality along the lines of "Wool mark" for which coordinated effort should be made by weavers, dyers, printers and makers up.

5. Instead of advertising, silk should be promoted by informing consumers about the qualities of silk, about its varieties, how to distinguish between various types of silk and how to clean and iron it.

III. PRODUCTION SCENARIO IN VARIOUS COUNTRIES

The silk production, consumption and situations prevailing in silk producing countries are not only distinct but unique as well. The production scenario in major silk producing countries (Tables 6 & 7) is briefly detailed as hereunder:

1. CHINA

China has been dominating in silk production for more than 3640 years. It served as second largest silk producing country during 1938-1970 when Japan was the largest silk producer in the world (Table 6) and became the first largest silk producing country subsequently. During 1983 and 1984 its production was 28,140 metric tonnes, which increased upto 77,900 metric tonnes during 1995 accounting for approximately 74% of the world silk production (Table 7). Over the past 40 years, the foreign exchange earned from silk export (90%) has accumulated to more than US $ 30 billions and silk is being sold to more than 130 countries and regions world wide. The ancient silk road has regained its former glory (Youzhe 1996).

The Chinese silk production consistently increased after 1984 every year barring 1988, which experienced a slight decrease of 1420 tonnes from that of 1987 (35,800 mt.). In the first phase of 1990s, mulberry silk production showed an upward trend till 1994 -95 when the land under mulberry (1.25 million ha.), *Bombyx mori* egg production (26 million boxes), mulberry cocoon output (7.7 lakh tons) and cocoon price (16.44 Yuan/Kg.) reached its highest historical records. The production of cocoons surpassed the demand and they were overstocked which led to decline in cocoon price resulting in substantial decrease in cocoon production to 400 thousand tons in 1996 (Li Long *et al.* 2002). Hence, after reaching a peak during 1995, its production has been consistently decreasing and during 1998 the production (49,430 mt.) experienced a decline of 36.55% as compared to that of 1995 due to the following government policies for balancing the demand and supply situations (Youzhe 1996):

i) Deregulation in weaving sector, which resulted in piling up of considerable fabric stock.

ii) European Unions (EU) imposition of quotas to regulate low price fabrics and made ups.

iii) Deregulation in agriculture sector posed stiff competition from other cash crops.

iv) Industrialisation and urban migration.

In 1999, the land under mulberry shrunk to 6.0 lakh ha., the silkworm egg brushing to 13.4 million boxes, accounting only 50% of 1994 and cocoon production decreased by 40%. However, productivity per box of eggs increased gradually in 1990s as it was around 25 Kg. during 1991 which gradually increased to 30.5 Kg. national average in 1999. The highest productivity was recorded in Zheijiang Province where it was 38.8 Kg. of green cocoons per box of eggs while lowest productivity was recorded in Sichuan and Chungqing areas in 1999. Mulberry silk production also followed the same trend and according to another estimate it reached its highest record of 87,767 tons in 1994 and fluctuated around 56,000 tons after 1997 (Li Long *et al.* 2002).

Production of quality raw silk at a reasonable competent price instead of high quantity low quality silk is now emphasised by Chinese Government. Chinese silk production is expected to continue at the same level for some more years. The Chinese Government is adopting precautionary measures by shifting production base to counter the production slide that may erupt due to industrialisation and migration.

According to Dandin (1998), inspite of present temporary trends towards decline of sericultural activities in China, the strengths and constraints of its silk industry could be summarised as hereunder:

Strengths

• Regions and seasons specific highly productive and resistant silkworm races.

• Organic manuring and assured moisture supply to mulberry gardens for sustained leaf productivity and quality.

- Perfect disinfection and disease prevention/control measures by application of effective chemicals specific to each pest and disease.

- Production, quality control and supply system in batches are highly monitored as they are under complete government control.

- Maintenance of high quality and uniformity of silk by reeling factories controlled by corporations through mass cocoon reeling.

- Well developed and perfected post-yarn processing and finishing technologies to meet the specific market demand.

- Increase in the profit margin and improvement in economic viability of silk industry through effective utilization of bye-products right from mulberry cultivation to fabric finishing at various levels.

- Field testing of technologies and their transfer to the field with more of field oriented research through well established system.

- High level of technical knowledge of all the field extension staff and perfect coordination between research and extension activities.

Constraints

- Only export market oriented production and supply due to poor domestic consumption (10-15%)

- Pushing sericulture activities aside due to rapid industrialisation in major silk producing areas.

- Scanty linkages and coordination between various organisation *viz.*, Agriculture Ministry, Textile Ministry, Trade Promotion Ministry, Provincial Corporations and county level functionaries of local organisations.

- Extensive expansion of sericulture to non-traditional and new areas for which production is yet to be confirmed.

Thus, the major constraints being faced by Chinese silk industry may be of policy and market aspects and not of technological in nature.

According to Youzhe (1996) Chinese silk industry will surely have good prospects under following conditions:

- The Central and local governments at various levels should attach importance to the problems of the silk industry and adopt measures actively.

- The local governments at various levels and the different departments of the silk industry should correctly deal with their respective benefits.

- They should consciously safeguard the integral benefits of the silk industry.

- They should together constitute a favourable environment for healthy and stable development of silk industry.

- China National Silk group integrating agriculture, industry and trade, with trade as the dragon head should be established as the final objective of this reform to unify understanding and restore the prestige of Chinese silk.

Finally, while China is leading producer and exporter of silk, Europe with Italy in forefront is the leading processor of silk and the driving force in the development of global silk industry. The silk economics of both countries are complementary to each other hence, bilateral cooperation and exchange will help to restructure and enhance the silk industries not only in both the regions but shall also lead to overall development of the global silk industry. China has been actively seeking to join W.T.O. To make Chinese economy compatible with world economy, China is stepping up the effort on internal market liberalization. Control of silk materials export and export of grey fabrics have already been liberalized. Chinese silk industry is now operating with market rules and in accordance with internationally accepted practices. New mechanisms have been established to favour cooperation between China and the capital countries. The advance technology and the art designs in foreign countries may be brought to China by other countries, production and processing bases may be set up there and their products may be sold in the world market to mutually benefit each other (Indian Silk Dec. 2001, ISA Newsletter July-Aug. 2001).

2. JAPAN

Japan served as the largest silk producing country during 1938 to 1970 and subsequently became the second largest producer of silk. However, it became third since 1987 to 1997 and subsequently it has become fifth largest producer of silk in the world. During 1938, 1966 and 1970 Japan produced 43,152, 18,694 and 20,515 mt. silk respectively. During 1983, the raw silk production of Japan was reduced to 12,456 metric tonnes, which has further come down to 1,080 metric tonnes by cutting back its 91.33% production during 1998. Its production has been found to decline consistently on account of its rapid industrialisation, which has negative correlation with the silk production (Tables 6&7). Japan now produces silk for meeting its 7% domestic requirement while 93% of its silk requirement is fulfilled by import of raw silk or silk products. Besides industrialisation, other main reasons which caused decline of silk industry in Japan are high labour cost, urbanisation, reduced land availability and lack of interest among the younger generation in sericulture.

In Japan, the raw silk production in fact decreased by 17% in 1995 due to the fact that the domestic cocoon production, raw silk production, number of farms, mulberry field and the number of spinning factories in 1995 decreased by 31%, 17%, 28%, 24% and 20% respectively as compared to the previous year. During 1995, the import was increased by 29% over the previous year while total supply of raw silk including the imported raw silk dropped by 6% due to decrease in domestic production. Interestingly, the domestic demand/delivery has also decreased by 14% during 1995 compared to previous year, which resulted in more stock piling of silk by 23% compared to previous year. The production as well as consumption of silk in Japan continued to dip during 2000 also. The cocoon production which stood at 1244 m.t. recorded 20% drop as compared to 1999 due to slump in cocoon prices and ageing sericulture workers which also decreased number of sericulture farms by 20% in the country. Japan produced 556.8 m.t. of raw silk during 2000 which is less by 14.3% over that of 1999. In the year of 2000, the supply of overall raw silk including imports and opening stocks was 4502 m.t. which is less by 4.4% than that of 1999 owing to drop in domestic demands. The import of raw silk also dropped during 2000 and stood at 2298 m.t. China and Brazil exported silk (70%) to Japan (Indian Silk Nov. 2001, ISA Newsletter Sept. 2001). Hence, the Japanese silk industry also suffered damage because of the decline in consumption, sudden increase in import of rich and graceful silk made ups of European style for ladies as well as women and increasing financial loss to the domestic industry (Indian Silk March, 1987 and ISA News Letter). Owing to the import of less expensive raw silk, cocoons as well as made ups, the price of the domestic raw silk, cocoons and silk products remain low in comparison to other farm products. Moreover, Japenese sericulture traditionally had a close connection with paddy cultivation and was a means of absorbing surplus labour thereby providing additional source of income to the farmer/labourers. Unlike paddy and other agricultural crops, sericulture requires longer working period and makes it difficult for silkworm rearers to do other jobs, hence the number of regular sericulturists is also showing a downward trend. Now, Japan is making earnest efforts to innovate its sericultural technology further and to develop new silk materials such as hybrid silk for inner wear and spun raw silk for outer wear which will help the country to conserve and develop its traditional silk technology and culture to open up new areas as well (Kumar and Yamamoto 1996).

The cocoon production in Japan is decreasing in the order of 25-30% each year. Hence future trends in sericultural research in Japan are towards innovative ways to utilize the basic biological functions of the silkworm for development of new material, which will enhance the quality of human life and economic strength. Presently, scientists in Japan are engaged in development of low cost artificial diet for silkworms; manufacture of silk powder, paint, silk leather and food; silk film and regenerated fibres for pens and healing of wounds, injuries, burns etc. and paper silk for beautiful flowers, lamp shades etc. Generally silkworm cocoons possess 75% fibroin and 25% sericin. Silkworm races with more sericin are being developed which do not have fibroin. Sericin is widely used in cosmetics for manufacture of silk powder and has anti oxidizing property, hence it is also used in medical fields. Many coloured sericin cocoon races and paper silk have been synthesised on the basis of cocoon races like white, yellow, green etc. Silkworm breeds with superfine silk filaments and new silk materials have recently been developed in Japan.Three silkworm hybrids Akibono, Hakugin and Honobono with thin silk filaments have been developed by Dr. T. Yamamoto. Hakugin is characterized by 1.65 denier while Honobono is characterized by 2.2. denier of filament as compared to 2.8 to 3.0 denier of silk filament size of majority of silkworm races. The filament size of hybrids has been further brought down to 1.36 and 1.72 denier by inducing trimoulting through treatment of III instar larvae with triflumizole. Filament size of 0.99 and 1.03 denier could also be achieved by inducing trimoulting in these races through treatment of IV instar larvae. Silk produced from these races is utilized for inner wear. Hybrid silk where synthetic fibre is wrapped in cotton filament during reeling is utilized for pantyhoses, socks and underwear in western style. The coarse raw silk is utilized for outerwear like jackets, sweaters etc., silk tow is used for blouses and silk wave is used as bed clothes, pads etc. (Singh *et al.* 2002).

3. INDIA

Sericulture has indeed been a tradition in India and its history is lost in antiquity. During 1938 it ranked sixth, subsequently it has attained fourth position in 1966, fifth in 1970, third in 1980 and became second largest raw silk producing country in the world since 1987 (FAO 1976, CSB 1986, 1999, Kumar and Das 2000, Tables 6&7). As situation in India is entirely different from other countries, the silk industry has been slowly but steadily growing during the past several years (Tables 9-14, 16). Mulberry raw silk production increased from 625 m.t. in 1951-52 to 14260 m.t. during 1998-99 and 13944 m.t in 1999-2000, while tasar, eri and muga raw silk production increased from 124, 100 and 45 m.t. during 1951-52 to 211, 974 and 85 m.t. during 1999-2000. Total raw silk production increased from 894 m.t. during 1951-52 to 15544 m.t. during 1998-99 and 15214 m.t. during 1999-2000 which is 17.39 and 17.02 fold increase respectively (Table 10). The growth of silk industry in India is due to strong domestic market. The saree culture of India and usage of silk is related to the cultural heritage and traditions of India. No ritual is supposed to be complete unless and until silk garments and dresses are worn. It is anticipated that there would be no major market problems for silk industry in next few decades.

Despite of producing 2600 m.t. raw silk, India imported 26 m.t. silk during 1971-72 which rose upto 2346 m.t. during 1997-98 and 6936 m.t. (Provisionally) during 1999-2000 though 15,236 and 15,214 mt. of (mulberry and non-mulberry) raw silk was produced respectively (Table 12). Hence, the consumption of silk in India is more than that of its production and both demand as well as production have been steadily increasing. According to another estimate, the consumption of silk in India was about 16500 m.t. while export of silk goods was of 2700 m.t. during 1997-98. The trends show that gap between demand and supply of silk in India will continue to increase in future also due to customary saree culture which may be encashed by proper planning and management. The healthy and stable steady growth of Indian silk industry is also due to practice of sericulture as a small family avocation by the marginal and small farmers in the rural sector which was least affected by massive industrial development witnessed during 1990s (Kumar and Das 2000).

During 2001-02, the total export earnings of all silk items amounted to Rs. 2235.38 crore as against Rs. 2421.98 during 2000-01. Hence, silk export was reduced by 7.7% during this period in India. The quality of raw silk imported is grade 2A and above. During 2001-02, 6797 mt. raw silk valued Rs. 620.78 crore was imported into India as compared to 4713 mt. valued Rs. 475.15 crore imported during the corresponding year of 2000-01. 6306 mt. (92.8%) of raw silk out of 6797 mt. imported during April-March 2001-02 was imported from China only. The total import of raw silk including silk yarn and fabrics during the period April-

March 2002 was Rs. 791.94 crore as compared to previous year's (2001) which was 566.93 crore (Indian Silk, August, 2003).

4. BRAZIL

Inspite of entering recently into the global silk club, the silk production of Brazil has been increasing inconsistently between 1938 (33 mt.) to 1998 (1821 mt.) [Table 6&7]. Its silk production increased from 1362 mt. during 1983 to 1780 mt. during 1986, remained stabilized upto 1988 by producing 1740-1780 mt., declined during 1989 and 1990 by producing 1680 mt., subsequently again enhanced its production from 2100 mt. during 1991 to peak production of 2,520 mt. during 1994 and subsequently consistently declined its production up to the level of 1821 mt. during 1998 which shows a decline of 28% of its silk production as compared to 1994 but it actually increased its production by 54.18% as compared to base year 1938. Since domestic consumption is very less (10%), majority (90%) of the raw silk produced is mostly exported. Presently, only three corporate companies are engaged in silk industry though earlier eight companies were coordinating sericulture and allied activities (Kumar and Das 2000). According to a study, out of 51000 people (9500 families) said to be working in silk sector, 4100 are children aged between 15-17 years and about 89% of them are going to school. Nova Experanca, the biggest cocoon producing city in Brazil, has 7500 children of 7-14 years of age living on silk rearing but 93% of them go to school Similarly, 89% of the rural children from silk rearing families attend schools as against the national average of 80%. These children do assist their families without affecting their regular education and hence no exploitation, as described in the North American guidelines for Children's Agricultural Tasks issued by the National Childrens Centre for Rural and Agricultural Health and Safety and accepted by several NGOs world over including Council of Economic Priorities, USA, Fair Trade Labelling Organisation, Germany and Business for Social Responsibility, USA (Editor, Indian Silk March, 2001).

5. PAKISTAN

Sericulture is an age old industry in Pakistan as it falls in the 6000 miles Silk Road running from China to Europe. This has originally been in practice in the provinces adjoining Punjab and in Sind, where sericulture was introduced in 1975-77. In 1987 the cocoon production of Pakistan stood at 2,10,000 kg out of which 1,86,000 kg (about 89%) were produced in Punjab and adjoining provinces while 24000 kg cocoons (11%) were produced in Sind, where commercial production started in 1977 and about 500 families were trained in silkworm rearing by 1991 (Sengupta 1991). The rearing has been carried out by and large only once in a year during spring season. Therefore, sericulture could not find a firm footing on farmlands as farmers were reluctant to engage their land for single crop.

6. TURKEY

The raw silk production was 100 metric tonnes during 1993, 80 mt. during 1994 and 40 mt. during 1995. Status of sericulture in Turkey afterwards are not available but the trends show declining trend of sericulture (Ronald Currie 1997).

7. COLOMBIA

Sericulture in Colombia started about 30 years back and the entire project was focussed towards Japan who were sole buyers of dried cocoons at that time. However, in 1984 Japan stopped the imports and during 1991-92 the prices fell at the international market hence sericulture landed into crisis in Colombia. The initial enthusiasm, with the construction of reeling plants generated an interest for silkworm rearing with an area of 600 ha. of mulberry in 1991, however, now the hactarage has declined by 50% as plants have reduced from 50,000 to 25,000. After repeated good performance during 1992-1995, the production dropped drastically. However, since 1998 a come back in the cultivation of mulberry has been seen due to good sales of cocoons. A sign of hope and recovery has been seen last year (1999) with a production of 62 mt. of cocoons, which represents a 78% increase in comparison to 1998 when 35 mt. of silk was produced. (Indian Silk Sept. 2000)

8. USSR/UZBEKISTAN

Major republics of erstwhile Soviet Union *viz*. Azarbaizan, Georgia, Russia, Tazakistan, Ukraine and Uzbekistan are practising sericulture in large scale. The erstwhile USSR was fourth largest raw silk producing country during 1938 (19000 mt.) and subsequently became third largest silk producing country during 1966-70 (2644-2940 mt.). Despite of the fact that Uzbekistan is a traditional producer of silk, its raw silk production has been decreasing gradually due to reduction in domestic consumption besides quality constraints to meet the international requirements. During 1983, the raw silk production of USSR was 3899 mt. and it was fourth largest silk producing country till 1988-1991 by producing 4020 mt. However, afterwards the silk production of Uzbekistan has been consistently decreasing and during 1998 it has come down to the level of 1500 mt. only though it has still maintained its status at fourth level globally (Tables 6&7). In Republic of Ukraine also, sericulture is currently practised in 18 out of 25 regions with a production of 1500 mt. of raw silk. Although 40,000 ha. land is covered under mulberry and 1800 farmers have adopted sericulture who earn 1000-1200 roubles/crop/box (Mukherjee 1999).

According to ISA Monthly News letter No. 230 (Source: Indian Silk 41(3), pp2) sericulture industry in Uzbekistan was always one of the main contributing factors of Uzbek economy; but industry started sinking with the collapse of USSR. While in 1991, the country produced 2,418 mt. of raw silk and 79 million meters of silk textiles, in 1998 only 548 mt. of raw silk and 8.8 million meters of textiles were produced. However, since last two years, the state is aimed for revival of sericulture step-by-step. Today, sericulture of Uzbekistan is represented by the Association of "UzbekIpagi", which was created by Govt. in 1998. It includes holding company Pilla Holding and the joint-stock company Shoi, which comprises of all the firms related to silk industry of the republic. In total, there are 215 associations in the country.

Accordingly to the data of the association in 2000, it was expected that 22,500 of green (living) cocoons and 1100 mt. of raw silk yarn are produced; an increase of 1.2 and 1.5 times, respectively over the last years index. By 2005, the production of the high quality grain (eggs) of domestic and import selection, which is highly competitive with the best world analogues in quality, will reach one million boxes. Since April 2000, the export of raw silk and silk waste has been prohibited to help increased processing of raw material in the home market and the exports will be carried out only in hard currency.

Uzbekistan Government has made an agreement with South Korea on joint implementation of a project to re-equip all the filatures and silk spinning enterprises of the association at a total cost of US $ 15 million with the proposed modernisation, the total capacity of all the spinning firms will increase from 1800 mt. to 2700 mt. of raw silk yarn threads per year. It will also increase the export potential of the industry to $ 30 million by helping in the production of 2A-3A quality of raw silk.

The Uzbekistan government is encouraging privatisation of the silk sector to attract private and foreign investments. Today, UzbekIpagi had four joint ventures including largest ones like Uzbek-Korean Tonmen and Uzbek-Japanese Silk Road. The Tonmen, with a project cost of $ 500,000, plans to process defective cocoons and export silk yarn worth $ 3.5 million. The Silk Road on the other hand, at a project cost of $ 7 million, proposes for production of silk yarn.

9. ITALY

Italy was third largest silk producing country during 1938 by producing 2738 mt. but declined during 1966-70 by producing 593-310 mt. silk though its domestic consumption was 4-6 times more. However, now it produces only a negligible amount of silk.

10. KOREA

Sericulture is an ancient industry in the Republic of Korea (South Korea) where it was introduced from China about 3000 years ago and played an important role in Korean economy. Republic of Korea (South Korea) was fifth largest producing country during 1938 and 1966 when it produced 1824 and 1154 m.t. raw silk respectively. During 1970, it became fourth largest silk producing country by producing 2846 m.t. raw silk.

The silk production of South Korea has been declining consistently since 1983 (2292 m.t.) to 1998 (146 m.t.). Now, South Korea stands at tenth position level globally. On the contrary, North Korea produced 1000 m.t. during 1988 and 1989 which rose up to 1300 m.t. during 1991 and subsequently its raw silk production observed a declining trend. During 1998, North Korea produced only 150 m.t. and it was ninth globally.

The silk industry consistently exhibited an upward trend in 20th Century barring the period of second world war. During 1976, the production of dried cocoons reached a peak above 41000 m.t. at an average of 65 kg/100 dfls and income from sericulture accounted to approximately 18% of the total agricultural income. Since 1977 the production declined consistently due to rapid industrialisation, lesser income generation and decrease in agricultural population. The import of relatively cheap Chinese cocoon also affected the industry and it faced almost extinction during 1995 as dry cocoon production was merely 210 m.t. and size of mulberry garden also reduced to 3200 ha from 14,900 ha. in 1989. The production scenario between 1989 to 1999 indicate that number of sericulture households, mulberry hectarage, number of boxes of eggs, cocoon production have declined from 36000, 14900 ha. 187000, 5500 m.t. in 1989 to 3796, 1700 ha., 45,682 to 1.0 m.t. respectively (Sharma and Sinha 2000). The modest increase in sericulture was noticed after 1995 due to the finding of the glucose lowering effect of powdery silkworm in the year 1995 by National Institute of Agriculture Science and Technology, Suwon, South Korea indicating that the powdery silkworm found a large role as functional food for diabetics. Silkworms produce sericin in silk glands just after 3rd day of the 5th instar, hence they are dried to get silkworm powder containing sericin which is a non-digestible protein.

The silkworm powder prepared out of the larvae fed with mulberry leaves was more effective in lowering down the glucose level in blood as compared to that prepared from hungry larvae. This finding led to further research on the properties of mulberry leaves which revealed that regular use of mulberry leaves for about 90 days or more may help in controlling blood glucose level in the body. It has anti oxidation property also hence its regular use may result as ante ageing. Besides, mulberry leaves are also very effective in controlling high blood pressure, lowering cholesterol level, anti cancer effect, heavy metal absorption in the drink, relief of arteria sclerosis and prevention of hyperlipemia. Young mulberry leaves collected in spring season, washed, dried and powdered have been found more effective in medicinal properties and hence consumed as a drink in cold water without boiling. Now sericulture is preserved more for food than dress in South Korea (Sharma and Sinha 2000). In South Korea a paradigm shift from traditional fabric making to other uses as food, industrial and medicinal products has been noticed since 1995 due to which sericulturists are now earning 5 times more income than the conventional cocoon production alone. The dried hydrolysed silk protein could be used as coating over the foam to give a good touch of silk leather, the film of fibroin silk protein is greatly used as artificial skin in plastic surgery and fibrous sclero protein is used in making digestible surgical threads. Besides sericin is also used for making cosmetic products, soap and silk wine. Non-traditional sericulture products of South Korea have great demand all over the world.

Scientists at Natural Product Research Institute, Seoul National University and National Sericulture and Entomology Research Institute, South Korea have developed the methodology to grow Cordyceps, a rare miraculous fungi on the body of silkworms during 1996. Cordyceps, a plant of ergot family commonly known as

"Winter Bug Summer Fungus" naturally occurs in China, Tibet and Nepal at an altitude above 10,000 feet. The sourcing and gathering of this herb is very rare so its supply often falls short of its demand and its price is more than its weight in gold. Cordyceps is used against fatigue, ageing and cancer. Besides, the amino acids obtained by biochemical resolution have been found to terminate hang over and activate liver function. The Cordyceps production reached to 6 m.t. in 1997 and increased to 30 m.t. in 1999 in South Korea. The sericulture farmers in South Korea are now earning 5 times more for producing silkworm powder and 6.5 times more for producing Cordyceps as compared to traditional cocoon production (Sharma and Sinha 2000).

11. VIETNAM

Though Vietnam is a traditional producer of silk, yet data of its production have become available only w.e.f. 1995 when it produced 2100 m.t. Subsequently, its production declined consistently upto 862 m.t. during 1998. However, it is trying to stabilize its silk industry to increase the production level to 2000 m.t. in next five years since domestic demand is increasing and prospects for export are reasonably good.

12. THAILAND

Thailand is a traditional producer and since 1990 it has ranked fifth to seventh largest producer of silk but its production was merely 50 m.t. in 1966 and 200 m.t. in 1970 which rose to 1503 m.t. during 1990, 1612 m.t. during 1991 and 1589 m.t. during 1992. Subsequently, its production started declining and it produced 900 m.t. during 1998. It appears that its production would be stabilised around 1000 m.t. annually which can meet 60% of the domestic demand. Hence, it has to import raw silk for domestic requirements. The country cannot enlarge its silk production base due to lack of land, limited water availability, industrialisation and high labour cost.

13. IRAN

Iran started sericulture in eighties. Its production was 381 m.t. during 1990 which rose to 750 mt. during 1995 and subsequently declined to 400 m.t. during 1998. It hardly meets its domestic demand for the carpet industry and due to socio-economic and agroclimatic conditions it cannot play a vital role in global silk industry.

According to ISA News Letter No.230 (Source: Indian Silk 41(4), pp 28-29), Iran has a long history of sericulture of many years before Christ and this artistic skill is linked with the culture of Persian people as a sacrosanct, respected activity some archaeologists believe that Iran is the origin for yellow cocoon production. Silk production in Iran reached its peak about 400 years ago, during the Safavion dynasty with an annual 3000 mt. of raw silk. Later on due to Pebrine disease, sericulture almost became extinct and silk production remained at low level till the pre-revolution period (1979). Now the present government of Republic of Iran realised the importance and potentials of silk industry and established an Iran Sericulture Company (ISC) in 1980 to provide developmental support to the silk industry and carpet industry in particular, as a social welfare measure to improve the socio-economic conditions of the farmers.

The sericultural activities of Iran are broadly categorised into two— Agriculture site and Filature site. Agriculture site is under the supervision of the Ministry of Agriculture Jihad. With the Islamic revolution in 1979, farmers were put under more care. There are two governmental farms in Gilan province, one each in Mazandaran and Khorasan for production and supply of domestic silkworm eggs. These farmers rear the parent silkworms using about 700 ha of improved mulberry fields.

Filature site, under the Ministry of Industries include one Government and three private reeling factories. At present, there are no facilities for silk weaving in Iran. On the contrary, the raw silk produced is exclusively used for carpet weaving.

Mulberry varieties in Iran generally comprise of local varieties like *M. alba* and *M. nigra* species.

At present, the annual demand for hybrid silkworm eggs in Iran is about 2,00,000 boxes. ISC, succeeded in meeting this demand through seven equipped silkworm egg production centres and their total production capacity is 2,40,000 boxes of hybrid silkworm egg production/year. About 80% of total annual production of bivoltine silkworm eggs is obtained during spring rearing season while the rest is produced in autumn crop. Nearly, 80,000 families are engaged in silkworm rearing. Now the average cocoon yield/box has jumped from 25 kg to 27-30 kg. At present the total production is about 6000 mt. of fresh cocoons.

Majority of the cocoons produced in the country are used for carpet weaving. Total raw silk production of Iran is about 800-850 mt. which is used for weaving of about 160,000 sq.mtr of silk carpet. The international price for Iranian silk is about US $ 1000/meter.

14. BANGLADESH

Sericulture in Bangladesh has seen many ups and downs and in the middle of twentieth century, raw silk production declined to merely 0.05 m.t. Currently, the total raw silk production of Bangladesh is 30 m.t. per annum; whereas the consumption/demand is estimated to be 180 m.t. Bangladesh imports silk mainly from China, India and Vietnam (Indian Silk, February, 1994)

15. INDONESIA

Silk in Indonesia is known by its Sanskrit name Sutera and its production is concentrated in the islands of Sulaveshu and Jawa. Although sericulture in

Indonesia has age old history, modern mulberry sericulture started only in 1950 with the Japanese support by the war veterans and in last 50 years bivoltine sericulture has established itself in the country. During 2000-01, mulberry plantation in Indonesia was around 12,500 hectares, producing around 750 tonnes of cocoons and 110 tonnes of raw silk yarn. On the other hand development of commercial wild silk production has started only during 1994-95 in Yogyakarta province in Jawa island. Popular wild silkworms under commercial production are (a) *Attacus attalus* L. feeding on Avocado, Mahogany, Soursoup and Guava trees (b) *Cricula trifenestrata* help feeding on Mango and Cashew trees.

Cricula is a major attraction in Indonesia as it produced golden coloured net like perforated cocoons. Though the exact production of wild silk in Indonesia is not known, it is estimated that around 250 tonnes of cocoons are annually available from these defoliators in nature with a potentiality to produce about 15 tonnes of spun silk yarn or wild silk handicraft products, if properly cared. Wild silk cocoons produced in Indonesia are unreelable and hence are subjected to spinning of hand spun yarn from both *Attacus* or *Cricula* cocoons.

Cricula silks are excellent materials to produce silk handicrafts. The wild *Attacus* silk is used in manufacture of heavy weight furnishing material mainly for bed and cushion, while *Cricula* wild silk is used in golden fabric for manufacture of ladies dress materials of higher value. It is also used for preparation of silk sheets for interiors, used as wall papers, screens and handicrafts. The *Cricula* shell is used in production of handicrafts like artificial flowers in natural golden colour and various colour combinations. The cocoon ornaments for hairs and dresses are popular among women.

The sericulture industry (both mulberry and wild) in Indonesia has also generated employment for about 75000 persons in reeling, spinning, weaving, batik printing and handicraft production and is the source of earnings for 12,500 families. (Mishra 2002)

16. OTHER COUNTRIES

Bulgaria, France, Greece, Turkey, Italy are the other silk producing, Romania, Spain, Yugoslavia, Switzerland, Pakistan, Khmer Repbulic, Egypt and Lebanon are the other silk producing countries which once took active role in global silk production during 1938-1970. In recent years the silk production in these countries has declined drastically due to one or the other reasons.

While attempts are being made to widen production bases in Bangladesh, Thailand, Philippines, Uganda, Nigeria, Kenya, Ivory Coast, Madagascar, Ghana, Bolivia, Brazil and many other developing countries, serious efforts are also being made to introduce sericulture in Myanmar, Nepal, Africa and Latin American countries with the assistance of international organisations. During 1988, other countries produced only 4.34% raw silk which has now declined to only 2% of the global silk production.

Table 8: Import of raw silk by major countries

Year	India *	Italy	Japan	West/ united Germany	France	U.S.A	Switzerland	Total
1980	366	1946	2976	92	604	233	179	6396
1981	319	2136	915	70	493	302	149	4384
1982	611	3450	2295	106	726	302	153	7676
1983	1098	3558	2229	104	640	336	156	8121
1984	1573	3270	1521	121	598	338	147	7568
1985	1181	4528	2098	132	756	285	201	9181
1986	1767	4631	1957	112	442	260	161	9330
1987	1745	3890	1457	105	505	305	178	8185
1988	2069	4044	1957	145	523	191	184	9113
1989	1395	4425	2048	155	757	294	190	9264
1990	1380	2980	2122	683	498	178	175	8016
1991	1598	2432	2756	680	271	162	149	8048
1992	2076	2712	1560	789	347	153	126	7763
1993	2768	2824	1484	545	323	184	43	8171
1994	4892	3814	1543	NA	581	182	55	11067
1995	5403	3252	1992	NA	434	122	NA	11203
1996	4149	2928	2076	NA	422	276	36	9887
1997	2911	NA	2242	NA	423	NA	NA	5576 (p)
1998	2346							
1999	2827							
2000	6936 (p)							
% increase/ decrease during 1997 over 1980	695.35	50.46	-24.66	492.39	-29.97	18.45	-79.89	—

Source: International Silk Association, Sericologia 1997
* Refers to financial year 1997 implies 1st April 1996 to 31st March 1997
NA : Not Available   P : Provisional
Adopted from C.S.B. 1999

IV. FUTURE OF GLOBAL SILK INDUSTRY

The fact that silk has been around us and survived for more than 5000 years by passing through the crisis of competition, supply and price proves that silk will continue to have a lot of appeal in the consumers market in future.

A perusal of export vis-a-vis import data by various countries indicates that the total export by all silk producing countries increased from 6069 m.t. during 1966 to 8237 m.t. during 1970 which shows that the export increased by 35.72% during the said period (Table 6).

Table 9: Indian Sericulture/Silk industry at a glance

Particulars.	Unit	1991-92	1995-96	1997-98	1998-99	1999-2000
1. MULBERRY SECTOR						
a. Mulberry area	Hectares	331237	286496	282244	270069	227151
b. Production of:						
i. Silkworm eggs/layings	Lakh Nos.	3183.93	2856.2	3355.2	3318.62	3309.8
ii. Reeling cocoons	Tonnes	107153	116362	127495	126565	124531
iii. Raw silk	Tonnes	10658	12884	14048	14260	13944
iv. Silk waste	Tonnes	—	4022	4250	4250	4153
2. NON-MULBERRY SECTOR						
Production of:						
i. Tasar reel cocoon	Lakh kahan	—	1.78	2.93	2.17	1.84
ii. Tasar raw silk	Tonnes	329	235	312	242	211
iii. Eri reel cocoon	Tonnes	—	1248	1357	1529	1553
iv. Eri raw silk	Tonnes	704	745	814	970	974
v. Muga reel cocoon	Lakh Nos.	—	3720.84	3371.72	3540	4260
vi. Muga raw silk	Tonnes	72	86	62	72	85
vii. Non-mulberry silk waste	Tonnes	—	208	250	253	248
3. Production of:						
i. Spun silk yarn	Tonnes	276	375	358	429	505
ii. Noil yarn	Tonnes	252	219	273	160	224
4. SILK FABRICS						
i. Quantity (estimated)	Lakh sq.mtr	2105.01	2732.97	2654.01	2757.36	3256.14
ii. Value (estimated)	Crore Rs.	318.17	4257.99	5150.70	6475.18	8561.64
5. IMPORT OF RAW SILK						
i. Quantity	Tonnes	2076	4149	2346	2827	6936(p)
ii. Value	Crore Rs.	147.42	—	189.4	—	—
6.CERTIFIED EXPORT OF SILK GOODS						
i. Quantity	Lakh sq.mtr	386.88	492.56	402.23	363.24	405.03
ii. Value	Crore Rs.	670.98	845.16	904.42	1003.76	1239.53
7. FOREIGN EXCHANGE EARNINGS	Crore Rs.	675.56	920.03	1060.16	1250.55	1501.78
	Mn.US $	—	274.88	285.22	297.04	346043
8. INSTALLED/ LICENSED SPINNING CAPACITY						
i. Spun silk yarn	Spindleage Nos.	30900	46480	33520	37464	38728
ii. Noil yarn	Spindleage Nos.	5140	7450	4930	4930	3930
9. NO.OF REELING UNITS						
i. Filature/cottage basin	Nos.	—	28655	25648	25649	25785
ii. Charka	Nos.	—	30590	32697	33742	34794
10. NO.OF LOOMS						
i. Handlooms	Nos.	227701	227701	227701	227701	227701
ii. Powerlooms	Nos.	—	29340	29340	29340	29340
11. EMPLOYMENT	Lakh persons	—	59.50	60.57	62.14	66.11
12.FINANCIAL ALLOCATION						
i. State sector	Lakh Rs.	—	12152.19	8453.09	10211.16	10044.59
ii. Central sector	Lakh Rs.	—	91.6	50.0	62.0	66.0
13. EXPENDITURE						
i. State sector	Lakh Rs.	—	9351.07	6735.37	7102.93	6572.13
ii. Central sector	Lakh Rs.	—	66.70	49.98	62.99	72.45
14. PRICES:						
A. Mulberry:						
i. Reeling cocoons	Rs./kg.	—	108.97	110.46	119.58	108.26
ii. Raw silk	Rs./kg	—	1023	1037	1126	1015

Particulars.	Unit	1991-92	1995-96	1997-98	1998-99	1999-2000
iii. Silk waste	Rs./kg	—	55-120	70-320	100-350	120-250
B. Tasar:						
i. Reeling cocoons	Rs./kg	—	600-1250	875-1600	710-1425	730-1150
ii. Raw silk	Rs./kg	—	1000-1275	1250-1415	1100-1310	1000-1300
C. Eri:						
i. Reeling cocoons	Rs./kg	—	145-180	135-180	135-150	110-200
ii. Raw silk	Rs./kg	—	400-580	300-650	300-650	500-00
D. Muga:						
i. Reeling cocoons	Rs./kg	—	250-700	300-800	450-720	340-550
ii. Raw silk	Rs./kg	—	3000-3600	2900-4400	2900-4200	2600-4300

Source: Silkmans Companion 1992, CSB 1999, 2001, Indian Silk - June 2000

* Refers to 1987 census, P: provisional

Source: CSB 1999

Table 10: Production of raw silk in India (in lakh kg.)

Year	Mulberry	Tasar	Eri	Muga	Total
1951-52	6.25	1.24	1.00	0.45	8.94
1961-61	11.85	1.79	1.10	0.39	15.13
1965-66	15.45	2.62	2.01	0.57	20.65
1968-69	17.81	2.56	2.14	0.69	23.20
1971-72	20.46	3.14	1.68	0.72	26.00
1973-74	24.21	2.57	1.41	0.75	28.94
1975-76	25.41	4.60	1.23	0.43	30.67
1977-78	31.86	4.34	0.56	0.35	37.11
1978-79	37.52	2.81	1.20	0.24	41.77
1979-80	41.93	3.84	1.83	0.45	48.05
1982-83	52.14	2.84	2.13	0.37	57.48
1984-85	68.95	4.44	2.79	0.55	76.73
1985-86	70.29	4.64	3.52	0.52	78.97
1986-87	79.05	5.48	3.92	0.55	89.00
1987-88	84.55	4.63	5.22	0.58	94.98
1988-89	96.83	3.58	5.65	0.45	106.51
1989-90	109.05	4.65	5.89	0.57	120.16
1990-91	114.87	4.84	6.24	0.70	125.50
1991-92	106.58	3.29	7.04	0.72	117.63
1992-93	130.00	3.82	7.26	0.60	141.68
1993-94	125.50	2.99	7.66	0.76	136.91
1994-95	134.50	2.57	7.98	0.74	145.79
1995-96	128.84	1.94	7.45	0.86	139.09
1996-97	129.54	2.35	8.64	0.73	141.26
1997-98	140.48	3.12	8.14	0.62	152.36
1998-99	142.60	2.42	9.70	0.72	155.44
1999-2000	139.44	2.11	9.74	0.85	152.14

Adopted from Silkmans Companion 1992, CSB 1999, 2001, Saratchandra 2000 a,b , Thangavelu 2000, Somashekhar 2000

An analysis of import of raw silk by major silk importing countries *viz.*, India, Italy, Japan, Germany, France, USA and Switzerland (Table 8) reveals that India, Italy and Japan serve as major importers of raw silk. Though data is not completely available, yet it is observed that imports in India, Italy, Germany and USA increased by 695.35, 50.46, 492.39 and 18.45 percent respectively while decrease in imports was observed in Japan, France and Switzerland by 24.66, 29.97 and 79.89 percent respectively during 1997 as compared to base year 1980. The total import by all the major countries from 6396 m.t. in 1980 to approximately 12000 m.t. in 1995 shows that consumption of silk has increased by 87.62% approximately during the said period despite of the fact that this increase was not steady (Table 8). When complete data regarding import by Japan, Germany, USA and Switzerland are known for 1997, picture would be discernible more clearly, yet present trends indicate that consumption of silk has been increasing since 1980 in the world. Hence, future of global silk industry is bright despite the fact that raw silk production after reaching a peak in 1995 (104203 m.t.) started to decline sharply for which many reasons have been assigned (Ronald Currie 2000, Kumar and Das 2000).

Ronald Currie (2000) has suggested following two areas in which steps should be taken for improving the future of silk marketing.

Information:　The unique characteristics of silk fibres should be informed to future users of silk from school age onwards.

Products:　The products on offer are seen as conservative, old fashioned and unimaginative, hence serious efforts should be made to innovate products adapted to the needs of today's consumers, i.e. younger, more dynamic, more mobile but attached to some solid values of quality and tradition positioning. According to him, if the style, colours, construction and prices are right, then silk will be a preferred clothing item of the younger generation.

4

Sericulture in India:
Yesterday, Today and Tomorrow

Indian silk industry has directly and indirectly been providing gainful employment to over six million people in rural and semi urban areas. The silk industry indeed consists of four independent parts *viz.*, mulberry cultivation and silkworm rearing, silk reeling, throwing and manufacturing. The word "Sericulture" has been derived from the word "Su" (Si) which means silk. The Indian silk industry during 1991-92, 1997-98 and 1999-2000 may be seen at a glance through Table 9. Sericulture is indeed an art and practice of cultivation of silk which involves a number of activities starting from mulberry production, development of breeding stocks of silkworms, egg production and rearing of silkworms. It is exclusively a labour intensive agro-based cottage industry taken up as a subsidiary avocation along with agricultural and horticultural activities by Indian farmers. Unlike mulberry sericulture, which is completely indoor, non-mulberry sericulture is practised outdoor as tasar and muga culture involve more or less similar activities in different manners for cultivation of silk. Hence, production scenario of mulberry and non-mulberry sericulture needs to be discussed individually. The production of total raw silk at the end of each plan period and variety wise contribution has been given in Table 11. (Thangavelu 2000).

MULBERRY SILK

The mulberry hectarage increased from 56,732 ha. to 2,70,069 ha., silkworm seed production increased from 862 lakh dfls to 3319 lakh dfls, total cocoon production increased from 16,287 m.t. to 1,26,565 m.t., cocoon production/ha/year increased from 287 kg. to 468.64 kg and cocoon production/100 dfls increased from 18.9 to 38.9 kg during the period of 1950-51 to 1998-99 which shows that mulberry hectarage increased by 4.8 times, silkworm seed production increased by 3.8 times, total cocoon production increased by 7.8 times, cocoon production/ha/year increased by 1.6 times and cocoon production/100 dfls increased by 2.0 times in our country during the said period. The total mulberry raw silk production has been found to increase from 957 m.t. to 14260 m.t., unit productivity of raw silk increased from 16.9 kg to 52.8 kg/ha/yr. and renditta declined from 17 kg to 8.9 kg during 1950-51 to 1998-99 which shows the total raw silk production has increased by 14.9 times, raw silk productivity/ha has increased by 3.1 times and renditta has improved by 1.91 times during the said period. The foreign exchange

Table 11: Production of raw silk at the end of each plan period and variety wise contribution over total raw silk

Plan	Mulberry		Tasar		Eri		Muga		Non-Mul.		Total
	Qnty. (MT)	% Contri.	Qnty. (MT)	% Contri.	Qnty. (MT)	% Contri.	Qnty. (MT)	% Contri.	Qnty. (MT)	% Contri.	raw silk
I Plan (1951-56)	1098	76.41	141	0.981	127	08.84	71	4.94	339	23.59	1437
II Plan (1956-61)	1185	78.32	179	11.83	110	7.27	39	2.58	328	21.68	1513
III Plan (1961-66)	1545	74.82	262	12.69	201	9.73	57	2.76	520	25.18	2065
Tra.period (1966-69)	1781	76.76	256	11.03	214	9.22	69	2.97	539	23.23	2320
IV Plan (1969-74)	2421	83.66	257	8.88	141	4.87	75	2.59	473	16.34	2894
V Plan (1974-78)	3186	85.85	434	11.69	56	1.51	35	0.94	525	14.15	3711
Tra.period (1978-80)	4193	87.26	384	7.99	183	3.81	45	0.94	612	12.74	4805
VI Plan (1980-85)	6895	89.86	444	5.79	279	3.64	55	0.72	778	10.14	7673
VII Plan (1985-90)	10805	90.68	465	3.90	589	4.94	57	0.48	1111	9.32	11916
Tra.period (1990-92)	10658	90.61	329	2.80	704	5.98	72	0.61	1105	9.39	11763
VIII Plan (1992-97)	12954	91.70	235	1.66	864	6.12	73	0.52	1172	8.30	14126
IX Plan (1997-2002)	14260	91.74	242	1.56	970	6.24	72	0.46	1284	8.26	15544
ñ Year change % over I Plan	1198.72		71.63		663.78		1.41		278.76		
% change over V Plan	347.58		- 44.24		1632.14		105.71		144.57		
% change over VI Plan	106.82		- 45.50		247.67		30.91		65.04		
% change over VII Plan	31.98		- 47.96		64.69		26.32		15.57		
% change over VIII Plan	10.10		2.98		12.27		-1.37		9.56		

Source: Thangavelu 2000

Table 12: Raw silk availability in India (Tonnes)

Year	Non-Mulberry				Mulberry	Grand	Raw silk imports	Total availability of raw silk
	Tasar	Eri	Muga	Total		Total	(Mul + Non-Mul)	
1971-72	314	168	72	554	2046	2600	26	2626
1972-73	361	143	66	570	2215	2785	8	2793
1973-74	257	141	75	473	2421	2894	13	2907
1974-75	402	115	41	558	2434	2992	29	3021
1975-76	360	123	43	526	2541	3067	89	3156
1976-77	423	106	53	582	2686	3268	62	3330
1977-78	434	56	35	525	3186	3711	162	3873
1978-79	281	120	24	425	3752	4177	176	4353
1979-80	384	183	45	612	4193	4805	366	5171
1980-81	265	135	48	448	4593	5041	319	5360
1981-82	257	147	44	448	4801	5241	641	5890
1982-83	284	213	37	534	5214	5748	1098	6846
1983-84	418	270	54	742	5681	6423	1573	7996
1984-85	444	279	55	778	6895	7673	1181	8854
1985-86	464	352	52	868	7029	7897	1767	9664
1986-87	548	392	55	995	7905	8900	1745	10645
1987-88	463	522	58	1043	8455	9498	2069	11567
1988-89	358	565	45	968	9683	10651	1395	12046
1989-90	465	589	57	1111	10805	12016	1380	13296
1990-91	380	624	70	1074	11486	12560	1598	14158
1991-92	329	704	72	1105	10658	11763	2076	13839
1992-93	382	726	60	1168	13000	14168	2768	16936
1993-94	299	766	76	1141	12550	13691	4892	18583
1994-95	257	798	74	1129	13450	14579	5403	19982
1995-96	194	745	86	1025	12884	13909	4149	18058
1996-97	235	864	73	1172	12954	14126	2911	17037
1997-98	312	814	62	1182	14048	15236	2346	17582
1998-99	242	970	72	1284	14260	15544	2827	18371
1999-2000	211	974	85	1270	13944	15214	6936(p)	22150 (p)

P : Provisional

Adopted from Silkmans Companion 1992, CSB 1999, 2001, Saratchandra 2000, b, Thangavelu 2000

earning during 1998-99 from mulberry as well as non-mulberry silk increased from Rs. 16.0 lakh to Rs. 1003.762 crores, which is 6273.5 fold increase (Table 14). The foreign exchange earnings from mulberry silk rose from 46.7418 crores in 1980-81 to Rs. 1142.8757 crores during 1999-2000 which amounts to 24.45 fold increase (Table 37). However, the total export earnings of India from silk items has been to the tune of Rs. 1501.7844 crore during 1999-2000 while during 1998-99 silk items of Rs.1250.5473 crore were exported which shows that earnings through the export of silk items registered an increase of 20.1% during this period (Table 38). During 1959-60 a total of 20 lakh persons were engaged in mulberry silk industry which has increased to 60 lakhs during 1994-95 which shows 3.0 fold increase in the field of employment generation in the country during the said period (Table 14). An analysis of mulberry sericulture (Samson 2000) reveals that during 1999-2000, area under mulberry (ha), dfl production, reeling cocoon production, raw silk production, renditta, raw silk productivity/ha, cocoon productivity/ha and cocoon production/100 dfls have increased by 81.85, 93.91, 238.96, 5387.6, 62.92, 201.82, 86.4 and 74.86 percent respectively basing the year of 1975-76. The corresponding increase/decrease during 1999-2000 has been found to be 28.25, 43.48, 11.52, 21.39, 14.61, 69.21, 48.78 and 2.131 percent with respect to the base year of 1990-91 (Table 13).

Table 13: Mulberry sericulture in India

Year	Area under mulberry (ha)	DFLs prodn. (Lakh Nos.)	Reeling cocoon Prodn. (MT)	Raw silk production (MT)	Renditta	Raw silk productivity Per ha. (Kg)	Cocoon productivity Per ha. (Kg)	Cocoon production/ 100 DFLs (Kg)
1975-76	124913	1706.90	36739	2541	14.5	20.34	294.12	21.52
1980-81	170000	2189.51	58208	4593	12.7	27.02	342.40	26.58
1985-86	217839	2628.79	76717	7029	10.9	32.27	352.17	29.18
1990-91	316610	3171.88	111663	11486	10.2	36.28	368.48	36.78
1991-92	331237	3183.43	107153	10658	10.1	32.18	323.49	33.66
1992-93	342764	3402.44	129685	13000	10.0	37.93	378.35	38.12
1993-94	319215	3034.32	117268	12550	9.3	39.32	367.36	38.65
1994-95	283093	3189.70	123115	13450	9.2	47.51	434.89	38.60
1995-96	286496	2856.20	116362	12884	9.0	44.97	406.16	40.74
1996-97	280651	3199.49	115655	12954	8.9	46.16	412.10	36.15
1997-98	282244	3355.20	127495	14048	9.1	49.77	451.72	38.00
1998-99	270069	3318.62	126565	14260	8.9	52.80	468.64	38.14
1999-2000	22715	3309.80	124531	13944	8.9	61.39	548.23	37.63
% increase/ decrease w.r.t								
1975-76	81.85	93.91	238.96	5387.60	62.92	201.82	86.40	74.86
1990-91	28.25	43.48	11.52	21.39	14.61	+69.21	+48.78	2.31

MT: Metric Tonnes

Source: Samson 2000

A. IMPACT OF MULBERRY VARIETIES AND SILKWORM BREEDS

It may be visualised that the progress made by India after independence in the field of sericulture has been due to horizontal expansion as well as enhancement of productivity per unit area vertically. The phenomenal vertical growth has been made possible due to evolution of high yielding superior mulberry varieties and silkworm breeds. Benchamin (2000b) critically analysed the impact of mulberry varieties and silkworm breeds on productivity and cost of silk production with the help of K2 and V1 mulberry varieties and PM x NB4D2 (Multi x Bi hybrids) and

Table 14: Horizontal and vertical growth of mulberry sericulture in India

Growth parameters	During 1950-51 (A)	During 1959-60 (B)	Gain (%)	During 1985-86 (C)	Gain (%)	During 1994-95 (D)	Gain (%)	During 1998-99 (E)	Gain (%)	Increase E/A
Mulberry hectarage	56,732	82,954 (46.22)	46.22	2,17,839 (283.98)	162.60	2,88,510 (408.55)	32.44	2,70,069 (376.04)	-6.39	4.8 times
Silkworm seed production (Lakh No.)	862.0	NA	—	2628.79 (204.96)	—	3189.70 (270.03)	21.34	3319.0 (285.03)	4.05	3.8 times
Cocoon production (kg./ha/year)	287.0	240.0 (-16.38)	-16.38	352.17 (22.71)	46.78	434.89 (51.53)	23.49	468.64 (63.07)	7.76	1.6 times
Cocoon production per 100 dfl (kg)	18.9	NA	—	29.18 (54.39)	—	38.60 (104.23)	32.28	38.9 (105.82)	0.777	2.0 times
Total cocoon production (M.T)	16,287	17,310 (6.281)	6.281	76,717 (371.032)	343.19	1,16,362 (614.48)	51.68	1,26,565 (677.09)	8.79	7.8 times
Raw silk production (M.T)	957	1,185 (23.82)	23.82	7,029 (634.48)	493.16	14,094 (1372.73)	100.51	14260 (1390.07)	1.18	14.9 times
Silk production (kg/ha/year)	16.8	14.29 (-14.94)	-14.94	32.27 (92.08)	125.82	45 (167.86)	39.45	52.8 (214.28)	17.33	3.1 times
Renditta (kg)	17.0	16.8 (1.19)	1.19	9.2 (84.78)	82.61	10.91 (55.82)	-15.67	8.9 (91.01)	22.58	1.91 times
Foreign exchange** earnings (Lakh Rs.)	16*	103 (543.75)	543.75	15982 (9787.5)	15416.51	92696.17 (579251.06)	480.0	100376.2 (627251.25)	8.28	6273.5 times
Employment generation (persons in lakh numbers)	—	20	—	51.52	157.60	60	16.45	60	0.0	3.0 times

* Corresponds to 1951 data. ** includes tasar also. ( ) Figures within parenthesis indicate percent gain over base year 1950-51.

Source: CSB 1986, 1999, Datta 2000a, Geiger 2000, Benchamin 2000b, Indian Silk - June 2000

CSR2 x CSR4 (Bi x Bi hybrid) silkworm breeds. While K2 yields 30,000 kg, V1 yields 52,500 kg. leaf/ha. Enhancement in leaf productivity not only reduces cost of production of one kg. of leaf from Rs. 1.17 to 0.78 but also increases silkworm rearing capacity/ha from 3000 to 4800 dfls. The cocoon production/ha increases from 1875 kg to 3900 kg., cocoon yield increases from 62.5 kg to 81.25 kg/100 dfl, cost of production of one kg. of cocoon reduces from Rs. 54.05 to 40.70 only when V1 variety of mulberry is fed to CSR2 x CSR4 silkworm breed as against feeding of K2 mulberry variety to PM x NB4D2 silkworm breed. The shell conversion rate (%) of V1, S36 and K2 mulberry varieties have been found to be 12.08, 7.98 and 6.04 with multi x bi hybrids of silkworm which increases to 14.8, 9.84 and 7.09 percent respectively when bi x bi silkworm hybrids are reared on these mulberry varieties (Table 15).

Table 15: Impact of mulberry varieties and silkworm breed change on productivity and cost

Sl. No.	Particulars	Variety/Breed	
1.	Mulberry variety	V1	K2
2.	Leaf yield/ha (kg)	67524	30000
3.	Cost of leaf production (Rs)	52500	35000
4.	Cost of production/kg of leaf (Rs)	0.78	1.17
5.	Silkworm hybrid	CSR2 x CSR4 (Bi x Bi)	PM x NB4D2 (Multi x Bi)
6.	Rearing capacity/ha (dfls)	4800	3000
7.	Cocoon yield/100 dfl (kg)	81.25	62.5
8.	Cocoon production/ha (kg)	3900	1875
9.	Cost of silkworm rearing (Rs.)	106156	66340
10.	Total cost of cocoon production (Rs.)	158656	101340
11.	Cost of production of 1 kg. of cocoon (Rs.)	40.70	54.05
12.	Shell conversion rate of mulberry varieties with Multi x Bi hybrid	V1 : 12.08, S36 : 7.98, K2 : 6.04	
13.	Shell conversion rate (%) of mulberry varieties with Bi x Bi hybrid	V1 : 14.8, S36 : 9.84, K2 : 7.09	

*Source:*Benchamin 2000b

The findings show that vertical enhancement of production in less cost may revolutionise sericulture in India with the help of superior varieties of mulberry and silkworm breeds in future also. According to Miyashita (1986) quoted by Ahsan (2000) maximum contribution for successful bivoltine crop is made by mulberry leaves (38.2%) followed by climate (37.0%), rearing technique (9.3%), silkworm race (4.2%), silkworm eggs (3.1%) and other factors (6.6%). Hence, quality of mulberry varieties is most important factor for silkworm rearing whereas silkworm race/breed is fourth in ranking.

B. SERICULTURE IN CHINA, JAPAN AND INDIA: A COMPARISON

Benchamin (2000 b) has made an attempt to analyse sericultural scenario of China, Japan and India prevailing during 1993-94, (Tables 16&17). Japan represents temperate sericulture while India represents tropical sericulture yet

India practices both temperate as well as tropical sericulture. Due to high investments in R&D over a long period of approximately 100 years, temperate sericulture technology is more advanced. Besides, productivity and quality in temperate sericulture have natural advantage over those of tropical sericulture. For comparison of productivity in Japan, China and India, production during 1993 is considered.

Table 16: Production and productivity of sericulture in China, Japan and India

Sl. No.	Particulars	1993-94			1998-99	
		China	Japan	India	China	India
1	Mulberry area (ha)	800000	485000	341799	597600	270069
2	Seed consumption/ha (No.)	1550	750	900	(-25.3%)	(-21%)
3	Cocoon production (MT)	619000	22750	123507	4090000 (-34%)	126565 (2.4759)
4	Raw silk production (MT)	71845	4260	13418	38700 (-46%)	14260 (6.275%)
5	Silkworm seed production (Lakh No.)	12380	354	3014	—	—
6	Cocoon production/ha (kg)	774	469	361	684 (-13.16%)	469 (29.92%)
7	Cocoon production/100 dfls (kg)	49.9	62.5	40.1	NA	NA
8	Raw silk production/ha (kg)	89.8	87.8	39.3	64.8 (-27.83%)	52.8 (34.35%)
9	Renditta (kg. cocoons required to produce 1 kg raw silk)	8.6	5.3	9.2	10.6 (-23.25%)	8.9 (3.26%)

Figures within parenthesis indicate percent increase/decrease with respect to base year 1993-94.
Source: Benchamin 2000 b

Table 17: Land, labour, input efficiency and cost of cocoon production in China, Japan and India (unit = 1kg cocoon)

Sl. No.	Particulars		China	Japan	India
1	Land requirement	(Ha.)*	0.001292	0.003192	0.00213
		(acre)	0.00526	0.00277	0.0068
2	Labour (hours)		4.14	2.22	7.66
3	Labour cost	(Rs.)	17	(0.378)	149.85
		(US $)	(3.4)	57.20	(1.28)
4	Input cost	(Rs.)	102.25	213.75	23.40
		(US $)	(2.3)	(4.75)	(0.52)
5	Total cost	(Rs.)	119.25	366.75	81.00
		(US $)	(2.65)	(8.15)	(1.80)

· Calculated from 1993-94 data
Source: Benchamin 2000 b

Despite the fact that China was only 2.34 times bigger in mulberry area, its cocoon and raw silk production is 5.0 and 5.3 times higher as compared to India (Table 16) due to quantitative as well as qualitative advantages of China over India in mulberry cultivation, silkworm rearing and reeling. The consumption of silkworm seed was 72% more and cocoon and silk production per unit area of land (1 ha.) is also higher by 114% and 128% respectively in China. The cocoon production/100 dfl and renditta are also better by 24.5 and 6.52% respectively. While India produces mostly yellow polyvoltine cocoons (95%), China produced more bivoltine and white polyvoltine (70%) and less yellow polyvoltine (30%) cocoons, hence cocoon and raw silk quality is also different in China. An over view of sericulture industry during 1995-96 in China and India is given in Table 18.

Table 18: Sericulture industry of China and India at a glance during 1995-96

Sl. No.	Particulars	China	India
1	Sericultural provinces/states	24 (mainly 5 areas)	24 (mainly 4 states)
2	Sericultural counties/ districts	1300	NA
3	Sericultural families	200 lakhs	About 12 lakhs
4	Area under mulberry	12.5 lakh hectares (including trees)	2.89 lakh hectares
5	Mulberry cultivation	All irrigated (80%)	Irrigated (23-25%) + rainfed
6	Avg. leaf yield (kg/ha/yr)	40,000	25,000
7	No. of crops/year	South China: 6-8 Taihu lake belt: 4 Sichuan province: 4-5	South & North East: 5-6 North : 3-4 (Temperate)
8	Seed production (1 box = 50 layings)	27 million boxes (=135 crore layings)	28.56 crore layings
9	Silkworm seed producers	All Govt. Centres	Pvt. (70%) + Govt. (30%)
10	Silkworm races	All bivoltines	Multi and Bivoltines
11	Silkworm rearing	Batch rearing (seasonal)	Continuous
12	Leaf cocoon ratio (by wt.)	18-20:1	26-28:1
13	Average yield/100 dfls	50-60 kgs	35-40 kgs
14	Cocoon production	7,59,850 tonnes	1,16,362 tonnes
15	Cocoon price fixation/kg	Based on quality (12-15 Y = Rs.50-70)	Open auction (Rs.90-120)
16	Marketing and reeling	All Govt. Centres	Pvt.(90%) + Govt. 10%
17	Raw silk production	66400 tonnes	12884 tonnes
18	Silk production centres	400 (All Govt.)	1045 (Pvt.70%) Govt.30%
19	No.of silk enterprises	3500	NA
20	No.of reeling ends	38 lakhs	31000 F.L.Basin + 34159 charkha
21	Renditta	6-8	8-10
22	Filament length	1000-1200 mts	700-900 mts
23	Silk content (Shell %)	21-23%	18-20%
24	Raw silk goods	>1A, 2A, 3A	B-E
25	Fabric production capacity	3000 lakh sq.mts	1883* lakh sq.mts
26	No.of weaving machines/loom	2.1 lakhs	2.56 # lakhs
27	Domestic consumption (%)	About 10-15%	90%

* Data refers to 1990-91   # As per 1992 Statistical Biennial

Source: Dandin 1998, compilation : KSSRDI, Bangalore

Table 19: Progress in improvement of multivoltine and bivoltine hybrids

Sl. No	Hybrid		Year	Yield/100 dfls (kg)	Cocoon shell ratio (%)	Renditta
1	PM x C.nichi	(Multi x Multi)	1950-70	15 - 20	14	14 - 16
2	PM x HS6	(Multi x Multi)				
3	PM x J112	(Multi x Multi)	1960-70	20 -25	15	13 - 15
4	PM x C 108	(Multi x Multi)				
5	PM x C 110	(Multi x Multi)				
6	PM x KA	(Multi x Bi)	1970-75	25 - 30	16	11 12
7	PM x NN6D	(Multi x Bi)				
8	PM x NB4D2	(Multi x Bi)	1975	30 - 35	17	10 -11
9	PM x NB18	(Multi x Bi)	1979	35 - 40	18	9 - 10
10	PM x NB7	(Multi x Bi)				
11	KA x NN6D	(Bi x Bi)	1970-74	30 -35	18	10
12	KA x NB4D2	(Bi x Bi)	1975-79	40 -45	19	8 -9
13	NB7 x NB18	(Bi x Bi)	1979	45 - 50	20	7 - 8

Source: Jolly 1981 - adopted from Datta 2000 c.

Table 20: Promising bivoltine breeds evolved at CSR&TI, Mysore (1970s)

Breed	Yld/10000 larvae (kg)	Cocoon Wt. (g)	Shell Wt. (g)	Shell ratio (%)
NB7	18.1	1.90	0.42	21.6
NB18	15.1	1.78	0.30	21.7
NB4D2	13.7	1.73	0.36	20.8

Source: Datta 2000c

Table 21: Performance of bivoltine breeds evolved at Pampore and Dehradun (1970s)

Breed	Survival (%)	Cocoon Wt. (g)	Shell Wt. (g)	Shell ratio (%)
DEHRADUN				
ID6	93.4	1.57	0.31	19.8
SH6	91.1	1.53	0.33	21.7
YS3	93.2	1.49	0.29	19.0
SF19	89.0	1.35	0.29	21.1
PAMPORE				
KY1	72.0	1.83	0.37	19.6
PLF	81.0	1.55	0.36	23.0
BL1	97.7	1.88	0.40	21.0

Source: Datta 2000c

Table 22: Performance of new multivoltine x bivoltine hybrids at CSR&TI, Mysore

Hybrid	Survival	Cocoon	Shell	Shell	Filament	Filament	Neatness	Reelability	Renditta
Rainfed									
BL23 x NB4D2 (Multi x Bi)	94.5	1.762	0.331	18.3	824	2.6	84.5	91.5	8.31
PM x C.nichi (Multi x Multi)	93.0	1.171	0.15	12.9	450	2.0	80.4	85.5	14.7
Irrigated									
BL24 x NB4D2 (Multi x Bi)	92.0	1.897	0.350	18.5	844	2.75	85.2	92.1	8.24
PM x NB4D2 (Multi x Bi)	91.4	1.724	0.310	17.9	732	2.55	81.6	91.3	10.08

Source: Datta 2000c

Table 23: Performance of bivoltine hybrids at CSR&TI, Mysore (1985)

Hybrid	Larval period (hr)	Yield/10000 larvae (No.)	Yield/10000 larvae (Kg.)	Cocoon wt (g)	Shell (g)	Shell ratio (%)
CC1 x NB4D2	571	9518	20.51	2.19	0.48	22.2
CA2 x NB4D2	571	9451	20.54	2.18	0.50	23.0 .
KA x NB4D2 (C)	567	8770	18.65	2.10	0.43	20.6

C: Control
Source: Datta 2000c

Table 24: Statewise bivoltine raw silk production in India

State	1998-99 DFLs Lakh No.	1998-99 Reeling cocoon (Tons)	1998-99 Raw silk (Tons)	1999-2000 DFLs Lakh No.	1999-2000 Reeling cocoon (Tons)	1999-2000 Raw silk (Tons)
Karnataka	57.25	1573.00	197.00	54.41	1692.00	212.00
Jammu & Kashmir	35.58	829.00	86.00*	36.16*	825.00*	85.00*
Uttaranchal	13.38	161.00	21.50	13.09	80.71	10.09
Madhya Pradesh	6.26	58.25	5.83	6.62	70.50	4.38
Assam	8.32	149.91	16.06	14.51	379.39	35.04
Manipur	4.75	178.00	17.90	5.52	229.00	22.70
Himachal Pradesh	4.61	118.00	11.80	4.58	108.80	10.80
West Bengal	2.21	62.00	6.00	0.93	32.00	4.00
Meghalaya	1.97	12.70	0.52	2.40	15.48	Negligible
Arunachal Pradesh	0.65	15.00	1.00	0.35	30.00	3.00
Orissa	0.39	11.70	1.30	0.25	6.38	0.68
Kerala	0.27	5.10	0.52	0.43	22.95	1.02
Nagaland	0.11	3.50	0.32	0.08	2.68	0.03
TOTAL	135.75	3177.16	365.75*	139.32	3494.89*	388.74*

* Ownership of mulberry to the farmer in whose land the tree exist

* Practicing of reeling in private sector permitted

* Establishment of cocoon markets to allow the free sale of cocoons.

Source: Central Silk Board, adopted from Ahsan 2000

A comparison with Japan indicates that seed consumption/ha. in India is higher by 20% which is probably due to higher leaf out put and use of silkworm breeds that consume less leaf but cocoon and raw silk productivity are lower by 35.8% and 55.2% respectively. Hence, a wide gap exist between India and other two countries *viz.*, China and Japan in sericulture and sericultural technology.

Table 25: Performance of Robust bivoltine hybrid CSR18 x CSR19 in Southern plateau during adverse summer and rainy seasons

Place	No.of DFLs	Yield/100 dfls (kg)	Rate/kg (Rs.)
1. RSRS, Salem	95	71.80	180.00
2. RSRS, Anantapur	20	57.88	—
3. RSRS, Kodathi	100	60.06	—
4. CSR&TI, Mysore (commercial rearing)	500	66.37	157.20
5. Gudimangalam (TN)	5780	51.38	140.00
6. Farmers (Karnataka)	4110	61.25	186.00
Total/Average	10605	61.45	166.00

Source: Datta 2000b

Japan has produced merely 108 M.T. raw silk during 1998-99 and now it stands at fifth position in the world. Sericulture has suffered in China also due to various reasons. Mulberry area has reduced by 25.3%, Cocoon production is lower by 34% and raw silk production has decreased by 46% in China in 1998-99 as compared to 1993-94. The productivity of China has also declined with respect to cocoon production/ha. and raw silk production/ha. by 13.16% and 27.83% respectively, while silk recovery renditta noticed a decrease by 23.25%. On the contrary, sericulture in India has noticed a positive trend during the said period despite the fact that 21% reduction has been observed in hectarage of mulberry. The total cocoon production and raw silk production slightly increased by 2.4759% and 6.275% respectively while productivity increased significantly with respect to cocoon and raw silk production/ ha. by 29.92% and 34.35%. Renditta has also been found to increase by 3.26% in India during the said period (Table 16). Though progress achieved by India in sericulture during last 50 years is significant, yet this progress has been achieved more by horizontal expansion and less by improvements in productivity. An analysis of gain in productivity in 1998-99 reveals that cocoon production/ha. increased by 1.6 times, cocoon output/100 dfls increased by two fold, raw silk production/ha. increased by 3.14 times and silk recovery (renditta) increased by 1.91 times as compared to base year of 1950-51 (Table 14). Yet cocoon and raw silk production/ha. in China is sill 46% and 23% higher than that of Indian productivity which may further be enhanced by 65% and 75% at par with the productivity level of China achieved during 1993-94. Hence, comparable to the achievements made by Japan and China there is lot of scope to improve productivity in India. At the same time decline in raw silk production by Japan and China is a signal to India to plan R&D in sericulture to improve productivity and quality besides reducing the cost of production.

Benchamin (2000b) also made an attempt to estimate the land, labour and input requirements and cost to produce 1 kg of cocoon in China, Japan and India. (Table 17). Labour efficiency has been found to be lowest in India requiring 7.66 labour hours as compared to 4.17 hours in China and 2.22 hrs in Japan to produce 1 kg of cocoon. The cost of labour for production of 1 kg of cocoon has been found to be lowest in China (0.378 US $ or Rs. 17) followed by India (US $ 1.28 or Rs. 57.20) and highest in Japan (US $ 3.4 or Rs. 149.85). The input cost per kg. of cocoon are highest in Japan (US $ 2.3 or Rs. 102.25) followed by China (US $ 4.75 or Rs. 213.75) and lowest in India (US $ 0.52 or Rs. 23.40). The total cost of cocoon production/kg. has also been found to be highest in Japan (US $ 2.15 or Rs. 366.75) followed by China (US $ 2.65 or Rs. 119.25) and lowest in India (US $1.80 or Rs. 81.00). Hence, India appears to have an edge over Japan as well as China but since Japan and China produce more than 70% superior bivoltine and white polyvoltine cocoons as compared to India which mainly produces yellow polyvoltine cocoons, India is at a disadvantageous condition because of the fact that bivoltine silk of Japan and China is far superior than polyvoltine Indian silk. The price of raw silk ranged between 26-28 US $ per kg. to China Silk, 37 US $ to Japan, 23 US $ to Europe and 20-22 US $ to Indian Silk. The current selling rate of Indian silk is US $ 30 whereas imported yarn costs only US $ 22 which indicates that raw silk reaching Indian market from China may be of low quality. Hence, India will have to shift from polyvoltine to bivoltine white silk production for improving quality of cocoon as well as raw silk. Such shift will also reduce the cost of production since the productivity through better conversion rate of leaf to cocoon and cocoon to silk is higher in bivoltine silkworms.

C. DEVELOPMENTAL PHASES IN INDIAN SERICULTURE

Multivoltine culture was practised to rear pure races in 19th and early 20th centuries. During 1920s Pure Mysore in Karnataka, Nistari in West Bengal and Sarupat and Moria multivoltine silkworm breeds were reared in Assam in India. During 1950s multivoltine breeds Nistid, Nismo, Ichot and Itan were evolved by CSR&TI, Berhampore and HS6 was evolved by Karnataka State, which yielded 18-20 kg cocoons/100 dfls with renditta of 16-18. Subsequently, during 1960s to 1990s many multivoltine breeds were evolved in India (Tables 43&44). On the contrary, the era of bivoltine breeds started with evolution of KA and KB in Kalimpong during 1968 from Chinese and Japanese introductions and after 1970s many bivoltine breeds have been evolved in India in different institutes for different regions (Tables 20, 21&44). While NB7, NB18 and NB4D2 were evolved at CSR&TI, Mysore, RSRS, Dehradun evolved ID6, SH6, YS3 and SF19 and RSRS, Pampore evolved KY1, PLF and BL1 bivoltine breeds. The performance of these breeds revealed that shell weight ranged from 0.29-0.42 gms and SR% ranged from 19-23 in these breeds. During 1950 to 1970 multivoltine hybrids of PM x C.nichi, PM x HS6, PM x J112, PM x C108 and PM x C110 were evolved and reared in India. While former two multi x multi hybrids yielded 15-20 kg cocoons/100 dfl with SR% of 14 and renditta of 14-16, latter three multi x multi hybrids exhibited yield of 20-25 kg cocoons/100 dfls, SR% of 15 and renditta of 13-15 only (Table 19). PM x C.nichi was probably the first hybrid exploited in Karnataka. The exploitation of hybrids in West Bengal and J&K came much later during 1956 and 1959 respectively (Thangavelu 1997). Unfortunately the multivoltine breeds which were evolved to increase the cocoon production and the quality were not successful in the field for long time. Hence, the mutant strains of Nistari were utilised for development of the cross breeds viz. CB2 x CB5. During 1970-79 multi x bi hybrids of Nistari x NB4D2, PM x KA, PM x NN6D, PM x NB4D2, PM x NB18 and PM x NB7 as well as Bi x Bi hybrids of KA x NN6D, KA x NB4D2 and NB7 x NB18 were also evolved. PM x NB4D2 and Nistari x NB4D2 crosses excelled over the other combinations since survivality of commercial hybrids is considered as the top priority of the industry instead of quality of silk. Hence, introduction of bivoltine in gene pools from China and Japan did not help much except for improving of breeds. The two indigenous races of PM in Karnataka and Nistari in West Bengal are found moderately tolerant to hazards like environmental stress and pathogen load. The bivoltine breeds, multivoltine x bivoltine hybrids and bivoltine x bivoltine hybrids could not reach the survival rate of the multivoltine breeds in India.

In fact, CSR&TI, Mysore introduced bivoltine production in south India during early 1970s from newly evolved hybrid of KA x NB4D2 due to which renditta came down to 10 as against 16 for the existing PM x C.nichi multivoltine hybrids and farmers earned good amount of money. The cocoon yield/100 dfls jumped from 25 kg in latter hybrid to 40 kg. in bivoltine hybrid and large quantity of bivoltine cocoons were reeled in many charka and cottage basins. However, from survivality point of view during 1970s the farmers preferred the cross breeds (PM x NB4D2 and PM x NB18 etc.) against bivoltine hybrids. Slowly, bivoltine hybrid rearing declined and farmers switched over to cross breed rearing. The cocoon yield of 30-40 kg./100 dfls and renditta of 9-11 in these multi x bi are also found good though they were not higher than those of bivoltine hybrids (Table 19). During 1990s, BL23 and BL24 polyvoltine breeds were evolved through hybridisation and selection from exotic and indigenous silkworm germplasm resources. The hybrids of BL23 x NB4D2 and BL24 x NB4D2 exhibited survivality of 92-94.5% and renditta of 8.31 and 8.24 was realised in both rainfed and irrigated areas respectively (Table 22).

During 1980s attempts were again made to revive and popularise bivoltine hybrids instead of multi x bi hybrids and CC1 and CA2 bivoltine breeds were evolved at CSR&TI, Mysore for commercial rearing in spring and autumn seasons in J&K and erstwhile U.P (Uttaranchal). CC1 and CA2 were utilised to produce CA2 x NB4D2 and CC1 x NB4D2 hybrids which yielded 20.51-20.54 kg cocoons/ 10000 larvae and shell ratio of 22.2-23% as against the yield of 18.65 kg/10000 larvae and shell ratio of 20.6% of KA x NB4D2 at CSR&TI, Mysore (control) (Table 23). However, the cocoon shell ratio realised in bivoltine hybrids at commercial level was approximately 18% only. In Northern India, bivoltine breeds like JD6 x SF19 and YS3 x SH6 which were evolved earlier became popular since 1980. At the end of National Sericulture Project (1989-96) the production of bivoltine raw silk was limited to a maximum of 500 MT. Hence, it was realised that the bivoltine production cannot be introduced in a big way by utilising the old silkworm breeds *viz.* KA, NB4D2 and NB18 etc. Even now the total bivoltine raw silk production during 1998-99 and 1999-2000 has been limited to only 365.75 and 388.74 M.T respectively (Table 24). According to Datta (2000 b), the main reasons for poor bivoltine sericulture in the country are:

a.   Inappropriate rearing house, poor disinfection, low maintenance of sanitary conditions in rearing areas and absence of young age silkworm rearing facility at constant temperature and humidity.

b.   Existing mulberry variety (K2) and the cultivation practices are qualitatively not suitable for bivoltine breeds.

c.   The cocoon shell ratio in earlier evolved breeds of silkworms is only 18-20% which does not fetch high price in the market.

d.   Bivoltine breeds introduced in India have their origin in China, Japan and Europe which have temperate climate. Tropical climatic conditions are considered as adverse for bivoltine rearing yet, these breeds were introduced in India without developing appropriate technologies or package to make them sustainable in tropical conditions.

Hence attempts were made since 1991 to make bivoltine rearing successful in the following ways:

1. BREEDING FOR HIGH YIELDING BIVOLTINE BREEDS

Japanese hybrids were crossed with the Indian bivoltine breeds like NB4D2, KA and NB7 and 18 breeding lines were evolved. Subsequently, 161 hybrids were evolved and evaluated at CSR&TI, Mysore. Seven hybrids *viz.*, CR2 x CSR5; CSR2 x CSR4; CSR3 x CSR6; CSR12 x CSR6; CSR13 x CSR5; CSR16 x CSR17 and CSR20 x CSR19 were identified as highly productive hybrids (Tables 45&46). Significant improvement was observed in the raw silk recovery, filament length and renditta in these hybrids which recorded pupation rate of 89.7-94.7%, cocoon shell ratio of 23.0-24.4.%, raw silk recovery of 18.9-19.9%, filament length of 1176-1328 m and renditta of 5.0-5.6. Except CSR13 x CSR5 and CSR20 x CSR29, above five productive hybrids were authorised by CSB for commercial exploitation (Basavaraja *et al.* 1995, Datta *et al.* 2000 a, b).

CSR2 x CSR4 and CSR2 x CSR5 hybrids were recommended for rearing during favourable months (Aug-Feb.) by selected farmers in South India, particularly Karnataka under Popularisation of Bivoltine Sericulture Technology programme. The performance of these hybrids is quite encouraging since the yield gap noticed earlier for bivoltine crops in the field has been very much reduced. Field trial of these hybrids established that Indian farmers are capable of producing cocoons of high quality which will produce raw silk with a renditta of 5.0 -6.1 and 2A - 4A grade. The average cocoon yield recorded by the farmers was 66.82 kg/100 dfls from these hybrids (Table 26). CSR2 x CSR4 was also reared at farmers level in Kashmir valley where it exhibited yield of 51.16 kg.cocoons/oz., shell wt. of 0.365

gm and SR% of 21.785 against 38.69 kg cocoon/oz., shell weight of 0.325 and SR% of 19.455 of NB4D2 x SH6 popular hybrid (Table 27).

Table 26: Performance (Avg.of 8 crops) of CSR bivoltine hybrids (CSR2 x CSR4 and CSR2 x CSR5) in the field under PPPBST.

Name of TSCs	No. of DFLs	Yield/100 dfls (kg)	Rate/kg (Rs.)
1. Srirangapatna	21375	62.23	189.00
2. Halagur	12000	61.87	201.00
3. Sira	22850	63.00	195.00
4. Chitradurga	7850	58.89	201.00
5. Halasahalli	2950	70.00	197.00
6. Kengeri	3100	59.00	191.00
7. Varthur	3950	81.30	204.00
8. Turuvekere	7195	78.27	197.00
Total / Average	81270	66.82	197.00

PPPBST : Promotion and Popularisation of Practical Bivoltine Sericulture Technology
TSC : Technical Service Centre
Source: Datta 2000b.

Table 27: Comparative rearing performance of CSR bivoltine hybrids vis-a-vis popular hybrids of North Western temperate regions in Kashmir valley at farmers level.

Combination	Cross	Avg.yield/ OZ. (kg)	Avg.Single Cocoon Wt. (g)	Avg. Shell Wt. (g)	Shell ratio (%)	Remarks
Spring 1999						
CSR2 x CSR5 and reciprocal	(Bi x Bi)	41.33	1.70	0.37	21.76	I Trial
NB4D2 x SH6 and reciprocal (Control)	(Bi x Bi)	35.37	1.75	0.34	19.42	do
Spring 2000						
CSR2 x CSR5 and reciprocal	(Bi x Bi)	61.0	1.65	0.36	21.81	II Trial
NB4D2 x SH6 and reciprocal (Control)	(Bi x Bi)	42.01	1.59	0.31	19.49	do
Average						
CSR2 x CSR5 and reciprocal	(Bi x Bi)	5.16	1.675	0.365	21.785	
NB4D2 x SH6 and reciprocal (Control)	(Bi x Bi)	38.69	1.670	0.325	19.455	

Source: Ahsan 2000

2. BREEDING FOR ROBUSTNESS

The tropical hot climatic conditions prevailing during summer are not conducive to rear high yielding productive bivoltine hybrids. Many quantitative traits such as viability, cocoon traits and reproductive rate sharply decline when temperature exceeds 28°C. Hence, development of robust hybrids for rearing throughout the year is very much required. CSR&TI, Mysore has been able to develop four compatible hybrids *viz.*, 1 HT x 2 HT, 1HT x 4 HT, 3 HT x 2 HT and 3 HT x 4 HT for rearing throughout the year by utilising Japanese thermo-tolerant hybrids as breeding resource material on the basis of their survival at 36 ± 1°C and 80 ± 5% RH (Datta *et al* 1997, Table 28). 1 HT x 2 HT (CSR18 x CSR19) hybrid was tested in Tamil Nadu, Andhra Pradesh and Karnataka where on an average it yielded 61.45 kg. cocoons/100 dfls during adverse summer and rainy seasons (Table 25). CSR18 x CSR19 is characterised by shorter larval duration (22 days) and consumes less leaf/100 dfls (1100 kg) compared to that of productive hybrids (CSR2 x CSR4, CSR2 x CSR5).

Table 28: Comparative performance of HT hybrids at 25 ± 1°C and 36 ± 1°C

Hybrid	25 ± 1°C				36 ± 1°C			
	Pupation (%)	Cocoon Wt. (g)	Shell Wt. (g)	Shell Ratio (%)	Pupation (%)	Cocoon Wt. (g)	Shell Wt. (g)	Shell Ratio (%)
1 HT x 2 HT (CSR18 x CSR19)	91.63	1.743	37.7	21.7	83.38	1.453	29.3	20.2
1 HT x 4 HT	84.15	1.776	38.3	21.6	69.50	1.441	29.4	20.7
3 HT x 2 HT	93.95	1.743	38.0	21.8	78.76	1.486	30.5	20.5
3 HT x 4 HT	87.75	1.753	37.8	21.6	69.71	1.471	28.5	19.4

Source: Datta 2000c

Table 29: Assumed proportions of silkworm *Bombyx mori* seed of different breeds

Year	Particulars of production	Total seed production	Bivoltine	Assumed breedwise proportion*		
				Pure Mysore	Nistari	Lemirin
1995-96				2262.2	532.58	12.66
(a)				79.2	18.65	
1996-97	Actual %	3199.49	53.34	2584.67	548.23	13.25
(b)	to total		1.67	80.78	17.13	0.41
1997-98	Actual %	3355.21	50.49	2690.57	600.23	13.92
(c)	to total		1.51	80.19	17.89	0.42
			Simple % annual growth			
	Against total (a) to (b)		2.02			
	Seed produced (b) to (c)		4.87			
	% proportion of bivoltine 1995-96		1.74			
	1996-97		1.69			
	1997-98		1.53			

*Values in lakh numbers

Source: Pavan Kumar 2000 b, Central Silk Board 1999*

Table 30: Sectorwise contribution in production of silkworm *Bombyx mori*

Values in lakh numbers

Year	Particulars	Production of silkworm seed				
		Total	State Govt	CSB	LSP	Others
1995-96	Actual %	2856.16	765.08	218.92	1766.49	105.71
	to total		26.79	7.67	61.85	3.7
1996-97	Actual %	3199.49	797.08	224.98	1765.77	411.66
	to total		24.91	7.03	55.19	12.87
1997-98	Actual %	3355.21	840.65	240.09	1872.06	402.41
	to total		25.06	7.16	55.8	11.99

Source: Central Silk board 1999, Pavan Kumar 2000b

Table 31: Tasar seed production at national level and supplied by BSM&TCs and their growth index

Year	Dfls production (lakh)		Dfls supplied by BSM&TC	
	Quantity	Index	Quantity	Index
1989-90	130.28	100.00	16.85	100.00
1990-91	154.69	119.00	14.22	84.00
1991-92	135.69	104.00	13.49	80.00
1992-93	133.85	103.00	15.19	90.00
1993-94	121.23	93.00	17.42	103.00
1994-95	98.19	75.00	13.07	78.00
1995-96	90.34	69.00	13.40	80.00
1996-97	102.51	79.00	13.75	82.0
1997-98	123.35	95.00	15.50	92.00
1998-99	89.00	68.00	11.11	66.00

Source: Thangavelu and Rai 2000.

Table 32: State wise indicators of tasar development in India during 1998-99.

State	Seed production Dfls (lakh)	Cocoon production (Lakh Kahan)	Silk production (MT)	Percentage of share of raw silk production
Bihar	50.00	1.20	120	49.59
M.P. (Chattisgarh)	9.67	0.54	68	28.1
Orissa	5.67	0.24	33	13.64
W.B.	11.14	0.13	16	6.61
U.P	1.72	0.03	3	1.24
A.P	0.90	0.02	1	0.41
Others	9.90	0.01	1	0.41
Total	89.00	2.17	242	100.00

Source: Thangavelu and Rai 2000.

Table 33: Non-Mulberry cocooon and raw silk production in India

Year	Cocoon production			Raw silk production				Estimated prodn. based on linear trend				
	Tasar (Lakh Kahan)	Eri (M.T)	Muga (Lakh No.)	Tasar (M.T)	Eri (M.T)	Muga (M.T)	Total raw silk (M.T)	Tasar (M.T)	Eri (M.T)	Muga (M.T)	Total raw silk (M.T)	Annual increase in production
1980-81	246	200	29	265	135	48	448	407	142	44	596	NA
1981-82	239	220	26	257	147	44	448	400	187	46	636	40
1982-83	300	326	14	284	213	37	534	394	233	47	676	40
1983-84	497	368	13	418	270	54	742	387	278	49	716	40
1984-85	485	401	21	444	279	55	778	380	324	51	756	40
1985-86	468	555	27	464	352	52	868	374	369	53	796	40
1986-87	430	631	26	548	392	55	995	367	415	55	836	40
1987-88	465	772	30	463	522	58	1043	360	460	57	876	40
1988-89	338	942	22	358	565	45	968	354	506	59	916	40
1989-90	450	984	28	465	589	57	1111	347	551	61	956	40
1990-91	392	1019	42	380	624	70	1074	341	597	62	996	41
1991-92	339	1169	40	329	704	72	1105	334	642	64	1037	40
1992-93	395	1187	31	382	726	60	1168	327	688	66	1077	40
1993-94	442	1254	39	299	766	76	1141	321	733	68	1117	40
1994-95	300	1297	39	257	798	74	1129	314	779	70	1157	40
1995-96	178	1248	37	194	745	86	1025	307	824	72	1197	40
1996-97	230	1422	37	235	864	73	1122	301	870	74	1237	40
1997-98	293	1357	34	312	814	62	1188	294	915	75	1277	40
1998-99	217	1529	35	242	970	72	1284	287	961	77	1317	40
compound growth rate	-1.77	8.65	3.60	01.94	9.42	2.96	4.19					
Co-efficient of determination	0.118	0.969	0.519	0.118	0.969	0.519	0.796					

Source: Thangavelu 2000

3. DEVELOPMENT OF SUITABLE MULBERRY VARIETIES

Requirement of mulberry in different regions/zones of India *viz.*, hills and plains of North Eastern regions, eastern regions, North temperate region, irrigated and rainfed southern regions, semi-arid southern coastal rainfed regions and saline regions are different due to varied inherent regional problems (Table 39, Sarkar 2000).

Table 34: Variety wise growth index of non-mulberry cocoon and raw silk production and its simple and compound growth rate in India

Year	Cocoon production			Raw silk production			
	Tasar (lakh kahan)	Eri (MT)	Muga (Lakh No.)	Tasar (MT)	Eri (MT)	Muga (MT)	Total raw silk (MT)
1980-81	100.00	100.00	100.00	100.00	100.00	100.00	100.00
1981-82	97.15	110.00	89.66	96.98	108.89	91.67	100.00
1982-83	121.95	163.00	48.28	107.17	157.78	77.08	119.20
1983-84	202.03	184.00	44.83	157.74	200.00	112.50	165.63
1984-85	197.15	201.00	72.41	167.55	206.67	114.58	137.66
1985-86	190.24	278.00	93.10	175.09	260.74	108.33	193.75
1986-87	174.80	316.00	89.66	206.79	290.37	114.58	222.10
1987-88	189.02	386.00	103.45	174.72	386.67	120.83	232.81
1988-89	137.40	471.00	75.86	135.09	418.52	93.75	216.07
1989-90	182.90	492.00	96.55	175.47	436.33	118.75	247.99
1990-91	159.35	510.00	144.83	143.40	462.22	145.83	239.73
1991-92	137.80	585.00	137.93	124.15	521.48	150.00	246.65
1992-93	160.57	594.00	106.90	144.15	537.78	125.00	260.71
1993-94	179.67	627.00	134.48	112.83	567.41	158.00	254.69
1994-95	121.95	649.00	134.48	96.98	591.11	154.17	252 01
1995-96	72.36	624.00	127.59	73.21	551.85	179.17	228.79
1996-97	93.50	711.00	127.59	88.68	640.00	152.00	250.45
1997-98	119.11	679.00	117.24	117.74	602.96	129.17	265.18
1998-99	88.21	765.00	120.69	91.32	718.52	150.00	286.61

*Source:*Thangavelu 2000

During 1950s, Mysore Local cultivar with a yield potential of 4-5 m.t./ha/yr. under rainfed conditions and 10-15 m.t/ha/yr under irrigated conditions was the popular variety in south India (Dandin 1997). In temperate zone of Kashmir valley Goshoerami, Tsukasaguwa and Ichinose were introduced from Japan while Kajali, a local indica type mulberry variety was popular in West Bengal (Table 40). In North Western states S-146, Chinese White, Himachal local, Tr-10, Chak Majra were recommended for plantation under sub-tropical regions whereas Chinese white, Goshoerami, Ichinose, KNG, Kokuso etc., were advocated for temperate agroclimatic conditions (Ahsan 2000). During mid 1960s, Kanva-2 and Mandalaya became popular varieties as Mysore-5 (M5) and Ber-S1(S1). Subsequently in southern

Table 35: Statewise production trend of non-mulberry raw silk (MT) in India in the recent years

State	1996-97				1997-98				1998-99				Basing 96-97 NM increase/decrease (%)
	Tasar	Eri	Muga	Total	Tasar	Eri	Muga	Total	Tasar	Eri	Muga	Total	
Andhra Pradesh	3	—	—	3	1.8	—	—	1.8	1	—	—	1	-66.67
Assam	—	439	72	511	—	406.61	59.56	466.17	—	493	70	563	10.18
Arunachal Pradesh	—	10	—	10	—	12.08	0.16	12.24	—	12	0.17	12.17	21.7
Bihar* (Jharkhand)	135	24	—	159	144	30	—	174	120	22	—	142	-10.69
Madhya Pradesh* (Chattisgarh)	33	—	—	33	96.73	—	—	96.73	68	—	—	68	106.06
Maharashtra	—	—	—	—	—	—	—	—	0.19	—	—	0.19	—
Mizoram	—	1	—	1	—	3.69	—	3.69	—	—	—	—	—
Meghalaya	—	194	1	195	—	159.96	1.52	161.48	—	234	1.76	235.76	20.90
Nagaland	—	23	—	23	0.05	24.40	—	24.45	0.12	24.62	—	24.74	7.57
Orissa	43	1	—	44	47.40	3	—	50.40	33.27	0.50	—	33.77	-23.25
Uttar Pradesh* (Uttaranchal)	2	—	—	2	2.50	—	—	2.50	2.74	—	—	2.74	37.00
West Bengal	18	3	—	21	18	4	0.09	22.09	16	6	0.13	22.13	5.38
Total	235	864	73	1172	312.48	813.74	61.53	1187.75	242	970.12	72.06	1284.18	9.57

Tasar includes tropical as well as temperate oak tasar

* Bihar, Uttar Pradesh and Madhya Pradesh have been divided into additional states of Jharkhand, Uttaranchal and Chattisgarh respectively.

Source:Thangavelu 2000

parts of India S-30, S-36, S-41, and S-54 varieties with yield potential of 30-35 m.t/ha/yr were evolved and popularised under recommended package of practices. In West Bengal, C776, C763, C799 and B2-259 with a yield potential of 25 m.t/ha/yr. were evolved, recommended and popularised. During 1980s for rainfed conditions of south India, RFS-135 and RFS-175 with a yield potential of 10-12 m.t/ha/yr were found promising. Recently evolved V1 variety has been found to yield 70 mt. leaf/ha/yr. under irrigated and recommended package of practices (Sarkar *et al.* 1999). Mulberry varieties presently under cultivation in different regions of India under various agroclimatic conditions, breeding methods used to develop them, place of their development and their leaf yield in respective areas has been given in Table 41 (Sarkar 2000).

Genotype related micronutrient deficiencies have been noticed widely in farmers field. Major incidences include Zn deficiency in DD (or Vishwa), Mn deficiency in V1, Mn and S deficiencies in M5 and P, K and Mg deficiencies in few other varieties need adequate attention. Nutrient use efficiencies and nutrient budgeting in respect of different genotypes is hence imperative for harnessing the genotypic potential (Bongale 2000).

Table 36: Wild silk (Tasar) export and its share in natural silk exports

Unit: Lakh sq.metre; Value Rs. in lakh

Year	Wild silk (Tasar) export				Total silk export				Wild silk share %	
	Qty.	Index	Value	Index	Qty.	Index	Value	Index	Qty.	Value
1980-81	15.02	100.00	563.92	100.00	125.82	100.00	5295.11	100.00	11.94	10.65
1981-82	18.29	121.77	686.03	121.65	142.99	113.65	6890.45	130.13	12.79	9.96
1982-83	17.07	113.65	661.55	117.31	140.90	111.99	7930.94	149.78	12.11	8.34
1983-84	21.38	142.34	855.72	151.74	147.75	117.42	9630.32	181.87	14.47	8.88
1984-85	23.42	155.93	1025.97	181.94	170.70	135.67	12532.84	236.69	13.71	8.18
1985-86	16.23	108.26	815.43	144.60	194.15	154.31	15921.14	300.68	8.36	5.12
1986-87	12.66	84.29	654.57	116.07	243.53	193.55	20000.69	377.72	5.19	3.27
1987-88	16.12	107.32	841.98	149.31	326.56	259.55	25179.49	475.52	4.94	3.34
1988-89	14.81	98.60	884.46	156.84	361.42	287.25	32791.76	619.28	4.09	2.70
1989-90	11.24	74.83	807.31	143.16	358.34	284.80	39247.77	741.21	3.13	2.60
1990-91	11.78	78.43	908.51	161.11	324.54	257.94	43593.75	823.28	3.62	2.08
1991-92	11.93	143.94	1141.93	202.50	386.88	307.49	67098.48	1267.18	3.08	1.70
1993-94	21.62	196.94	2881.13	510.91	466.32	370.62	78622.25	1484.81	4.63	3.66
1994-95	29.58	103.46	4123.08	731.15	610.33	485.08	92696.17	1750.60	4.84	4.45
1995-96	15.54	67.91	2530.63	448.76	492.56	391.48	84516.45	1592.12	3.15	2.99
1996-97	10.47	68.51	1722.97	305.53	480.83	382.16	87646.53	1655.24	2.17	1.97
1997-98	10.20	67.91	1622.15	287.66	402.23	319.69	90442.44	17080.04	2.54	1.80
1998-99	10.29	68.51	1924.53	323.54	363.34	288.78	100376.20	1895.64	2.83	1.82

Among non-mulberry silk only tasar silk is exported.
Source: Thangavelu 2000

Table 37: Foreign exchange earnings of India through silk exports (Rs.in crore)

Year	Mulberry	Tasar	Mixed/ Blended	Total	Silk Waste	Grand Total
1950-51	—	—	—	0.16	0.37	0.53
1960-61	—	—	—	1.03	0.34	1.37
1965-66	—	—	—	2.28	0.41	2.69
1968-69	—	—	—	6.37	0.4	6.83
1973-74	—	—	—	12.37	2.09	14.46
1977-78	—	—	—	31.60	1.46	33.06
1979-80	—	—	—	47.57	1.26	48.83
1980-81	46.7418	5.6392	0.5701	52.9511	0.1694	53.1205
1981-82	61.3208	6.8603	0.7234	68.9045	0.8276	69.7321
1982-83	71.6920	6.6155	1.0019	79.3094	3.5373 0.0017*	82.8484
1983-84	85.9193	8.5572	1.8267	96.3032	15.3694	111.6726
1984-85	112.6223	10.2597	2.4464	125.3284	3.7230	129.0514
1985-86	148.0124	8.1543	3.0447	159.2114	0.6067	159.8181
1986-87	188.3903	6.5457	5.0709	200.0069	1.4792	201.4861
1987-88	236.4192	8.4198	6.9559	251.7949	3.1760	254.9709
1988-89	311.0046	8.8416	8.0714	327.9176	2.6210	330.5386
1989-90	375.3327	8.0731	9.0719	292.4777	8.1323	400.6100
1990-91	417.1967	9.0851	9.6557	435.9375	4.5932	440.5307
1991-92	646.9600	11.4193	12.6055	670.9848	4.5815	675.5663
1992-93	—	—	—	—	—	733.1217**
1993-94	738.4994	28.8113	18.9118	786.2225	2.8818	789.1043
1994-95	862.5069	41.2308	23.2240	926.9617	3.5048 7.0722*	937.5387
1995-96	793.3702	25.3063	26.4880	845.1645	0.4724 0.4392*	846.0761
1996-97	834.7167	17.2297	24.5189	876.4653	3.4402 0.5388*	880.4443
1997-98	856.3567	16.2215	31.8462	904.4244	20.8935 0.9673*	926.2852
1998-99	943.2350	18.2453	42.2817	1003.7620	32.3515 0.1621* 50.0601***	1036.1135
1999-2000	1142.8757	22.0186	74.6342	1239.5285	36.3561 11.35* 25.6818***	1275.8846

* Refers to silk yarn export   ** DGCIS data and break up not available   *** DEPB, not certified by CSB.

Adopted from Silkmans Companion 1992, CSB 1999, Indian Silk, June 2000

Table 38: Total export earnings of India from silk items (Value in lakh Rs. & Million US $)

Export	April to March				% increase over 1998-99	
	1999-2000		1998-99			
	Rs.	US $	Rs.	US $	Rs.	US $
1. Certified Export						
(i) Silk goods	123952.85	285.93	100376.20	238.42	23.5	
(ii) Silk yarn	11.35	0.03	16.21	0.04		
(iii) Silk waste	3635.61	8.39	3235.15	7.69	12.4	
Sub total (i+ii+iii)	127599.81	294.35	103627.56	246.15	23.1	
2. DEPB (without CSB certification)	2568.18	5.92	5006.01	11.89	-48.7	19.9
3. 100% EOUs +						
(a) Silk goods	17315.98	39.94	14571.98	34.61	18.8	9.1
(b) Silk yarn/waste	2694.47	6.22	1849.18	4.39	45.7	
Sub total (a+b)	20010.45	46.16	16421.16	39.00	21.9	19.6
Grand Total (1+2+3)	150178.44	346.43	125054.73	297.04	20.1	16.6

+ Refers to Tariff certificates issued by CSB against self declaration.

Source: Indian Silk, June 2000

Under BSTD (Bivoltine Sericulture Technology Development) project, new varieties were tested for suitability to bivoltine rearing. Varieties S36 and V1 were found suitable for young age while V1 variety was found suitable for the late age rearing of bivoltines. The characteristics of S36 and V1 varieties are given in Table 42.

4. CULTIVATION OF MULBERRY VARIETIES

For bivoltine sericulture Indo-Japanese paired row system of plantation (90+150 cm) x 60 cm and a fertilizer package of 350:140:140 (N:P:K) kg/ha/year with 20 mt.of FYM/ha/year have been recommended for cultivation of both S36 as well as V1 plantations. The mulberry leaf yield significantly increased from existing 35 mt/ha/yr. to 70 mt/ha/yr. due to use of newly evolved V1 variety and this type of package of cultivation in South India (Table 42).

5. DEVELOPMENT OF TROPICAL BIVOLTINE SERICULTURE TECHNOLOGY

Bivoltine Sericulture Development Project (BSTD) was launched in 1991 with the joint collaboration of Japanese and Indian specialists to develop appropriate technology package for bivoltine sericulture in the tropics. Besides development of suitable bivoltine hybrids, mulberry varieties and their cultivation practices discussed above, following technologies were also developed and field tested as a package since 1997. These technologies guarantee the success of bivoltine cocoon production and promise to support ushering of a new era in Indian Sericulture (Datta 2000b, Pavan Kumar *et al.* 2000 a, b).

(a) Disinfection

For successful bivoltine crop, disinfection of the rearing house, surroundings, appliances and practice of hygienic measures are very essential. Disinfectant formalin is replaced by chlorine dioxide mixed with slaked lime solution in appropriate quantity for rearing houses. As compared to formalin or bleaching powder solution, this chemical is less corrosive. The powder of bed disinfectant VIJETHA developed at CSRTI, Mysore should be dusted on the silkworms after each moult, half an hour to one hour prior to resumption of feeding and during the third day of fifth stage.

(b) Silkworm rearing

As chawki or young age rearing is most crucial period for ensuring the crop stability and higher cocoon yield, it should be done in a separate Chawki Rearing Centre (CRC) which should cater to the needs of 10-20 farmers and must have facilities for maintenance of constant temperature and humidity, absolute disinfection and use of bed disinfectants. Succulent leaves of V1 or S36 varieties with a protein content of 26% and moisture content of 80% should be fed to young age larvae. Instead of tray rearing, shelf or shoot rearing where 2-3 feedings are sufficient should be practised for late age rearing to reduce labour cost by 40%. Rotary mountages should be used for spinning of cocoons of uniform shape which fetch higher price of Rs. 20-25 over the usual cocoon price in the market.

(c) Seed crop rearing and seed production

In India sector wise approximate contribution in production of silkworm, *B.mori* seeds of all types has been found to be 25% from State Governments, 55% from private sector (Licensed Seed Producers : LSP), 13% from others and 7% from National Silkworm Seed Project (NSSP) of CSB (Table 30). During 1995-96, 1996-97 and 1997-98, the proportion of F1 bivoltine to multivoltine silkworm seeds was found to be merely 1.74%, 1.69% and 1.53% respectively (Table 29) which was produced by Govt. sector only excepting Shanshi Seritech Ltd. (A.P). The three Silkworm Seed Production Centres (SSPCs) at Udhampur, Dehradun and Bangalore with production of 25 lakh dfls/year contributed about 50% of bivoltine F1 (industrial) quality seed for commercialisation while Basic Seed Farms (BSF) contributed towards maintenance and multiplication of traditional and elite reproductive seeds. The participation of private sector is not noticed in J&K, H.P, Uttaranchal and Punjab which are core bivoltine areas in the country. In BSF the seed is multiplied true to mother type during August/Sept.-Feb./March on the basis of F1 requirement in field. Hence, development of protocol for certification of seeds is absolutely necessary to ensure quality maintenance (Pavan Kumar *et al.* 2000a).

In Karnataka, production pattern of Bivoltine Seed Cocoons (BSC) for production of F1 seed indicates that 67.93% of the seed cocoons is generated from non-seed area. Yet, Adopted Seed Rearer (ASR) and Adopted Seed Area (ASA) should also be considered since certain areas offer agroclimatic conditions more suitable for seed cocoon generation. *e.g.*, Channarayapatna of Karnataka for CSR breeds. Chawki (young age) worms of CSR breeds are distributed to selected farmers who can strictly adopt bivoltine technology right from mulberry cultivation to mounting facilities. For creating competition among farmers to produce good seed cocoons according to the set quality parameters price of the cocoons needs to be quality based as in other countries.

BIVOLTINE SEED PRODUCTION TECHNOLOGY

Salient features of recently evolved bivoltine seed production technology (Pavan Kumar 2000 b) are enumerated below:

I. Transportation, sorting and preservation of basic seed cocoons (BSC)

1. Loosely packed BSCs should be transported during cool hours (evening) in plastic crates (60 x 40 x 22 cm lbh) that accommodate 7-8 kg or 5000 cocoons approximately.

2. At SSPC, BSCs are spread in single layer in plastic trays (60 x 90 x 5 cm lbh) arranged in double rows in cocoon preservation stands (190 x 135 x 65 cm lbh).

II. Pre-emergence (eclosion) operations

i.   Seed cocoons are sorted out on the basis of breed characters like cocoon shape, cocoon colour, grains etc., before processing.

ii. Flimsy cocoons like pointed end, thin end, too small or big, deformed, open end cocoons are rejected.

iii. Selected cocoons are preserved at 24-25°C and 75-80% RH with circulation of fresh air.

iv. Selected cocoons are deflossed and cut and naked pupae are separated sexwise by visual examination and weighing.

Subsequently, 1100 male or 1000 female pupae are preserved in each tray in a single layer on corrugated paper (60 x 90 cm) and covered with double layered plastic moth picking net.

For regulating mating of desired races/breeds, pigmentation related to pupal development and emergence is followed as hereunder:

—   Beginning of pigmentation of compound eyes indicates completion of half period from mounting to emergence.

—   Blackening of compound eyes indicates that moth emergence may occur within 3-4 days.

—   Turning dark of antennae indicates that moth emergence may occur within 2-3 days.

Softening and loosening of pupal body indicates that moth will emerge next day in case of Chinese races and subsequent day in case of Japanese races.

Synchronization of moth emergence of different breeds is critical and hence planned from brushing. On the basis of larval period, the breeds of oval cocoons are brushed one day later to breeds that spin constricted cocoons or else the seed cocoons of different races of varying but matching spinning dates may be procured to synchronize moth emergence. Besides, the cocoons, pupae or moth may be preserved at low temperature whenever required. The emergence occurs after 12-13 days from spinning and continues for 4-5 days. The peak emergence occurs on third day while cumulative maximum emergence occurs between second to fourth day.

III. Post emergence operations

(i) Rejection and pairing: Defective females are rejected which are identified as females with compressed and elongated body segments, abnormal body shape with black specks on wings and body segments, abnormal or underdeveloped wings, naked body with scaly hairs, heavy and bulky body, fused terminal abdominal segments/ovipositors and protruded organs from hind end. The inactive, sluggish and incapable moths for copulation are also rejected.

For pairing healthy moths are collected, allowed to spread wings and 10-20% more male moths are mixed with female moths. Unpaired male moths are collected after 30-45 minutes and paired separately. The pairing room should be kept dark and temperature and relative humidity is maintained at 24-20°C and 70-80% respectively. After three and half to four hours female moths are decoupled, gently stimulated for urination and shifted for oviposition.

(ii) Oviposition: Large starched kraft sheets (60 x 90 cm) are used to place 250-300 female moths of F1 combination wise. Whole sheet is rejected if pebrine is noticed which causes loss in grainage economics. Hence, the use of megacellule with 100% moth testing was recommended by Pavan Kumar *et al.* (2000 c).

(iii) Mother moth testing and loose egg preparation: Dry moth testing is popular in North India while fresh moth testing is popular in South India. Microscopic examination of female moths is resorted to eliminate the diseased layings.

Loose egg production is now preferred over laying on sheet as it facilitates effective surface disinfection and acid treatment and elimination of defective and low weight eggs. It also helps in standardisation of number of eggs/gm and per box, improves grainage economics, saves space and labour during oviposition, ensures easy preservation and incubation of eggs and evaluation of breeds combination wise, region and season wise etc.

(d) Improvement of reeling technology

Sorting of bivoltine cocoon has been suggested prior to reeling to maintain uniform quality of the yarn. CSTRI, Bangalore has suggested use of multiend reeling machines having the facilities of hot air cocoon stifling, steam cooking (3 pan cooking), soaking of reeled silk in permeation chamber etc. to produce international quality silk 3A-4A grade.

(e) Maintenance of silkworm race

A new system of maintenance and multiplication of basic stocks (breeders stock to P2) has been implemented in all the selected Basic Seed Farms of NSSP and State Departments and systematic training was imparted to technical staff on knowledge of breed characteristics, its maintenance and multiplication, transportation of cocoons and egg production.

6. POPULARISATION OF BIVOLTINE TROPICAL TECHNOLOGY

The extension staff have to be trained to educate the farmers correctly about the bivoltine tropical technologies. The extension workers should also do a continuous follow up of the programme to popularise the technologies as well as to get the feedback and guide farmers timely. More than 100 farmers were selected and trained under Indo-Japanese collaboration project to popularise the technology at their places with the support of Dept. of Sericulture, Karnataka, Andhra Pradesh and Tamil Nadu. The field performance of CSR hybrids by these farmers (Table 26) clearly affirms that bivoltine hybrid sericulture can successfully be introduced in the southern peninsula if adequate care is taken from plantation of mulberry varieties to rearing of cocoon and its transportation to the market. This technology will certainly perform better in the Northern regions of India where climate is more congenial during spring and autumn. Although small attempt has been made and fairly good results have been obtained with CSR breeds in Kashmir and other northern areas yet a break through is still to be made (Table 27).

The fact that introduction of bivoltine breeds alone, did not yield expected results, cautions for creation of facilities like supply of quality incubated silkworm seeds to well equipped chawki rearing centres, with the facility to maintain required humidity and temperature, utilisation of independent rearing houses with adequate disinfection measures, cultivation of suitable mulberry varieties under irrigated conditions, bivoltine market facility for the sale of cocoons and silk reeling using multiend machines. The State Governments should create the above facilities first and then switch over to CSR breeds with required extension support.

D. CHALLENGES AHEAD FOR INDIAN SERICULTURE

Although progress made by India in sericulture during last two decades have been very significant since cocoon yield/100 dfls has increased from 25 to 35-40 kg, renditta has come down from 14.0 to 8.75 and silk filament length has increased from around 350-400 meters to 650-700 meters due to the introduction of multi x bi cross breeds involving male BV and multivoltine female in a big way. Yet recent surveys conducted by CSB indicate that lighter sarees are being preferred. Over 90% of silk consumption in India is in silk sarees. Unlike in Japan and China where traditional women's Kimono and Sanghai gowns are replaced with western dress which does not necessarily require silk, Indian women will continue to wear saree fabric due to reinforcement in custom and religion. Recently, shift is also perceived in the market gradually for increasing demand of finer and uniform deniers of yarns. The quality requirements of various deniers of silk with "ply" day "twist" details in various weaving clusters of the country have also been increasing. Made-ups and garments from silk have unlimited scope and ever-growing market. De-reservation of garment sector from small scale industry sector under recent Textile Policy is likely to push garment sector a big leap forward, hence, a lot of pressure will be exerted on the weaving sector to manufacture quality fabrics to meet the requirements at reasonable and acceptable prices. In the backdrop of this emerging scenario, the domestic consumption is estimated to cross 25000 m.t. in the next few years which amounts to doubling of the production from current levels of 13944 mt. during 1999-2000 and 15900 mt during 2000-01. The demand and supply gap that exists today is approximately 7000 m.t. which is being filled by imports and is superior to bulk of the raw silk produced in India. The entire imported silk is consumed in power loom sector which indicates that the demand and supply gap that needs to be filled is by high quality yarn and this gap would continue to widen unless efforts are made to produce high quality silk. Hence, quality improvement has become a dire necessity even for internal market (Somashekar 2000). With the World Trade Organisation liberalising the import restrictions by 2002, China's low quality yarn will be sold to India in a big way. Even now China's ungraded or 'C' grade silk is imported officially to the extent of approximately 5000 m.t. annually. Without restrictions, this will go up with Chinese silk also coming through Nepal and Bangladesh. Chinese silk is highly subsidised and its products are sold at a cheaper rate as compared to the silk goods originating from other countries. This situation has adversely affected the trade opportunities of other countries besides undermining the centuries old image of silk. The Indian silk yarn is mostly of "H" quality, while internationally traded silk yarn is 2A-3A grade. Powerloom weaving with poor quality yarn results in costly and defective cloth compared with other silk producing countries. Hence, the entire industry has remained in a syndrome of poor farmers, poor reelers with ancient machines, poor handloom and small power loom weavers controlled by master weavers in unorganised sectors. On the other hand, silk and silk items from other countries shall enter Indian market easily under the Agreement on Textiles and Clothing (ATC) of WTO. The incentives, under the Subsidies and Countervailing Measures (SCM), will have to be regularised on the basis of standard production cost and other parameters which would create a healthy trade situation and uniformity in price for silk in the international market. Besides, the imposition of tariffs cannot control the cost of silk beyond the bound rates and would be much lower than that stipulated now.

E. STRATEGIES REQUIRED FOR INDIAN SERICULTURE

On the basis of past experiences, current production scenario and national and international requirements there is a dire need to devise certain strategies for development of Indian silk industry to equip itself and focus more on quality, product attributes, market competitiveness, service match, pricing and trade promotion etc. as early as possible to counter the expected challenges arising out of World Trade Organisation (WTO) and General Agreement on Trade and Tariff (GATT). These strategies may be enumerated briefly hereunder:

1. Quality should be the prime criteria in all the segments of the silk industry.

2. More emphasis should be given on vertical growth rather than on horizontal expansion.

3. Bivoltine sericulture technology integrating different sets of genetic materials and practices should be standardised and instilled at the grass root level in different agroclimatic regions of the country. Major silk producing states like Karnataka, Andhra Pradesh and Tamil Nadu have already finalised their bivoltine production plans for up to the end of X Plan (Oommen 2000). Since, Bivoltine Sericulture Tropical Technology developed by CSR&TI, Mysore is fairing well, it should be implemented with appropriate modifications in Northern India where climate is more congenial (Datta 2000b).

4. The past experience of introducing bivoltine breeds alone met failures and not the success. Hence, State Govt. planning to switch over to CSR breeds should caution their extension staff not to go ahead unless they are equipped with the facilities for implementation of a package for promotion of new bivoltine sericulture (Datta 2000b).

5. Appropriate importance must be given to multivoltine sericulture with cross breeds to improve the productivity and quality in regions more suitable for multivoltine sericulture.

6. Improved and cost effective machineries affordable to the reelers should be developed to instil the quality as well as to appreciate higher value for quality products at grass root or primary producer level (Kumar & Das 2000).

7. The organised textile sector of India have not evinced any interest in the indigenous silk industry. Although few companies *viz.*, Tatas, Kirloskars and Thapars evinced some interest in integrated silk production from mulberry to weaving silk cloth, they erred to produce all the raw material of cocoons by themselves through contract farming and also by importing high cost reeling and weaving machines without ensuring the bivoltine raw materials. Probably due to the fact that under the existing laws they could not ensure that seri-farmers whom they wanted to assist in funding better inputs will certainly sell the bivoltine cocoons to them only because as the Law stands now all the reeling cocoons must be brought to markets and all the reelers can take part in the auction. Since contract farming is being practised in other agro-based industries *viz.*, Cotton, Sugar, Tobacco, Coffee etc., the same provisions/ laws should be regulated in sericulture also to facilitate entry of private companies in the field of sericulture. The companies should be allowed to construct a filature with multi-end or automatic reeling machines to purchase cocoons with its contract farmers from an area proximate to the filature. To protect the interest of farmers, the choice can still be left to the farmers

whether to sell the bivoltine cocoon produced by them to the company filature or take it to the cocoon markets (Balasubramaniam 2000).

8. The existing laws also prohibit a modern grainage in the private sector from selling its hybrid eggs outside its "licensed area". While farmers will buy good quality eggs from any grainage, such legal restrictions have only bred unhealthy practices, and unreasonable restrictions which hampered hybrid bivoltine egg production. Hence, CSB and State Departments may withdraw from commercial nature of activities such as hybrid egg production instead they should encourage organised companies to take it up even by leasing the large number of world's big grainages build under the World Bank Projects to the private sector retaining only the quality control aspects (Balasubramaniam 2000).

9. Soft loans should be given to the small reelers to convert their charkha (which accounts for over 45% of the raw silk produced today) and the cottage basins (which is obsolete technology) to multiend basins and automatic reeling filatures (Balasubramaniam 2000).

10. Regular market surveys should be conducted to understand the demand and pulls, changing fashions, product requirements etc. to formulate policies and programmes for implementation in future by various agencies.

11. Exclusive brand promotion should be encouraged in Indian Silk for better quality assurance to the consumers (Kumar & Das 2000).

12. Basic information about qualities of silk is lacking among the younger generation, hence a campaign should be started from school age onwards about the unique ecofriendly characteristics of the silk to the users.

13. Too often the products offered are conservative, unimaginative and old fashioned. Hence, serious efforts should be made to innovate new products according to the needs of younger consumers who are more dynamic, more mobile yet attached to some solid value of quality and traditions (Ronald Currie 2000).

14. A National Information System on sericulture and silk should be developed to create a sound database for better policy formulation, planning, management and implementation of the programmes.

15. Sericulture should be integrated with other allied ecofriendly remunerative ventures like silviculture, horticulture, agriculture, pisciculture and dairy to enable farmers to earn much higher income in less expenditures/investments.

16. Bye product utilization should be given greater thrust under R&D activities in sericulture.

17. Non-textile use of silk, silkworms and silkplants in pharmaceuticals and cosmetic industries should be promoted in the line of Korea and other countries to increase the income of farmers manifold.

2. TASAR SILK

Tropical tasar silk originated in India and practise of tasar culture is an important facet of the country's tribal tradition. It is an exclusive and centuries old craft of the hill folk and tribals of Central India. India has been the second largest producer of tasar silk in the world since a long time. Nearly 77% of the tribal population out of 38 million tribal populations are located in tasar cultivated states and 0.2 million tribal families are engaged in tasar culture. Hence, there has been a requirement of 20 million tasar dfls for tasar farmers @ 100 dfls/farmer in the country. But the organized sectors of CSB and State Sericulture Departments have not been able to meet the entire requirement due to one or the other reasons. The highest total dfls production was recorded as 154.69 lakh (15.47 million) during 1990-91, subsequently it declined upto 89.00 lakh dfls during 1998-99. The present growth rate of dfls production is 68.31% against the base year 1989-90) which is the lowest growth rate in the decade. Similarly, during 1989-90 the production of dfls by C.S.B. was 16.85 lakh, which has come down to 11.11 lakh during 1998-99 registering a decline of 34% against the base year (Table 31).

The cocoon and raw silk production during 1998-99 have registered 88.21% and 91.31% growth rate approximately since 1980-81. The cocoon production was 246000 kahan and raw silk production was 265 m.t. during 1980-81 which has come down to 217000 and 242 m.t. respectively during 1998-99. The compound annual growth indicates that cocoon and raw silk production are declining by 1.77% and 1.94% respectively every year (Thangavelu and Rai 2000). The cocoon and raw silk production from tasar sector since 1980-81 to 1998-99 are shown in Table 33).

The tasar raw silk contributed 18% of non-mulberry and 1.55% approximately of the total raw silk production in India during 1998-99 which is indeed very low. The data indicate that the contribution of tasar raw silk was 12.07% of the total raw silk production during 1971-72. Hence, the contribution of tasar raw silk to total raw silk has declined by 87.158% during the last thirty years (Tables 10, 11&12).

Tasar culture is practised in six major states of India *viz.*, Bihar, Orissa, Chattisgarh (erstwhile M.P), Andhra Pradesh, West Bengal and Uttar Pradesh etc. However, wide variations occur in the production of tasar dfls, cocoon production and raw silk production amongst these states. During 1998-99, Bihar (Jharkhand) was the highest producer of tasar raw silk contributing 49.5% production, followed by Chattisgarh (28.1%), Orissa (13.64%), W.B (16%) and A.P (%) (Tables 32&35).

The last twenty years export data show declining trend in quantity as well as in export value, which is significantly alarming. The contribution of wild silk (tasar) was 11.94% in quantity and 10.65% in value to total silk exports, which was 125.82 lakh sq.metre in quantity and Rs. 52.9511 crores in value during 1980-81. The contribution of tasar in silk export has come down to 2.83% in quantity and 1.82% in value during 1998-99 as compared to the base year of 1980-81 (Tables 36&37). Tasar being natural fabrics has great demands in USA, Germany and other European countries. Its wider export market has been explored since 1950s thereby finding way for almost 90% of its production for export which is in sharp contrast to eri and muga silk whose export is almost negligible (Koshy 1998).

3. MUGA SILK

Like tropical tasar, muga silk also originated in India. It is exclusively confined to India, hence it is a rich heritage of our country. During 1980-81, muga cocoon and raw silk production was 29 lakh and 48 m.t. which rose upto 35 lakhs and 72 m.t. respectively registering increase of 20.69% and 50% respectively (Tables 33&34) during 1998-99. The compound growth of muga cocoon and raw silk production has been found to be 3.6% and 2.96% respectively. The present share of 0.46% of muga silk in total raw silk production is lowest among all the four types of silks cultivated in India (Table 11). There is no export of muga silk from India because of the high consumption and less production within the country.

The production of muga cocoon and silk has been fluctuating during 1980-81 to 1998-99. The highest muga cocoon and raw silk production has been noticed during 1990-91, 1991-92 and 1998-99 (Table 33). Besides Assam, other states *viz.*, West Bengal, Arunachal Pradesh, Mizoram, Meghalaya, Nagaland and Manipur also practise muga culture on a very limited scale.

4. ERI SILK

Eri culture is a traditional practice among the Indo-Mongoloid and Tibeto-Burma sub-tribes and tribals of North-eastern India due to conducive environment for eri food plant cultivation as well as ericulture. Hence, this region accounts for more than 90% of the total eri silk production of the country. During 1980-81 the cocoon production and raw eri silk production was 200 m.t. and 135 m.t. which rose to 1529 m.t. and 970 m.t. respectively during 1998-99 registering an increase of 664.5% and 618.52% respectively. The compound growth rate of eri cocoon and eri silk has been found to be 8.65 and 9.42 respectively during the said period (Tables 33&34).

The production of eri cocoon and eri silk has been found to exhibit a consistent increasing trend during the last twenty years since 1980-81 to 1998-99.

The eri silk is fine, hygroscopic and its thermal qualities are unmatching. Assam Textile Institute in collaboration with Wool Research Association and SASMIRA, Bombay proved that eri could be a finer blend with wool, polyester and similar fibres for the manufacturing of baby blankets, children garments and the like. As eri silk is not reelable, eri cocoons go either for hand or machine spinnings and hence eri silk availability is mostly in the form of hand spun coarser yarn for making of chaddar, a traditional gown for the rural folk. Currently there is no export of eri silk but it has tremendous export potential due to finer counts of spun eri silk for carpet industry, cleaner and finer eri silk tops, thermal underwears, knitted dress materials and suitings (Koshy 1998).

5. TEMPERATE OAK TASAR SILK

Ever since inception of temperate tasar in 1969-70 in sub-Himalayan belt of India only spring to spring *i.e.* only one crop was assured. However, diapause in pupae could be broken by subjecting the cocoons at 16 hrs. photoperiodic treatment for 25 days and summer and autumn crops were also raised. Till 1983, spring and autumn crops were equally good and effective yield of cocoons ranged from 9.28-59.3 during spring and 2.59-35.6 cocoons/dfl during autumn. The cocoon production during 1976 was recorded as 0.23 to 2.44 lakh during spring and 0.3 to 3.34 lakh during autumn in Central Silk Board units of Manipur. During 1977 to 1979 Manipur produced more than 2 crore oak tasar cocoons, which came down to 50 lakh during 1987-88. Currently, Manipur is producing hardly 15-20 lakh cocoons yearly while Himachal Pradesh (HP), Jammu & Kashmir (J&K) and Uttaranchal together produce hardly 10-15 lakh cocoons/year. After 1983, a sharp decline was recorded in the effective yield of cocoons during autumn season and production of cocoons further slowed down. Since autumn crop was generally a failure, the brushing of dfls was avoided during autumn to preserve the seed cocoons for raising spring to spring crop. Hence, only univoltine cocoons were raised after 1986. The main draw back of univoltine stock is poor fecundity and fertility which led to poor recovery of cocoons from seed cocoons (cocoon:cocoon ratio). After 1999-2000 again some sign of stabilization of II-crop is being noticed; yet the total production of cocoons in India is seldom more than 30 lakh.

The Chinese oak tasar silkworm (*Antheraea pernyi*) was first discovered in Shandong Province during Han Dynasty more than 2000 years ago and it was then introduced after 1200 years in North-East China which produces nearly 70% of the total amount of oak tasar produced in China. There is no authentic record available in Chinese history regarding domestication of tasar in China. "Shing" dynasty first started domestication in 968 AD which continued upto "Ming" dynasty. Some records also show that tasar was domesticated in the year 26 AD in China. On the contrary, Indian oak tasar industry was started with synthesis of *Antheraea proylei* J. only recently in 1969. Hence, lot of R&D activities have to be undertaken on silkworm as well as on host plant aspects for overall effective growth of oak tasar industry in India since it has to compete with other remunerative crops also.

A total of 18,41500 ha. of land approximately is covered by various species of oaks in three North Western and six North eastern states of India out of which 46500 ha. falls under exploitable area. A total of 55000, 30500, 135500, 40000, 15000-20000, 24000, 1225000, 15000 and 23000 ha. oak flora are distributed in Jammu & Kashmir, Uttaranchal, Himachal Pradesh, Manipur, Nagaland, Assam, Arunachal Pradesh, Mizoram and Meghalaya respectively. The corresponding exploitable area in these states is 1000, 5000, 3000, 20000, 5000, 2000, 5000, 5000 and 500 ha. respectively. In Manipur, Nagaland, Assam, Meghalaya and Mizoram mostly *Quercus serrata* auct. non Thumb is used for oak tasar culture. Whereas, *Q. acutissima* Carr. (600-1800 m.ASL) exploitable for raising cocoon crop which causes mortality of worms in the later stages of II crop due to its fast maturation. *Q. dealbata* is also available in these states (600-1000 m ASL) it cannot be exploited due to low productivity of oak tasar cocoon on it. The population of *Q. grifithii* is very low and scattered hence it cannot be exploited commercially. In Arunachal Pradesh the oak plantation, mostly of *Q. grifithii* is generally available at higher altitudes (above 1000 m ASL). Due to nature of the food plant, high altitude low temperature and high velocity of wind the larval span prolongs too much even upto 90 days making, this avocation uneconomical in Arunachal Pradesh. In North Western India, *Q.incana* serves as host plant of *A. proylei* J. at low altitude (1200-2500 m ASL) and matures even faster than *Q.serrata* of North eastern region., while at higher altitude (2500-3500 mASL) more suitable oak species *viz.*, *Q. semecarpifolia* is found. The productivity of cocoons is uncertain in *Q.incana* and *Q. semecarpifolia*.

Reasons for declining trend of oak tasar cocoon production after 1983 in India may be one or many. Although no definite factor is discernible yet following main factors/reasons have been identified for causing decline in production.

(a) Physiological deterioration: Prior to 1983, there was no marked difference in the cocoon production and productivity of the I and II crops but 1984 onwards it appears that spring crop rearing was emphasised than second crop rearing which led to the gradual loss in the vitality of the breed itself. Generally, prolonged preservation of spring crop harvest (8 months) at room temperature makes the pupae to utilize the food reserve stored within them, which might in turn affect the fecundity and fertility. For better production and productivity either suitable preservation methods of univoltine seed cocoons has to be evolved or a bivoltine race should be evolved with stabilized summer/autumn crop rearing.

(b) Genetic deterioration: Earlier *A. pernyi* and *A. proylei* stocks were available which exhibited different characteristics but now no marked differences are discernible in both the breeds. Cytologically, morphologically and physiologically both the stocks appear to be the same and available materials of oak feeding silkworms are not found to have high productivity, which shows that genetic deterioration has occurred in *A. proylei*. Hence, we should have more breeds of silkworm for evaluation and *A.pernyi* should be replaced with fresh stock from China for raising fresh F1 hybrids. Recently B6 and BY1 two silkworm breeds with high fecundity and survival rate were introduced from China but they have brought tiger band and virulent pebrine diseases also to India. Although both the diseases have now been controlled, strict disease surveillance is still required. The PRP series of lines especially PRP12, PRP5, PRP3 and PRP2 evolved at Regional Tasar Research Station, Imphal have shown better productivity, yet further improvements are required through biotechnological innovations. Presently, PRP12 and PR5 seeds are at multiplication and multilocational level.

(c) Host plant deterioration: Due to regular exploitation of host plants (especially *Q.serrata* in Manipur) without any supplementation with chemical and/or organic fertilizers, manures etc. through the soil, the host plant leaves might have deteriorated nutritionally to support the viability of silkworms particularly during autumn season. However, fertilization of host plants has been found to improve cocoon production during spring crop whereas no improvements were noticed in II crop due to fertilization/fortification of the host plants. Hence, host plant factor seems to be less responsible for failure of second crop.

(d) Environmental deterioration: The environmental conditions during 1976-1983 were more congenial for second crop. Subsequently, the environmental temperature increased considerably due to global warming which affected II crop of oak tasar in Manipur adversely.

It is possible that either any one or combined effect of all the above factors is responsible for failure of second crop oak tasar silkworms which needs to be rectified immediately if oak tasar culture has to sustain.

F. NON-MULBERRY SILK: OVERALL PRESENT STATUS

The percentage share of mulberry and eri silk in total raw silk has increased whereas percentage share of tasar, muga and total non-mulberry silk has declined drastically after V Plan period (1974-78). The present share of mulberry, non-mulberry, tasar, eri and muga silk are 91.74%, 8.26%, 1.56%, 6.24% and 0.46% respectively (Table 11).

During 1980-81 to 1998-99 tasar raw silk production decreased by 7 m.t. approximately per year which is alarming. On the contrary, total non-mulberry, eri and muga silk production has been increasing annually by 40 m.t., 45 m.t. and 2 m.t. respectively on the basis of actual and estimated raw silk production data. Eri silk has shown having highest growth rate even in comparison to total non-mulberry silk. The compound annual growth rate of tasar, eri, muga and non-mulberry silk has been found to be 1.94%, 9.42%, 2.96% and 4.19% respectively (Table 33)

The overall trend of non-mulberry silk production is not encouraging particularly that of tasar silk which supplies most of the raw materials for export production in the non-mulberry silk category. During 1980-81 to 1998-99 over a period of about twenty years, eri silk production has increased from 135 m.t. to 970 m.t. whereas muga silk production increased from 48 m.t. (1980-81) to 86 m.t. (1995-96) and subsequently decreased to 72 m.t. Likewise, tropical tasar has also shown fluctuating trend by increasing from 265 m.t. (1980-81) to 548 m.t. (1986-87) and subsequently again declining upto 242 m.t. (1998-99) (Table 33).

State-wise production trend during last three years 1996-97 to 1998-99 indicate that tasar raw silk production in Bihar, Orissa and West Bengal has declined by 11.11%, 22.63% and 11.11% during 1998-99 in comparison to 1996-97 and the highest increase of 106.06% was registered in Madhya Pradesh (Chattisgarh). Muga silk has also shown a declining trend in Assam. During 1998-99 in comparison to 1996-97, muga silk has declined by 2.78% in Assam though its production increased by 17.53% as compared to the year 1997-98. On the contrary, eri silk production has shown significant achievement in most of the states. The state-wise increase during 1998-99 in eri silk production is registered 12.30% in Assam, 20.62% in Meghalaya, 7.04% in Nagaland 20% in Arunachal Pradesh and 3.82% in Bihar against the base year 1996-97 (Table 35).

G. CONSTRAINTS IN NON-MULBERRY SERICULTURE

Thangavelu (2000) and Srivastav *et al.* (2000d) have categorised constraints facing non-mulberry sericulture in India into two types *viz.*, man-made and inherent biological constraints which are unique to wild silkworms and their food plants.

(a) Man-made constraints

1. Non-mulberry sericulture is the inherent strength of Indian sericulture yet undue importance given to mulberry silk overshadows it since the planners and administrators often consider the immediate economic returns to the farmers from mulberry sericulture while the long term effect of non-mulberry sericulture on ecological and environmental balance is seldom considered. Hence, opportunities, funds and infrastructure made available to scientists for research and developmental activities in non-mulberry sericulture are highly inadequate to commensurate the numerous and complex problems unique to non-mulberry silk.

2. During early eighties India had 11.16 million hectare of forest land under tasar host plants, out of which 9.7 m.h. accounted for *Shorea robusta* alone while 1.46 m.h. land accounted for *Terminalia arjuna, T. tomentosa* and other host plants. Reduction of forest cover from 23 to 12% during late eighties indicates that 50% of tropical tasar food plants might have been eliminated. Althrough 7845 ha. of land was brought under tasar host plants under Inter State Tasar Project (ISTP) in tropical states during 1981-86, yet it was not more than 0.15% of what we have lost. By then more than 15 years have further passed away and we might have further lost considerable forest area not only due to grazing and illegal felling of trees but also due to dreaded pests and diseases (Srivastav *et al.* 2000 d).

   Likewise during early nineties, Manipur, North East region and whole country had 20000, 37500 and 46500 ha. of land under accessible oak forests but rate of loss of forests has been 7000 ha./yr, 15800 ha./yr and 774100 ha./ year in India after 1995, which is very much alarming (Anonymous 1999).

   As estimated by Planning Commission, the total muga plantation area (10m x 10m spacing) has declined to 2500 ha. in 1998 from 5765 ha. in 1980-81. Against a total requirement of 6420 ha.only 2500-2730 ha. of land is under muga food plantations. While 3920 ha. of additional plantation at the existing rate of utilization is required, only 531 ha. has been covered under Augmentation of Muga Food Plants (AMFP) scheme. Currently, only 7793 ha. of land is under eri food plantation out of which 296 ha. was raised during Augmentation of Eri Food Plants (AEFP) scheme during 1993-99 in the North East region.

   Hence, scarcity of non-mulberry food plantations is a very acute problem. The systematic bloc plantations introduced with the support of state and central sector/schemes form less than 1% in tropical tasar and less than 20% in muga and eri sector which is very inadequate not only from non-mulberry sericulture but also from the point of view of environmental protection.

3.   Excessive use of chemical fertilizers renders the soil unfertile in long course of time which is highly detrimental. The initial increase in yield observed after fertilisation has been decreasing over the years. Moreover, most of the rearers depend on forest plantation where they do not want to apply costly fertilizers. Further, by application of all recommended doses of fertilizers/manures productivity of leaf/ha. could be enhanced from 13000 to 18000 kg/ha in tropical tasar, 8000 to 13000 kg/ha in oak tasar, 16000 to 26500 kg./ha in muga and 3000 to 4000 kg./ha approximately per crop in eri which could enhance rearing capacity from 450 to 700 dfls, 1000 to 1500 dfls, 1000 to 1500 dfls and 300 to 400 dfls approximately per crop respectively which is less economic to farmers.

4.   Non-mulberry farm cannot be raised by farmers due to non-availability of land, less income through non-mulberry sericulture and acute competition with other cash crops.

5.   The forest and economic plantations have been raised through ordinary heterogeneous seeds produced due to open pollination which resulted into heterogeneous population bearing different degrees of productivity levels, quality and susceptibility to various pests and diseases.

6.   The current spacing of 4'x4' being followed in case of tropical and temperate tasar host plants is unsuitable for intercropping and also enhances pest infestation and disease infection.

7.   Although *T.tomentosa* has been found better than *T. arjuna* for rearing of *A. mylitta* as well as for commercial purposes, yet most of the economic plantations have been raised with *T. arjuna* alone.

8.   Non-availability of sufficient quantity of quality silkworm seeds in right time has compelled traditional rearer to switch over to other avocations.

(b) Inherent biological constraints

1.   Non-mulberry sericulture has remained as a subsidiary crop maintained at subsistence level for a low supplementary income. The maximum range of gross annual income/ha/yr. has been found to be Rs. 12000-15000/-, 6000-10000/-, 35000-45000/- and 25000-35000/- from tropical tasar, temperate (oak) tasar, muga and eri culture respectively. Hence, it is not treated at par with other cash crops.

2.   The recommended pest and disease control measures are not only toxic and uneconomical but also impracticable for scattered and mixed forest plantations due to which tribals do not want to adopt them.

3.   Non-mulberry host plants (except Tapoica and Castor) are generally slow growing.

4.   Well identified high yielding qualitatively superior pest and disease resistant varieties of non-mulberry food plants are lacking.

5.   Non-mulberry silkworms are wild in nature, hence less amenable to human handling.

6. Barring eri silkworms, non-mulberry silkworms are reared out door. Hence, they are open to adverse climatic conditions, pests, predators and parasites.

7. Non-mulberry silkworms are highly susceptible to diseases which are aggravated by adverse climatic conditions and hence prone to extremely low survival rate under human handling.

8. Non-mulberry silkworms exhibit unstable voltinism and unstable crop due to low fecundity and physiological and genetic degeneration of the races.

9. Non-availability of genetically distinct and character specific stable races/ strains and lack of high yielding cross breeds/hybrids of non-mulberry silkworms.

10. Low cocoon yield (low effective rate of rearing) and low multiplication rate of non-mulberry silkworms.

11. Despite of occurrence of extensive natural variants in the form of ecoraces and physiogenetic forms, maintenance of germplasm is very difficult in wild silkworms.

12. Genetics of non-mulberry silkworms is not studied in detail in a systematic way to evolve specific purelines and determine racial characters.

13. Pure lines isolated from existing population of mosaic forms (with different genetic linkage groups) exhibit very low survival rate, hence it is very difficult to maintain pure lines for evaluation and evolution of breeds.

14. The non-mulberry silkworms often exhibit loss of robustness of cocoons, health of worms and other quantitative characters in successive generation due to loss of physiological vigour.

15. Biotic potential of the non-mulberry silkworms in respect of fecundity and silk yield could not be studied.

16. Nutritional requirement of the non-mulberry silkworms has not been yet understood.

17. Causes of low fecundity, retention of eggs in the ovary staggered oviposition and consequently the hatching of eggs over 3 to 4 days and low rate of fertilisation which are responsible for low multiplication rate (seed to seed/ dfl to dfl) have not been properly understood.

H. STRATEGIES REQUIRED FOR NON-MULBERRY SERICULTURE

On the basis of foregoing discussion, if we want to make non-mulberry sericulture a sustainable and profitable venture to our farmers besides protecting the environment, following strategies should be adopted in the near future to make India a global power in sericulture in post GATT and WTO era.

(a) Host Plant Improvement

Generally, host plant sector in sericulture as a whole and non-mulberry sericulture in particular need more attention. The non-mulberry sericulture could not make much progress and therefore, following strategies should be adopted for improvement of non-mulberry host plants.

1. New ventures should be launched for raising economic plantations of all non-mulberry food plants with elite genotypes having high nutritive value in the near future to tackle the problem of shortage of food plants.

2. Poly culture system of plantation should be adopted to avoid losses caused due to pests and diseases. Superior genotypes of *Terminali tomentosa* needs to be evolved for incorporation in plantations along with *Terminalia arjuna* and their spontaneous hybrids. Similarly, elite genotypes of *Litsaea polyantha* should be planted along with *Persia bombycina* and *Heteropanax fragrans/Mannihot utilitissima* should be planted along with *Ricinus communis* in new plantations.

3. Different types of sericulture should be integrated with each other as well as with silviculture, agriculture, horticulture, pisciculture, mushroom culture and dairy to attract and benefit farmers more and more.

4. The spacing between the plants and rows should be enhanced from 4' x 4' to 8' x 8' or 9' x 9' to facilitate intercropping of leguminous plants which will not only fix nitrogen in the soil, but also fetch vegetables/pulses and also serve as green manure.

5. Ecofriendly cost effective biofertilisation and organic manuring should be encouraged. Research on use of phylloplane microflora for plant protection as well as leaf production should be intensified.

6. Easy cost effective and efficient package of practices like foliar spray of urea, DAP, hormones, minerals and botanicals etc. should be encouraged for adoption by farmers.

7. Use of ecofriendly neem derivatives should be encouraged under integrated pest management programmes to control pests of non-mulberry host plants.

8. Survey, collection, maintenance, characterization and evaluation of all available natural variants of non-mulberry food plants should be carried out.

9. Fast growing elite genotypes of non-mulberry host plants with high nutritive value and resistance to drought, pests and diseases should be isolated for plantations in future.

10. Attempts should be made to isolate dwarfing gene and fertilizer using efficient genotypes of host plants in the pattern of wheat and rice.

11. Molecular breeding of host plants for introgression of genetic markers associated with disease/pest resistance or tolerance may be attempted through marker assisted selection by involving the most prevalent class of resistance genes which encode a Nucleotide Binding Site (NBS) and a Leucine Rich Repeat (LRR) region.

12. *In-vitro* culture and propagation methods through cuttings should be developed for all main non-mulberry food plants for raising homogenous plantations.

13. Molecular characterisation of non-mulberry host plant genotypes should be carried out through RAPD (Random Amplified Polymorphic DNA), RFLP

(Restriction Fragment Length Polymorphism), AFLP (Amplified Fragment Length Polymorphism), SSR (Single Sequence Repeats), ISSR (Inter Single Sequence Repeats) and EST (Expressed Sequence Tagging) to avoid duplications in Germplasm Banks since they are more reliable as compared to morphological and physiological characters which may be affected by environment also.

14. Subsequently, map based cloning and transposon tagging methods may be employed to isolate genes corresponding to desirable traits in long course of time.

15. Genetic engineering tools may be employed for incorporation of Nif (Nitrogen fixing) genes into non-mulberry food plants in long course of time.

(b) Non-Mulberry Silkworm Improvement

The conventional and traditional research approach adopted for improvement of non-mulberry silkworms for three to four decades has not yielded the much desired results barring some aspects of rearing technology, silkworm seed production and seed organisation. Hence, following strategies should be adopted for improvement of non-mulberry silkworms in future.

1. Survey, collection, maintenance, characterisation and systematic evaluation of all available natural variants of wild silkworms taken up earlier should be intensified urgently to evolve genetically stable pure lines which may be used immediately for evolution of specific lines based on genetic and biochemical parameters. These specific/pure lines should subsequently be used for synthesis of high yielding cross breeds/hybrids in the pattern of cross breeds of mulberry silkworm.

2. Suitable rearing techniques should be developed for maintenance and evaluation of wild silkworm germplasm to assist the breeders to evolve suitable high yielding and stable races and cross breeds/hybrids.

3. Much detail is not available on the linkage groups and gene maps of any of the wild silkworms. Hence, genome of wild silkworms should be deciphered using biotechnological techniques.

4. The biomolecular characterisation of non-mulberry silkworms through RFLP, RAPD, SSR, ISSR, AFLP and EST should be urgently taken up to avoid duplications in the germplasm since they are more reliable.

5. Rearing technologies specific for production of seed cocoons during adverse climatic conditions and production of commercial cocoons during favourable seasons should be developed with specific approaches and methods. Similarly, silkworm rearing techniques for different rearing seasons (monsoon, spring, autumn etc.) should be developed.

6. Wild silkworm rearing technique should aim at minimising the effect of temperature and humidity prevailing during adverse climatic conditions by restructuring the host plant canopy by increasing aeration in rearing site to

prevent excess moisture and reduce high temperature which will greatly reduce incidence and spread of diseases in non-mulberry wild silkworms.

7. The silkworm crop loss due to diseases all times reach more than 60%. This loss is 20-40% even in the regular favourable season. Hence, following strategies are urgently required:

    a. Studies on pathogen specificity, identification of pathogens at species and strains level, epizootics and etiology are required.

    b. Identification of cost effective, ecofriendly disinfectants and antibiotics and development of technology to apply them in open habitat on dwarf bushes as well as large trees for not only prevention of diseases but also for curative effect.

    c. Organic farming to produce quality leaves with high nutritional value for feeding the silkworms which will impart resistance to them is required.

8. Nutritional requirement, growth promoting and deterrent status of biochemicals (allelo chemicals, semi chemicals alkaloids, tannins, fibres and polyphenols) present in the leaves of host plants should be studied, which will facilitate identification of the most suitable elite host plants of tasar, eri and muga silkworms containing the desired nutrients and feeding stimulants for healthy growth, development of disease resistance and promotion of uniform development of eggs in the ovary of these silkworms.

9. Reproductive biology (development and maturation of oocytes, transfer of spermatozoa, fertilisation and oviposition) and physiology of diapause of non-mulberry silkworms should be investigated through integrated studies on the role of brain hormone, insect nutrition and interaction with environment especially altitude, photoperiod, temperature and relative humidity.

10. Artificial diet should be developed to facilitate indoor rearing of wild silkworms atleast during the young stages to produce robust and healthy young silkworms resistant to diseases.

11. Indoor rearing methods should be developed to domesticate wild silkworms especially tropical tasar and muga silkworms to facilitate maintenance of genetically pure lines.

The first phase (1991-96) of "Bivoltines Sericulture Technology Project", a collaborative programme of C.S.B. with Japan International Cooperation Agency (JICA), has already resulted in the development of bivoltine CSR hybrids suitable to the tropical climate of India, evolution of high yielding mulberry varieties like V1, development of suitable cultivation technology, seed technology, rearing and reeling technologies. The second phase (1997-2002) which has just concluded, verified the technologies with 142 selected farmers in three Southern States of Karnataka, Andhra Pradesh and Tamil Nadu has turned out to be a big success as the average yield shot up from 28 kg. to 68 kg./100 dfls while the renditta reduced from 8.5 to 6.0 at farmers levels which has improved socio-economic levels of adopted farmers significantly since some of them are earning a record income of

Rs. 1.2 lakh per year. Large number of farmers inspired by them are keen to switch over to new bivoltine sericulture technologies in tropical states. The success of bivoltine technologies in temperate-subtemperate states is likely to be more encouraging due to more suitable climatic conditions since new "Dun" breeds and their hybrids are at par with the races/hybrids developed in other countries for different economic traits,uniformity in cocoon shape, cocoon grain, larval marking and grainage performance. Cocoon yield of approximately 80 kg./100 dfls, filament length between 1028-1182 mts., raw silk recovery between 17-19%, neatness above 90 points and renditta between 6-7 recorded in "Dun" hybrids *viz.*, Dun 6 x Dun 22 and Dun 12 x Dun 19 indicates that these hybrids are capable of increasing cocoon productivity by two fold in the years to come (Siddiqui *et al.* 2002).

The third phase of the project commencing shortly (2002-07) will focus on its large scale dissemination and socialization and hence calls for meticulous planning and close coordination amongst concerned agencies and seri-extension networks (Editor, Indian Silk, April 2002).

China has controlled the global silk trade for more than 4000 years. Now, the time is ripening to make India a global power in silk trade through slow yet steady improvement in its quality of silk and enhancement in production of raw silk vertically as well as horizontally. Now, India has to get ready and bring silk revolution in the country in the line of "green revolution", "white revolution" and "blue revolution" through scientific know how and technical manpower.

5

Utilisation and Conservation of Seri-Biodiversity

Total sum of variability present in a particular taxa, accumulated through years of evolution under domestication and natural selection, has been referred to as "gene pool" while biological diversity refers to the variety of distinct ecosystem or habitats, the number and variety of species within them, and the range of genetic diversity within the population of each of those species. Hence, variety of living things (beings) and their richness in the world as a whole or in any location within the world have been termed as biodiversity. There is increasing realisation that at least 10 and possibly 40% of the world's currently available species may become extinct within next 50 years (Mukherjee 2000). As creation of gene and/or genotype is much cumbersome, time consuming and expensive, the sensible use of existing natural variability as the basis for improvement as well as resources for future needs is much easier. The rate of species loss may be minimised by immediately adopting appropriate measures against the factors, which cause genetic erosion. The well being of current and future generations may be best served if genetic erosion, the decline and extinction of species and the degradation of ecosystem are kept at minimum level for supporting national development and further sustainable use of biodiversity. It should be our endeavour to preserve the biodiversity available in the nature for our posterity. Any detrimental action of today on the ecosystem will have far reaching consequences on the species composition and diversity and any damage to the ecosystem will be irreversible.

Conservation of biodiversity has been a burning issue of global importance. The issue gains great importance for our country which is one of the 12 identified mega biodiversity country of the world and one of the 12 centres of origin of cultivated plants with two out of 18 identified hot spots *viz.*, Western Ghats and the Eastern Himalayas. India is an abode of more than 17,500 species of higher plants, including 168 major and minor crop species and 334 wild relatives of these crop species. Bulbs, flowers, fruits, seeds and nuts of 1532 wild edible species are used by native tribes of India and 7500 plants of ethno botanical importance have been reported by NAAS (1998). It also harbours 26 breeds of cattles, 40 breeds of sheeps, 20 breeds of goats, and 18 breeds of poultry and is quite rich in faunal biodiversity as well with 81,000 species of animals thereby constituting 6.4% of recorded faunal species of the world (Gautam 1997, Dandin 2000).

International undertaking on Plant Genetic Resources was developed by FAO in 1983 on the concept that "Plant Genetic Resources are the common heritage of mankind". This undertaking was signed by 112 countries including India. Subsequently, historic "Earth Summit" held at Rio de Janeiro, Brazil adopted a "Convention on Biological Diversity" (CBD) in 1992 which was signed by 174 countries and India adopted the same in 1994. In order to prevent exploitation of developing countries which are rich in biodiversity and traditional knowledge by the techno-economically rich developed countries; suitable provisions were made under General Agreement on Tariff and Trade (GATT). The "Rio Earth Summit" rightly recognised the intrinsic values of biological diversity and the "sovereign rights of the nations (States) over their biological diversity (resources)". The convention addresses:

* The fundamental requirements of *in situ* conservation of ecosystems and natural habitats.
* The supporting role of *ex situ* conservation.
* The vital role of local communities and women in the conservation and sustainable use of biological diversity.
* The desirability of sharing equitably the benefits arising from the use of traditional knowledge, skill, innovations and practice along with the biodiversity.
* The importance and need to promote regional and global co-operation for conservation.
* The requirement of substantial investments to conserve biological diversity.

Under Article-42, provisions have been made to address the aforesaid issues in a broader sense and all the signatories have been entrusted to take appropriate measures to derive the benefits of CBD and develop suitable strategies for conservations and sustainable use of their biodiversity for improvement of society.

Under Article-6, a set of norms were established for developing National strategies/plans and programmes for the conservation and sustainable use of biological diversity as well as equitable sharing of benefits derived from such wealth and associated traditional knowledge.

1: SERI-BIODIVERSITY IN INDIA

India is the only country of the world which still harbours all the five known types of commercial silkworms and their host plants in wild/semi-domesticated/domesticated conditions and hence the country is very rich from seri-biodiversity point of view and also therefore has to face the greatest challenge of their conservation as well as sustainable utilization to support the national development.

A brief description of seri-biodiversity with respect to various host plants as well as silkworms has already been made in Chapters I and IV. Hence, only additional pertinent features are being dealt hereunder:

A. HOST PLANT GENETIC RESOURCES

The genetic resources of host plants of silkworms are discussed under mulberry and non-mulberry.

(a) Genetic Resources of Mulberry

A brief description of genetic resources of mulberry has been made in Chapters-I and IV. Hence, only additional pertinent features are dealt in this chapter.

Sericulturally, India may be divided into six areas for mulberry cultivation viz., hills and plains of North Eastern regions, eastern region, temperate (Northern hills), southern region (irrigated), southern coastal region (semi-arid rainfed) and saline regions (Table 39). Accordingly, the problems and requirements for cultivation of mulberry in different zones with respect to agro climatic conditions, local preferences, the extent of water resources, the distribution of diseases/pests/ nematodes and similar other factors are different and have a bearing on the cultivation of mulberry varieties.

Mulberry is believed to have originated on the lower slopes of Himalayas (Sarkar 2000) and wild mulberry (*M.nigra* and *M.laevigata*) are available in the North West and North Eastern India. *M.nigra* is endemic in the north western sector. *M. indica* and *M.alba* are indigenous and found extensively. *M. laevigata* is widely distributed in India along the Himalayan belt, and Andaman and Nicobar islands. There are developed gene pool of *M. laevigata* in India, particularly at Yercaud in Tamil Nadu and its source and origin are unknown. Introduced mulberry gene pool is available in several parts of India, particularly in the tea and coffee estates and some pockets in South India, Madhya Pradesh, Rajasthan and Gujarat.

Between 1991-1999, Central Silkworm Germplasm Resources Centre (CSGRC) erstwhile Silkworm and Mulberry Germplasm Station (SMGS) at Hosur, Tamil Nadu has collected and assembled 908 mulberry resources from different sources after elimination of duplicates from estimated 1600 mulberry genotypes available in the country. The current stock of CSGRC represents more than 80% of the entire institutional resources of 15 species of mulberry from 26 countries (Table 43) viz. *M. australis* Poirch., *M. multicaulis* Poirch., *M. nigra* L., *M. cathayana* Hensi, *M. bombycis* Koidz., *M. alba* L., *M. latifolia* Poirch; *M. serrata* Roxb., *M. rotundiloba* Koidz., *M. thou* Seringe, *M. rubra* L., *M. tiliaefolia* Matino, *M. sinensis* Hort., *M. indica* L., *M. laevigata* Wall. (Mukherjee 2000, Tikader *et al.* 2002).

In early phase of research, selection from natural variability, plant introductions and domestication were considered to be the important steps in identifying a suitable variety for a particular area. The popular mulberry variety K2 was selected from local cultivars in South India, while S1, a genotype from Mandalaya

(Burma) and Goshoerami a genotype collected from Japan were considered as the best varieties for Southern India, West Bengal and North-East India and Kashmir areas respectively. Subsequently, mutation, hybridization and polyploidization technologies were used to develop varieties like S13, S34, S36, S41, S54, V1 and Tr-10 and S1635. The leaf yield in these mulberry varieties ranges from 12 MT/ha/yr in S13 and S34 to 70 MT/ha/yr in V1 varieties. S13 and S34 were developed for rainfed condition, whereas S36, S54 and V1 were evolved for irrigated condition. Pr minent mulberry varieties presently under cultivation in different regions of India, methods used for their development, period of evolution and productivity in respective areas are given in Tables 40&41. The important features of V1 and S36 varieties recommended for bivoltine sericulture are detailed in Table 42.

Table 39: Requirement of mulberry in different zones of India

Areas	Problems	Requirements
1. Hills and plains of North Eastern regions	a. Acid soils	Tolerant to acidity.
	b. More acceptability of other crops like ginger etc.	Suitable plant type for intercropping.
	c. Higher incidence of mildew	Tolerant to diseases.
2. Eastern region	a. Medium yield	High yield
	b. Low yielding during winter season	Stability of yield during winter season.
	c. Higher incidence of leaf spot diseases in rainy season and mildew in winter.	Tolerant to diseases.
	d. Soil salinity in coastal areas.	Tolerant to salinity
3. Temperate (Northern hills)	a. Soil acidity	Tolerant to soil acidity
	b. Low rooting	Superior genotypes with high rooting.
	c. Cold	Tolerant to frost and cold.
	d. High dormancy	Less dormancy.
4. Southern region (irrigated)	a. Medium yield	High yield
	b. Competition with sugarcane and other horticultural crops.	Varieties favouring mechanisation.
	c. High incidence of leaf spot and leaf rust.	Disease tolerant.
	d. Nematode infestation.	Resistant to nematodes.
	e. High incidence of mealy bugs and tukra.	Resistant to sucking pests.
5. Southern region (rainfed and alkaline)	a. Scarcity of water	Tolerant to water stress
	b. High atmospheric temperature, low humidity and high wind velocity.	Drought tolerant.
	c. Erratic rainfall	- do -
	d. Poor annual rainfall with uneven distribution.	- do -
	e. Poor yield	Comparatively high yield
	f. Soil alkalinity	Tolerant to alkalinity
	g. Saline soil	Tolerant to salinity.
6. Coastal saline regions	a. Saline soil	Tolerant to salinity.

Source: Sarkar 2000 and personal observations

Table 40: Evolution/Adoption of mulberry varieties in India

Period	Regions of cultivation and mulberry varieties		
	South India	North East India	North India (J&K)
1950s	Mysore Local, Japanese grafts	Kajjali (local variety)	Goshoerami*, Shatut, Tsukasaguwa*
1960s	Mysore Local, Japanese grafts	Kajjali, Matigara black (local variety)	Goshoerami*, Shatut Sujanpuri
1970s	Kanva-2, Mysore Local	Kajjali, Matigara black, Mandalaya (S-1)	Goshoerami*, Shatut, Tsukasaguwa*, Sujanpuri, Chak majra.
1980s	Kanva-2, Mysore Local, S54, S36, MR2, S30	Berhampore S1, Tr-4, Tr-10, BC259, Kosen*, C776, C763, S1635	Selection-146, Chinese White**
1990s	Kanva-2, S34, S36, S54, S13	TR10, BC259, S766, S799, S1635, Kosen*, Tr-4, C776, C763,	S146, Ichinose*, Kenmochi*, Chinese White**
1995s	Kanva-2, S36, S31, S13, MR2, Visva, V1, S1635, RFS175, RFS 135.		

* Japanese and **Chinese mulberry varieties were introduced into India for the temperate region.

Source: Dandin 1997 and personal observations

(b) Genetic resources of tropical tasar food plants

Genetic resources of primary as well secondary host plants are described in Chapters-I and only additional details are furnished in this chapter.

During 1987 to 1993 explorations were conducted in Madhya Pradesh/ Chattisgarh, Jharkhand/Bihar and Orissa and 130 genotypes were identified as plus trees belonging to various taxa in section "Pentaptera" of genus *Terminalia* which comprises of *Terminalia arjuna* and *T. tomentosa* species complexes (Table 44). The distinguishing features of various taxa of both species complexes are given in Table 45. Altogether four taxa/morphotypes *viz.*, *T. berryi W.&A.*, *T. glabra W.& A.*, small leaf form (minutifolia), putative hybrids closer to *T. arjuna* in *T.arjuna* species complex and eight taxa/morphotypes *viz.*, *T. alata* var. *alata* Heyne ex. Roth., *T. alata* var. *nepalensis* (Haines) Fernandes; Kahvi, big leaf forms and small leaf forms of *T. crenulata* Heyne ex. Roth; big leaf forms and small leaf forms of *T. coriacea* (Roxb.) W. & A. and putative hybrids closer to *T. tomentosa* in *T. tomentosa* species complex were earmarked (Srivastav and Thangavelu 1995, 2000, Srivastav *et al.* 1996, Thangavelu *et al.* 2000). Spontaneous hybrids of both the species were reported as early as in 1925 by Parker.

I. Geographical diversity and gene centres

Conventional methods for delimitation of taxa of both the species complexes is difficult and controversial. While Parkinson (1936) used foliar, bark and fruit characters, Bahadur and Gaur (1980) used geographical distribution pattern also for species delimitation. According to them *T. crenulata* is widely distributed in south west India along Western Ghats, whereas *T. coriacea* is more common in South East India along Eastern Ghats and the typical *T. alata* is restricted to Central India although its variety *nepalensis* is also distributed in sub-mountain Himalayan tracts in the foot hills of Northern India. Distribution of these taxa overlaps in several marginal areas of Central India where their occurrences converge from all directions. Central India has been as such regarded as the "Centre of origin" of section 'Pentaptera' of genus *Terminalia* and 'Indo-Malayan' as well as Australian regions may be regarded as the ancestral home of genus *Terminalia* which has bitopic/polytopic rather than monotopic centres of origin (Srivastav *et al.* 1997a, Srivastav and Thangavelu, 1997). The aforesaid 130 genotypes exhibit maximum variability and include representatives of all the taxa and/biotypes of *Pentaptera* section.

Table 41: Mulberry varieties presently under cultivation in India

Sl. No.	Variety	Region	Developed at	Breeding methods used	Leaf yield in respective areas (MT/ha/yr)
1	Kanva-2	South India (Irrigated)	CSRTI, Mysore	Selection from natural variability.	30 - 35
2	S-36	- do -	- do -	Developed through EMS treatment of Berhampore Local.	40 - 45
3	S-54	- do -	- do -	- do -	65 - 70
4	Victory-1	- do -	- do -	Hybrid from S30 x Ber.C776	65 - 70
5	DD	- do -	KSSRDI, Bangalore	Clonal selection	40
6	S-13	South India (Rainfed)	CSRTI, Mysore	Selection from polycross (Mixed pollen) progeny.	12 - 15
7	S-34	- do -	- do -	- do -	12 - 15
8	MR-2	- do -	- do -	Selection from open pollinated hybrids.	35
9	S-1	Eastern and North-Eastern India (Irrigated)	CSRTI, Berhampore	Introduction from Mandalaya (Burma)	30 - 35
10	S-799	- do -	- do -	Selection from open pollinated hybrids.	35 - 40
11	S-1635	- do -	- do -	Triploid selection	40 - 45
12	S-146	N. India and foot hills of J&K (Irrigated)	- do -	Selection from unknown source.	30 - 35
13	Tr-10	Hills of Eastern India	- do -	Triploid of Berhampore S-1	20 -25
14	BC-259	- do -	- do -	*M.indica* (Local) crossed with Kosen and backcross twice with Kosen.	18 - 20
15	Goshoerami	Temperate	CSRTI, Pampore	Introduction from Japan	15 - 20
16	Kosen	Temperate	CSRTI, Pampore	Introduce from Japan	15 - 20
17	Chak Majra	Sub-temperate	RSRS, Jammu	Selection from natural variability	25 - 30
18	China White	Temperate	CSRTI, Pampore	Clonal selection	15 - 20

Source: Sarkar 2000 and personal observations

Table 42: Important features of V1 and S36 mulberry varieties recommended for bivoltine sericulture in Southern Plateau

Particulars	V-1	S-36
Plant type	Erect	Spreading
Leaf	Ovate, large, entire, dark green, smooth glossy and thick	Cordate, entire, large pale green, smooth and glossy.
Rooting ability	94%	48%
Leaf yield	60-70 MT/ha/yr.	40-45 MT/ha/yr.
Soil	Red loamy and black soil pH upto 8.0	Red loamy soil
Notice	Necessity of more fertilizer and irrigation.	—
NPK requirement	N350, P140, K140 kg/ha/yr	N350, P140, K140 kg/ha/yr
FYM requirement	20 MT/ha/yr	20 MT/ ha/yr
Spacing in paired row Indo-Japanese system of plantation	(90+150 cm) x 60 cm	(90+150 cm) x 60 cm
Suitability to bivoltine silkworms	For young and late age silkworms	For young age silkworms.

Datta : 2000b

II. Morphogenetic diversity

Morphogenetic diversity revealed following features with respect to fruit, seedling and foliar diversity in *Pentaptera* (species complex of *T. arjuna* and *T. tomentosa*)

(i)   24 pure populations were grouped into 10 clusters for 4 fruit characters *viz.*, length, weight, breadth of wings and germination percentage on the basis of genetic divergence measured by D^2 statistics and canonical analysis. Fruit weight, fruit length, breadth of wing and germination percentage were found to contribute 32.97, 32.60 16.30 and 18.11 percent respectively towards divergence (Srivastav and Goyal 1991).

(ii)   The genetic divergence as measured through D^2-statistics and canonical analysis revealed that half-sib seedlings raised from above 24 populations of seeds from 24 biotypes showed grouping into 8 clusters and seedling height, leaf length/leaf breath ratio, leaf width and leaf length have contributed 88.04, 5.8, 3.62 and 2.54 percent respectively towards the divergence (Srivastav *et al.* 1992).

(iii)   Genetic divergence as measured through D^2 statistics in 39 genotypes for 9 foliar characters revealed that 39 genotypes fell into 7 clusters. Maximum contribution (60.46%) towards divergence was made by leaf weight followed by stomatal frequency (14.44%) and stomatal length (12.55%) (Srivastav *et al.* 1997a).

Such studies on morphogenetic diversity support the view that separate taxonomic status is required for *T. glabra* and *T. berryi* within Arjuna complex and *T. crenulata*, *T. coriacea* and *T. alata* within *T. tomentosa* complex.

Half-sib progenies of 22 accessions belonging to *T. glabra* and putative hybrids were tested for bioassay along with 2 typical local controls of *T. arjuna* and *T. tomentosa* at Central Tasar Research and Training Institute, Ranchi and D, DS1, N5, DS2, B2, S1 and B6 accessions were found superior for tasar silkworm rearing. Nine accessions *viz.*, DS4, N6, DS2, O1, O2, S1, S2, N3 and PBG19 were ascertained as high leaf yielding genotypes by raising their half-sib progenies. Four accessions *viz.*, S1, O1, PBG12, PBG33 and PBG13 exhibited nutritional superiority through biochemical assay (Srivastav *et al.* 1998b, 2002b). Thus S1 exhibited qualitative as well as quantitative superiority constantly. Physio genetically, PBG11 and PBG48 exhibited highest response to rooting through air layering and PBG15, PBG16, PBG19, PBG24, PBG25, PBG26 accessions exhibited drought resistant characteristics through stomatal studies (Srivastav and Thangavelu 1997, 2000, Srivastav *et al.* 2000b).

(c) Genetic resources of temperate tasar food plants

The preliminary explorations conducted in West Imphal, East Imphal, Senapati, Thoubal, Churachandpur, Bishenpur and Ukhrul districts in Manipur revealed that *Quercus serrata* auct. non Thunb. syn *Q. acutissima* Carr. exhibits continuous rather than discontinuous variations at the levels of shape, texture, margin, apex, base, size, number of lateral veins and weight of single leaf as well as size, weight, shape, texture, and involucre of fruits. On the basis of foliar variations laceoalata, lanceo-linear, elliptica, oblonga, obovata, oblanceolata, spatulata and cuneata varieties may be isolated but detailed investigation is yet to be undertaken. Further, *Q. acutissima* Carr. has been considered as the valid name of *Q. serrata* Thunb. (Singh and Das 1991, Srivastav and Singh 1997a, Srivastav *et al.* 2000c) since *Q. serrata* Thunb. should infact be awn less in the margins of leaves and its involucres should be very short around fruits but awn less specimens have not been encountered so far though some of the genotypes have been found to exhibit intermediate characters between both the species. Hence, occurrence of true to type *Q. serrata* cannot be completely ruled out. Furthermore, some morphotypes are identical to *Q. semiserrata* Roxb. but they do not exactly represent the true to type *Q. semiserrata* because they exhibit coarsely serrated or undulated margins whereas true to type *Q. semiserrata* should have coarsely serrated or semiserrated margins towards leaf apex only (Brandis 1906, Kanjilal *et al.* 1940). There is every likelihood that *Q. dentata* Thunb. and *Q.variabilis* Bl. reported from Korea, China and Japan may also be forming a complex in North-eastern region of India because of their allied nature and geography. In *Q. grifithii* Hook. & Thom.ex Miq. obovate-oblong and oblanceolate morphotypes with either dentate or serrate margins have been observed. Lot of natural variability occurs in seeds of *Q. grifithii* also. *Q. grifithii* of India is identical to *Q. aliena* Bl. and *Q. acutidentata* (Maxim.) koidz of Japan and china. Hence, extensive explorations for collection and enrichment of genetic resources of *Quercus* should be taken up in future.

Although genetic variability has not been studied in *Q. leucotrichophora* śyn. *Q. incana* and *Q.himalayana* syn. *Q. dialata* syn. *Q. floribunda* so far, in North

Western region of Himachal Pradesh, seven morphotypes of *Q. semecarpifolia* were identified. 10 morphotypes in *Q.serrata* and 7 morphotypes in *Q. semecarpifolia* were subjected for bioassay in natural/economic plantations which revealed that lanceolate-oblanceolate or lanceolinear morphotypes (C,B/E,G) in *Q. serrata* and thick non-spiny varieties (M2, M5) in *Q. semecarpifolia* behaved as superior varieties for oak tasar rearing (Srivastav *et al.* 20002a Raja Ram *et al.* 1998). The distribution of various oak species in 29 lakh acres at the altitude of 2000-9000 ft. AMSL in the sub-Himalayan regions from Jammu & Kashmir in North-east, Garhwal and Kumaon hills of Uttarakhand and Himachal Pradesh in Central zone to Manipur and Arunachal Pradesh in North-east indicates their rich natural diversity in India (Misra *et al.* 2000). Negi and Naithani (1995) reported 35 species of oaks in India, out of which 15 species belong to genus *Quercus* while other species belong to *Lithocarpus, Castnaea* and *Castanopsis*.

(d) Genetic resources of muga food plants

In Soalu, (*Litsaea polyantha*) earlier two varieties *viz.*, oblong and obovate were identified earlier on the basis of leaf shape, size and texture. Recently, 10 morphotypes *viz.*, F1-F6 and M1-M4 have been identified (Benchamin 2000a, Das *et al.* 2000), out of which F1, F5, M1 and F6 have been found qualitatively superior through biochemical assay (Srivastava *et al.*, Lucknow Univ., Pers. Comm.).

Jolly *et al.* (1979) first reported that Som (*Persea bombycina/Machilus bombycina*), can be classified on the basis of leaf shape into 16 morphotypes of which Naharpatiya, Azarpatiya and Jampatia are important. Subsequently, Yadav *et al.* (1985) reported four types *viz.*, Belpatia, Kathalpalia, Naharpatia and Ampatia. However, Rajaram *et al.* (1993a) and Sengupta *et al.* (1991, 1993) collected altogether only 8 morphotypes of which morphotypes I and II were obtained from Boko (Kamrup) while the remaining varieties were obtained from Borahibari (Sibasagar). Currently, CMER&TI, Jorhat and RMRS, Boko are maintaining altogether 14 morphovariants (S1-S8, A1-A2, R1-R2 and T1-T2) of Som, 10 mophotypes of Soalu (F1-F6, M1-M4) and 3 morphotypes of Gonsoroi (Benchamin 2000a, Das *et al.* 2000). Out of 8 morphotypes of Som subjected for bioassay S6, S3, S5 and S4 exhibited superiority for rearing (Siddiqui, Pers. Comm.). Bioassay of morphotypes of Som revealed range of ERR from 41.16% to 61.16% and range of absolute silk yield from 67.63 to 87.65 gm/per 300 larvae while, filament length was found to vary from 293.91 to 363.68 mtr. with 44.59% to 53.49% recovery. The chemoassay of 14 morphotypes of Som revealed that crude protein ranged from 8.52 to 11.32%, lipid from 5.42 to 7.10%, crude fibre from 19.62 to 28.07%, total ash from 3.72 to 5.09%, lignin from 7.84 to 15.99%, Cellulose from 20.27 to 35.93% and moisture ranged from 31.06 to 46.40%. Biochemical studies revealed qualitative superiority of S3, S2 and S1 morphotypes and drought resistant chracteristics were exhibited by S2, S1, S6, S3 morphotypes through stomatal studies (Das *et al.* 2000, Siddiqui Pers. Comm.). However, Singh *et al.* (2000) reported that S5 and S7 morphotypes are more suitable for muga silkworm rearing on the basis of joint assessment of morphological and biochemical constitution. Studies on 8 different collections of Som at Regional Research Laboratory, Jorhat revealed that growth and life cycle of silkworms were influenced unevenly by plant type. The sugar percentage of leaves positively influence growth of worms while crude fibre negatively influenced growth of worms and cocoon quality (Choudhury *et al.* 1998).

(e) Genetic resources of eri food plants

Among the eri food plants, 41 accessions of *Ricinus communis* (Castor) and 12 accessions of *Manihot utilitissima* (tapioca) collected from Meghalaya, Assam, Manipur, Hyderabad and Laos are being maintained at Regional Eri Research Station, Mendipathar. 10 accessions of *R. communis* were evaluated through palatability test, survivality and leaf yield/plant which revealed that Local red(non-powdery), Hojai, local green and Damalgiri are superior in descending order (Sarmah 2000). Evaluation of 24 castor genotypes collected from Regional Research Station, Raichur at U.A.S., Dharwad, Karnataka revealed that RG-323 is best variety for commercial rearing (Patil *et al.* 1998). Evaluation of 4 castor varieties (Aruna, Assam local, bilara local and Udaipur local) revealed superiority of Aruna in semi-arid climate of Rajasthan due to highest SR% (14.23%), fecundity (350) and hatachability 72% (Dookia 1986) while evaluation of 10 castor varieties in Jorhat exhibited superiority of red petioled variety which produced highest cocoon weight (2.79g), shell weight (0.35g) and SR% (14.06) even by consuming least amount of foliage (Hazarika and Hazarika 1996). Local red (non-powdery) and Hojai are high leaf yielding varieties also besides being qualitatively superior (Sarmah 2000). Correlation studies on bioassay and chemoassay of castor genotypes indicated that moisture, amino acids, carbohydrates and mineral contents have significant positive correlation with larval duration, weight, survivality, ERR%, cocoon, shell and pupal characters and also with grainage parameters, hence, selection of genotypes is an important aspect for eri silkworms (Jayaramaiaha and Sannappa 1998).

Evaluation of 9 out of 12 varieties of *M.utilitissima* revealed that H-1423, H-97 and H-972 are much better than other cultivars of Tapioca for eri silkworm rearing (Saratchandra and Joshi 1988).

A. SILKWORM GENETIC RESOURCES

(a) Genetic resources of mulberry silkworm

Even though silk is mentioned in the *Rig-Veda*, the *Ramayana*, the *Mahabharata* and *Abhigyana Shankuntalam* etc., we do not know anything about ancient Indian silkworms. The oldest known mulberry silkworms are Ichot, Nismo, Itan, Barapolu, Cheenepolu, Chotapolu, 'C'Nichi, Kashmir race, Nistari, Pure Mysore and Sarupat (Table 46). While cheenapolu was introduced from China in West Bengal, 'C' Nichi was introduced from Japan in Mysore and remaining races were indigenous. Barapolu and Kashmir races were univoltine while rest of the races were multivoltine. Now-a-days Barapolu, Ichot, Nismo, Itan and Chotapolu races have become extinct. Later Tamil Nadu White and Moria races were evolved.

Performance of indigenous breeds shows that Tamil Nadu white was the best multivoltine race with 70% pupation rate, 19 kg/100 dfl yield, 14.35 avg. shell percentage and 517 mtr. filament length in South India while Moria was best in Assam and some parts of North-east India (Table 47). Prior to 1922, only pure races were reared and subsequently exploitation of hybrids became popular. Pure Mysore x C. Nichi was probably the first hybrid exploited in Karnataka and exploitation of hybrids was resorted much later during 1956 in West Bengal and J&K in 1959 (Thangavelu 1997).

Systematic breeding of mulberry silkworms indeed started in 1950s at Central Sericultural Research & Training Institute, Berhampore which evolved Nistid, Nismo, Ichot and Itan multivoltine breeds and Department of Sericulture, Karnataka evolved HS6 which exhibited bivoltine character initially but later became multivoltine. Before 1970s, the popular bivoltine exotic races *viz.*, C-110, C-108, C-122, J-112, J-122 and C. Nichi were exploited.

During 1960s to 1980s, Central Sericultural Research and Training Institute (CSR&TI), Berhampore evolved MBD-IV, MBD-V, A4e, D14b, G, CB2, CB5, Nistari (SL), A-23, A-25, whereas CSR&TI, Mysore evolved Kollegal Jawan, Kolar Gold, Mysore Princess, Hosa Mysore, MY-1, P2D1, RD1, PCN and Director of Sericulture, Tamil Nadu evolved Tamil Nadu white multivoltine breeds. Regional Sericultural Research Station, Kalimpong evolved KA, KPG-B, KB; RSRS, Pampore evolved PLF, BL1, SH6; RSRS, Dehradun evolved S21, YS3, SF19 and Karnataka State Sericultural Research & Development Institute, Bangalore evolved NP2 and SP2 bivoltine breeds. During 1990s, Central Sericultural Research & Training Institute, Mysore developed BL-23 and BL-24 multivoltine breeds and CSR series, 1HT and 2HT bivoltine breeds, whereas University of Mysore, Mysore developed MU-1, MU-11, MU-303 multivoltine breeds and MG-408, MG-414, MG-852 and MG-854 bivoltine breeds. During the same period CSR&TI, Pampore evolved PAM-101, PAM-111; KSSRDI, Bangalore evolved KSO1, KSO2; CSR&TI, Berhampore evolved SK3, SK4 and RSRS, Coonoor evolved CNR3, CNR4, CNR14 bivoltine breeds (Table 48). Characteristics of eggs, larvae and cocoons and/or commercial features of some of these important bivoltine breeds *viz.*, KA, KPG-B, NB4D2, NB7, NB18, P5, CC1, SH6, CSR2, CSR4, CSR5, CSR18, and CSR19 have been given in Tables 49 & 50. CSR hybrids have indeed the potentiality to produce more than 60-70 kg/100 dfls with a shell ratio of 21.5-24.4%, filament length of 1176-1328 meter and renditta of less than 5.1-5.6 which has rekindled the national hope to produce international grade (2A-4A) silk with the help of multiend reeling machinery package along with process technology developed by Central Silk Technological Research & Training Institute, Bangalore (Tables 51 & 52) (Kawakami 2001).

Central Sericultural Germplasm Resources Centre, at Hosur (Tamil Nadu), has so far collected 355 silkworm (*Bombyx mori* L.) races from 12 countries and 7 states of India between 1991-1999, which include 136 exotic, and 219 indigenous races belonging to 292 uni/bivoltine and 63 multivoltine breeds. These 63 multivoltine and 292 bivoltine silkworm races have been collected from different sources after elimination of duplicates from 775 silkworm germplasm accessions available in the country (Table 60, CSGRC, 2000, Jayaswal *et al.* 2000). The silkworm gene pool includes indigenous univoltine races "Boropolu" collected from Majoli, Assam, perhaps the only place where it is available today and commercially exploited. Some of the breeds which are cultivated since long and have high adaptability like "Lamarin" from Manipur and "Jam" races from Jammu and "Mirgund" races from Kashmir are also collected and maintained. The exploration trips revealed availability of *Theophila* species and *Bombyx* mandarina in North-Eastern parts of the country and *Ocinara* species in Andman & Nicobar islands (Anonymous 1999b).

Table 43: Country / State wise mulberry base collection at CSGRC, Hosur

SI. No.	Country-wise mulberry collections			State-wise mulberry collections		
	Code	Country name	Collections	Code	Country name	Collections
1	AFG	Afghanistan	2	A&N	Andaman & Nicobar	15
2	AST	Australia	2	APR	Andhra Pradesh	3
3	BGD	Bangladesh	5	ARU	Arunachal Pradesh	5
4	CHI	China	52	ASM	Assam	10
5	CYP	Cyprus	1	BIH	Bihar	18
6	EGY	Egypt	3	GUJ	Gujarat	1
7	FRA	France	32	HAR	Haryana	2
8	HUN	Hungary	1	HPR	Himachal Pradesh	9
9	IND	India	647	J&K	Jammu & Kashmir	22
10	IDA	Indonesia	6	KAR	Karnataka	99
11	ITL	Italy	7	KER	Kerala	66
12	JPN	Japan	69	MPR	Madhya Pradesh	12
13	BUR	Myanmar	6	MAH	Maharashtra	6
14	PAK	Pakistan	8	MEG	Manipur	12
15	PAP	Papua New Guinea	1	MEG	Meghalaya	21
16	PAR	Paraguay	4	NAG	Nagaland	1
17	PHI	Phillippines	1	NED	New Delhi	2
18	PRT	Portugal	1	PUN	Punjab	5
19	SUN	Russia	2	RAJ	Rajasthan	5
20	SKR	South Korea	6	SKM	Sikkim	1
21	ESP	Spain	2	TNU	Tamil Nadu	53
22	THI	Thailand	11	UPR	Uttar Pradesh	129
23	TUR	Turkey	1	UTT	Uttaranchal	7
24	VEN	Venizuela	1	WBL	West Bengal	143
25	VTM	Vietnam	3			
26	ZIM	Zimbabwe	11			
27		Unidentified	23			
	Total Collections		908	Total Collections		647

Source: Mukherjee 2000, Tikader *et al.* 2002

Table 44: Species-wise composition of *Terminalia* (Pentaptera) genotypes available at Central Tasar Research & Training Institute, Ranchi, India

Year of exploration	Taxa	No. of accessions	Location
1987	1. *T.glabra* syn. *T.arjuna* var. *arjuna*	12	Dhamtari, Jagadalpur, (Chattisgarh), Sundergarh, Nowrangpur (Orissa)
	2. Putative hybrids	10	Sundergarh, Nowrangpur (Orissa) and Jagdalpur (Chattisgarh)
1989	1. Putative hybrids	2	Nowrangpur (Orissa)
1990	1. *T. glabra*	12	Ranchi (Jharkhand)
	2. *T. berryi* syn. *T. arjuna* var. *angustifolia*	2	- do -
	3. *T.alata* var. *alata* syn. *T.tomentosa* var. *alata*	8	- do -
	4. *T.coriacea* syn. *T.tomentosa* var. *coriacea*	4	- do -
	5. *T.crenulata* syn. *T.tomentosa* var. *crenulata*	2	- do -
	6. Putative hybrids	11	- do -
1993	1. *T.glabra*	20	Ranchi (Jharkhand), Balaghat, Bilaspur and Jagdalpur (Chattisgarh)
	2. *T.berryi*	5	Ranchi (Jharkhand), Balaghat (Madhya Pradesh)
	3. *T.alata* var. *alata*	2	Jagdalpur and Bilaspur (Chattisgarh)
	4. *T.alata* var. *nepalensis*	2	Ranchi (Ranchi)
	5. *T.coreacea*	1	Balaghat (Madhya Pradesh)
	6. *T.crenulata*	24	Ranchi (Jharkhand)
	7. Putative hybrids	13	Bilaspur and Raigarh (Chattisgarh) Ranchi (Jharkhand), Balaghat (M.P.) and Jagdalpur and Bilaspur (Chattisgarh)
	Total	130	

Srivastav & Thangavelu 2000, Thangavelu *et al.* 2000.

(b) Genetic resources of non-mulberry silkworms

Genetic resources of tropical and temperate tasar, muga and eri silkworms *viz.*, *Antheraea mylitta, Antheraea proylei/A. pernyi/A. frithii, A. assama* and *Samia cynthia ricini* syn. *Philosamia ricini/Samia cynthia* syn. *P. cynthia* are described in Chapter-I.

Presently Regional Eri Research Station, Mendipathar is maintaining seven eco-races *viz.*, Borduar, Titabar, Khanapara, Mendi, Sille, Dhanubhanga and Nongpoh in eri silkworm germplasm. Six pure lines, strains *viz.*, Yellow Plain (YP), Yellow Spotted (YS), Yellow Zebra (YZ), Greenish Blue Plain (GBP), Greenish Blue Spotted (GBS) and Greenish Blue Zebra (GBZ) were isolated from Borduar and Titabar ecoraces. On the basis of larval body colour and markings on the body cuticle (Table 3) since these two ecoraces have been found more productive amongst all the six ecoraces in different zones (Sarmah 2000, Benchamin 2000a).

Table 45: Distinguishing features of various taxa in genus *Terminalia* (section: Pentaptera)

Species	Taxa	Distinguishing features
1. *Terminalia arjuna* Bedd. (Species complex)		Fruits comparatively very big 1-2 inches, bark smooth, grey, flaking off in large thin layers. Wings of the fruit not very broad, their stratons curving upwards. Leaves glabrous.
	1. *T.berryi* W.&A. syn. *T.arjuna* var.angustifolia.	Leaves narrow, elongate - oblong, suddenly narrowed into petiole with drooping branches, L/B=4.
	2. *T.glabra* W.&A. syn. *T.arjuna* var. arjuna.	Leaves often cordate, obtuse or very shortly acute with horizontal branches. L/B=4.
	3. Small leaf form	Leaves smallest, elliptic-lanceolate, L/B= 3.0-35.
	4. Tutative hybrids closer to *T.arjuna*.	Bark grey, slightly rough, leaves broader than those of *T.glabra* and not perfectly smooth.
2. *Terminalia tomentosa* W.&A. (species complex)		Bark rough, grey-black, not flaking in thin layers. Wings very broad in fruits, striations of wings carried horizontally to the edge. Leaves tomentosa rarely glabrous.
	1. *T.alata* syn. *T.tomentosa*	Leaves cordate or oblong suddenly narrowed into the petiole. Adult more or less hairy beneath, bark grey to black. Young ovary villous, fruit glabrous.
	(i) *T.alata* var.alata Heyne ex. Roth. syn. *T.tomentosa* var. alata	Leaves oblong to elliptic or lancealate. 13-30 = 4-10 cm. strongly tomentosa beneath, distributes in Northern India.
	(ii) *T.crenulata* Heyne ex. Roth. syn. *T.tomentosa* var. crenulata.	Leaves narrowed into the petiole, obovate to elliptic, adult nearly glabrous beneath, young ovary and fruits glabrous, common in Western ghats.
	a. Big leaf forms	Obovate leaves
	b. Small leaf forms	Oblong leaves
	c. Kahvi	Elliptic leaves
	3. *T.coriacea* (Roxb). W.&A. syn. *coriacea*	Bark dark coloured like crocodile skin, leaves like *T.alata* but beneath with a close hard fluvous hard tomentosa rather than villous. Fruits pubescent between wings. Common along Eastern ghats.
	a. Big leaf forms	Obovate leaves
	b. Small leaf forms	Oblong leaves
	4. Putative hybrids closer to *T.termentosa*,	Bark greenish black, rough and slightly cracked. Larval bigger than those of *T.glabra*, nearly galbrous, veins protruding.

Source: Srivastav and Thangavelu 2000, Thangavelu *et al.* 2000

The six pure lines were crossed following diallel crossing technique and on the basis of combining ability and heterosis tests seven crosses were selected and evaluated. The YZxGBS(ES1) and GBSxGBZ(ES2) hybrids exhibited superiority for ERR%, cocoon weight, shell weight and fecundity, larval weight, SR% and absolute silk yield respectively (Table 53). The average performance of these elite ecoraces at field level revealed superiority of ES1 followed by ES2 for all commercial characters (Table 54, Debraj *et al.* 2001).

Table 46: Silkworm races of India.

Silkworm race	Original place of cultivation (home)	Characteristics
1. Barapolu	Malda, West Bengal	Univoltine, cocoon greenish white (extinct)
2. Bulupolu or Cheenapolu	Introduced from China, reared in West Bengal for a long period.	Multivoltine, cocoon greenish, small.
3. Chotapolu	Malda - Murshidabad, reared in West Bengal for more than 200 years.	Multivoltine, cocoon white, spindle shaped (extinct)
4. 'C' Nichi	Introduced from Japan in Mysore State	Bivoltine/Multivoltine, cocoon white, dumb-bell shaped.
5. Kashmir race	Kashmir	Univoltine, silkworm white, cocoon yellow.
6. Nistari (N)	Malda, West Bengal	Multivoltine, silkworm highly disease resistant (hardy), cocoon small, spindle shaped, golden yellow.
7. Pure Mysore (PM)	Mysore (Karnataka) and other South Indian states	Multivoltine, silkworm moderately disease resistant, cocoon small, yellow.
8. Sarupat	Assam, Meghalaya and Mizoram	Multivoltine, cocoon small, flossy, white or light green.
9. Ichot	—	— (extinct)
10. Nismo	—	— (extinct)
11. Itan	—	— (extinct)

Source: Sarker 1998, Anonymous 1999b

Table 47: Performance of indigenous multivoltine breeds during summer

Race/Breed	Pupation rate	Yield/100 layings(kg)	Avg.shell (%)	Filament length (mtr)	Region
Nistari	76.00	17.00	11.29	367	West Bengal
Sarupat	45.00	16.00	12.82	434	Assam and North eastern India
Pure Mysore	55.00	16.00	11.78	342	Tamil Nadu, Karnataka and Andhra Pradesh
Tamil Nadu White	70.00	19.00	14.35	517	- do -
Moria	58.00	18.00	14.34	523	Assam and some parts of NE regions of India.

Sengupta *et. al* 1997

Table 48: Multivoltine and Bivoltine breeds evolved/available in India

Year	Multivoltine		Bivoltine	
	Breed	Breeder	Breed	Breeder
1920s	Pure Mysore	Karnataka		
	Nistari	West Bengal		
	Sarupat and Moria	Assam		
1950s	Nistid, Nismo,Ichot,	CSRTI, Berhampore		
	Itan HS6	DOS, Karnataka		
1960s	Kollegal Jawan,	CSRTI, Mysore	KA,KPG-B, KB	RSRS,Kalimpong
	Kolar Gold,		PLF, BL1, SH6, S21	RSRS, Pampore
	Mysore Princess			RSRS,Dehradun
	MBD-IV, MBD-V,	CSRTI,Berhampore		
	A4e, D14b.			
1970s	Hosa Mysore,	CSRTI, Mysore	NB4D2, NB7, NB18	CSRTI, Mysore
	Tamil Nadu White	DOS,Tamil Nadu		
1980s	G,CB2,CB5,	CSRTI,Berhampore	CC1, CA2	CSRTI, Mysore
	Nistari(SL), A-23,		YS3, SF19	RSRS,Dehradun
	A-25		NP2, SP2	KSSRDI, Bangalore
	MY-1,P2D1, RD1,	CSRTI, Mysore		
	PCN			
1990s	BL23, BL24, MU-1,	CSRTI, Mysore	CSR series, 1HT, 2 HT	CSRTI, Mysore
	MU-11, MU-303	Mysore University	PAM-101, PAM-111	CSRTI,Pampore
			KSO1, KSO2	KSSRDI,Bangalore
			MG-408, MG-414, MG-852,	Mysore University,
			MG-854	Mysore
			SK3, SK4	CSRTI, Berhampore
			CNR3, CNR4, CNR14	RSRS,Coonoor

CSRTI : Central Sericultural Research & Training Institute

RSRS : Regional Sericultural Research Station

DOS : Department of Sericulture

KSSRDI : Karnataka State Sericultural Research & Development Institute

Source: Samson 2000

2. SURVEY, EXPLORATION, COLLECTION AND INTRODUCTION

For collecting the available genetic resources, survey and exploration form the basic input in germplasm activity. Hence, survey, exploration and collection activities are intensified in new areas. Survey is an attempt to make an inventory of the available germplasm through available literature before commencing the actual collection through exploration or by any other means to conserve the germplasm distributed in different geographic regions. Assessment of variation

Table 49: Characteristics of important bivoltine breeds

Breed	Egg	Larva	Cocoon
KA	Newly laid eggs light yellow, fecundity 470-650, chorion colour light yellow.	Newly hatched larvae blackish, larvae plain, bluish white, robust and active in feeding, larval duration 23-24 days.	Plumpy oval, medium to coarse grains, cocoon weight 1.7 to 1.9 gm, shell weight 0.31 to 0.36 gm, SR% 18-19, emergence after 11-12 days.
KPG-B	—	Plain	Elliptical, big, white
NB4D2	Newly laid eggs bright yellow, fecundity 475-550, chorion light yellow to deep yellows.	Newly hatched larvae deep brown, larvae plain, faint bluish and robust, larval duration 24-25 days.	Elongated constricted with round ends, grains medium, cocoon weight 1.6-1.9 gm, shell weight 0.32-0.40 gm, SR% 20-21, emergence after 11-12 days.
NB7	Newly laid eggs yellow, fecundity 450-550, chorion light yellow to deep blue.	Newly hatched larvae deep brown, larvae plain, matured larvae bluish with reddish tinge, larval duration 24-25 days.	Elongated ovals with medium grains, cocoon weight 1.6-1.8 gm, shell weight 0.31-0.36 gm, SR% 19-20, emergence after 11-12 days.
NB18	New laid eggs bright yellow, fecundity 450-550, chorion light yellow to deep yellow.	Newly hatched larvae deep brown, larvae plain, larval duration 25-26 days.	Elongated constricted with round ends, constriction shallow to deep, grains medium, cocoon weight 1.6-1.8 gm, shell weight 0.35-0.40 gm, SR% 21-22, emergence after 12-13 days.
P5	—	Marked	Elliptical, with construction, white, bigger than NB18.
CC1	Newly laid eggs yellow, fecundity 550-650, chorion light yellow.	Newly hatched larvae deep brown, larvae plain, bluish, white and plumpy, larval duration 24-25 days.	Elongated ovals with medium grains, cocoon weight 1.7-2.0 gm, shell weight 0.34-0.42 gm, SR% 20-21, emergence after 11-12 days.
SH6	—	Marked	Elliptical white
CSR2	Newly laid eggs yellow, fecundity 450-550, chorion light yellow to deep yellow.	Newly hatched larvae deep brown, larvae plain, larval duration 23-24 days under control conditions.	Ovals with fine to medium grains, cocoon weight 1.8-1.95 gm, shell weight 0.45-0.50 gm, SR% 24-26, emergence takes 13 days.
CSR4	Newly laid eggs yellow, fecundity 600-650, chorion yellows.	Newly hatched larvae blackish, larvae plain, mature larvae bluish with reddish tinge, larval duration 24-25 days under control conditions.	Elongated dumb-bell with round ends, grains fine to medium, cocoon weight 1.9-2.15 gm, shell weight 0.40- 0.45 gm, SR% 21-22 emergence takes 14 days.
CSR5	Newly laid eggs bright yellow, fecundity 475-525, chorion yellow.	Newly hatched larvae blackish, larvae plain, mature larvae bluish, with reddish tinge, larval duration 24-25 days under control conditions.	Elongated dumb-bell with round ends, grains fine to medium, cocoon weight 1.80 -1.95 gm, shell weight 0.45-0.50 gm, SR% 24-25, emergence takes 14 days.

Compiled by the authors

and degree of threat it faces in the ecosystem is also made through survey of literature and other means. CSGRC, Hosur has conducted forty survey and exploration trips during 1993 to 2000 within the country and assembled 908 accessions of mulberry (Anonymous 2000, Tikader *et al.* 2002) and 355 (63 MV and

Table 50: Characteristics of CSR breeds

Characters	Characteristics of productive CSR Breeds			Characteristics of robust CSR Breeds	
	CSR2	CSR4	CSR5	CSR18	CSR19
1. Fecundity	500-550	470-525	475-500	500-550	475-500
2. Larval marking	Plain, bluish white	Plain, bluish white	Plain, creamish white	Sex limited*	Sex limited*
3. Larval shape	Plumpy	Slender	Slender	Plumpy	Slender
4. Larval duration	23-24 days	24-25 days	24-25 days	21-22 days	21-22 days
5. Cocoon colour and shape	White, oval	White, dumb-bell	Creamish white, dumb-bell	Creamish oval	Creamish dumb-bell
6. Cocoon grains	Fine to medium	Fine to medium	Fine to medium	Medium to coarse	Medium to coarse
7. Cocoon floss	scanty floss	Scanty floss	Scanty floss	—	—
8. Pupation rate	85-90%	85-90%	85-90%	90-91%	92-93%
9. Cocoon weight	1.80-1.95 g	1.80-1.90 g	1.80-1.90 g	1.65-185 g	1.55-1.75 g
10. Cocoon shell wt.	0.45-0.50 g	0.38-0.43 g	0.42-0.48 g	0.35-0.40 g	0.33-0.38 g
11. Cocoon shell ratio	24-26%	21-23%	23-25%	21-22%	21-22%
12. Raw silk	19-20%	17-18%	19-20%	17-18%	17-18%
13. Filament length	1000-1100 m	900-950 m	900-1000 m	780-790 m	750-780 m
14. Neatness	85-90 points	90-92 points	90-92 points	90-92 points	90-92 points
15. Denier	3.00-3.15	3.00-3.19	3.00-3.20	2.65-2.75	2.65-2.75
16. Reelability	80-85%	75-80%	75-80%	80-85%	80-85%
17. Rearing season	Sept.-Feb.	Sept.-Feb.	Sept.-Feb.	Sept.-Feb.	Sept.-Feb.

*Marked larvae = Females, Plain larvae = Males
Source: Basavaraja *et al.* 2001

292 BV) silkworm germplasm accessions from different institutes, silk farms, explorations and introductions from other countries (Sinha *et al.* 2002). Before undertaking the survey and exploration, relevant information is collected on the occurrence of mulberry and silkworm germplasm from literatures sample specimens in Botanical Survey of India, Zoological survey of India, Forest Research Institutes, State Forest Departments, different Universities and National Bureau of Plant Genetic Resources. CTR&TI, Ranchi has conducted four exploration trips, while RMRS, Boko and RERS, Mendipathar have also conducted several exploration trips for collection of non-mulberry germplasm resources. It is proposed to conduct joint survey and exploration trips in collaboration with Regional Stations of CSB, State Sericulture Officials and Forest Department within the country. In the case of non-mulberry germplasm, systematic survey and exploration on sustained basis should be organised. Import of trait oriented valuable germplasm for production of quality silk needs to be made from sericulturally developed countries like China and Japan on mutual exchange basis through official channel after following strict quarantine measures to check introduction of dreaded pathogens/transgenic materials/ pests. At the time of collection, necessary information with respect to collection data, passport data and specimen data needs to be collected (Anonymous 1999b). Besides, how much germplasm collection should be made and maintained at one point of time should also be assessed properly because it involves high expenditure continuously towards conservation.

3. CHARACTERISATION AND CLASSIFICATION

Highly heritable phenotypic characters which can be easily seen by the eyes and are expressed in all environment are recorded as a part of preliminary evaluation. Recordal of data is started after establishing the plant material or the silkworm races in the "Core collection" after raising two disease free cycles. Morphological characters/data are recorded twice in a year as per set descriptor in case of mulberry and silkworms (Anonymous 1999b, Dandin and Jolly 1986, Jolly and Dandin 1986). Suitable modifications may be made for characterisation of non-mulberry host plants according to prevailing conditions (Srivastav *et al.* 1997b). Besides morphological characterisation, additional studies through biochemical, biomolecular and cytological methods etc., are also made to characterise the germplasm. Such studies help in identifying the markers and eliminate the duplicates besides establishing phytogenetic relationships between various taxa. For making rapid progress, characterisation of germplasm resources should be done with the help of computer aided image analyser. DNA finger printing and isozyme profile are also be used for establishing unique identity of genotype/variety/race and such informations along with specimen sample should be made available in computer data base for utilisation in future. A brief account of number of descriptors being followed by Central Sericultural Germplasm Resources Centre, Hosur is given in Diagram 2.

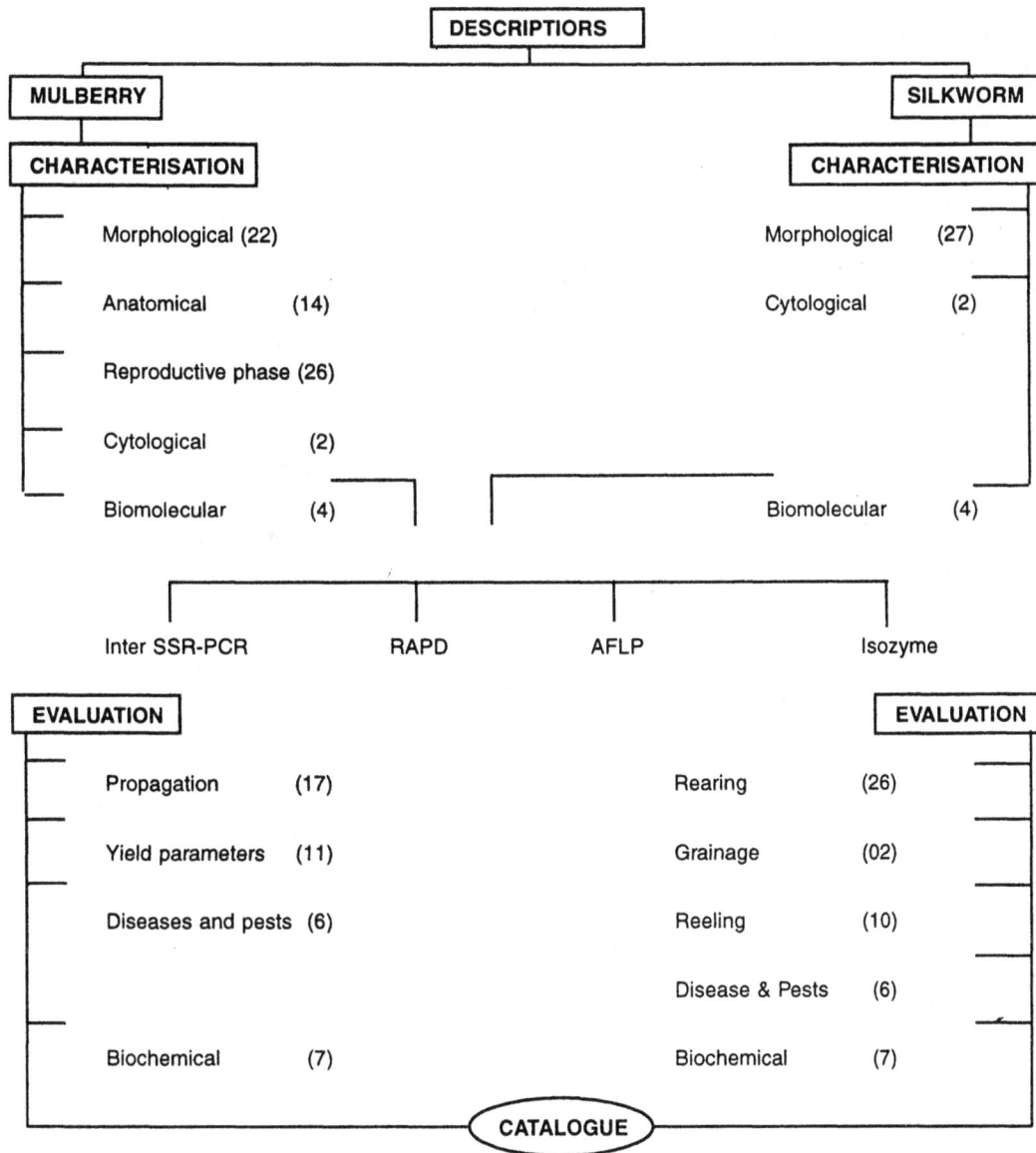

Diag. 2: No. of descriptors for characterisation of mulberry and silkworm germplasm

Source: Anonymous 1999

For promoting utilisation, classification of the genetic stock is essential to facilitate breed specific accessions for meeting objectives of breeding programme. Accordingly, Central Sericultural Germplasm Resources Centre, Hosur has classified its collection of 908 mulberry genetic resources into two major groups *i.e.*, indigenous and exotic followed by local/cultivars, wild, improved/evolved popular varieties, polyploids, mutants and others. Similarly, the silkworm genetic resources are also classified into bivoltine and multivoltine breeds, apart from evolved breeds, breeding stocks, popular breeds, geographical races and land races etc. which include breeds in current use, obsolete breeds, breeding materials and other collections (Anonymous 1995).

So far 628 mulberry and 355 silkworm germplasm accessions have been characterised by CSGRC, Hosur in two phases ending March 2000 (CSGRC 2000, Anonymous 2000, Tikader *et al.* 1997,2000, 2002, Sinha *et al.* 2002).

4. EVALUATION

Genetic resources are evaluated to know their genetic potential for their commercial exploitation directly as well as indirectly through breeding programmes. Hence, evaluation of germplasm is the key to its utilisation. On the basis of information gathered from genotype X environment interaction, multilocational tests are made to provide the objective assessment particularly in the *in situ* condition. As direction of the breeding changes with time, environment, agro-climate and varying needs of the industry, evaluation must be a never ending process. Data on preliminary evaluation facilitates selection of right material for further evaluation and breeding programme. Normally, measurement or rating on easy physiological traits is made in primary evaluation and subsequently accessions are selected for further evaluation with respect to additional descriptors useful in crop improvement *viz.*, stress tolerance, disease and pest tolerance and quality characters etc.

In the first phase, CSGRC, Hosur evaluated 328 mulberry accessions in two stages *i.e.* juvenile and adult for 33 characters and all the silkworm breeds assembled there have also been subjected for evaluation for 41 characters under different seasons which indicated wide range of diversity in the germplasm maintained there (Anonymous 1999b). The status of germplasm evaluation in other institutes of India is detailed in Table 55.

CSGRC, Hosur has evaluated 63 multivoltine and 292 bivoltine germplasm for 22 grainage and rearing parameters, 328 silkworm germplasm accessions for four reeling parameters (filament length, non-breakable filament length, denier and boil off loss). Besides, 56 multivoltine and 40 bivoltine germplasm have also been evaluated for post cocoon traits and quality test of raw silk based on 17 parameters (Sinha *et al.* 2002).

These germplasm were evaluated in the process of regular maintenance by research institutes and the information mostly remained with the institutes as very few got published in research journals. As such the information on evaluation could not be utilised by scientists and breeders due to following reasons (Anonymous 1999a).

1.   Uniform norms were not followed for evaluation by various research institutes.

2.   Evaluation was undertaken in the process of maintenance without aiming to meet the requirement of breeders.

3.   Informations are not available in a compiled form or document.

4.   Due to absence of proper evaluation network between the institutes, some aspects of germplasm evaluation, *viz.*, stress tolerance, diseases and pests resistance could not attract the attention of breeders.

5.   Good number of accessions have not been evaluated so far, for all the required parameters at the national level.

CSGRC, Hosur has taken up only one multi-locational trial comprised of 50 mulberry varieties through All India Mulberry and Silkworm Germplasm Evaluation Programme (AIMSGEP) which was mainly aimed at identifying yield potentialities at different agro climatic locations for identifying region specific germplasm. A modality has already been worked out and All India Mulberry and Silkworm Germplasm Evaluation Programme (AIMSGEP) has been implemented in collaboration with the CSB institutes and Universities in the first and second phases respectively (Dandin *et al.* 2000, Anonymous 1999b). Currently, CSR&TI, Mysore, Berhampore and Pampore; RSRS, Conoor, Kalimpong, Jammu, Sahaspur and Jorhat and CSGRC, Hosur are involved in the network units in AIMSGEP for evaluation of selected accessions of mulberry in different agro climatic conditions. In the case of silkworms, only 10 elite bivoltine silkworm germplasm are being evaluated through eight networking units under different agro climatic conditions of the country to identify region /season specific breeds/races (Sinha *et al.* 2002).

5. DOCUMENTATION AND CATALOGUING

Although various sericultural research institutes/stations/universities etc., generate huge amounts of data, it leads to data redundancy due to lack of scientific and systematic documentation. Hence, there is an urgent need for developing user friendly documentation system with standard format to facilitate the data exchange and dissemination of information on characterisation and evaluation for wider use.

Cataloguing is the systematic documentation of all recorded information on different aspects of germplasm to facilitate breeders in selecting the materials of their interest. CSGRC, Hosur has characterised and evaluated 328 mulberry and 225 silkworm accessions, compiled the data and published two catalogues, one each on mulberry and silkworm germplasm which is first of its kind in India (Anonymous-CSGRC. 2000). The second volume of catalogue on mulberry and silkworm contain data on 300 mulberry and 110 silkworm.

6. NATIONAL DATABASE

A database can collate all data on the characterisation and evaluation of germplasm resources for statistical analysis to classify the total diversity into groups on the basis of correlated characters, geographical pattern of characters such as disease resistance and stability of these characters in different agro climatic conditions. For easy retrieval of the required information, the recorded data are maintained in systematic manner which may also uniformly be followed by other Research Institutes to make the system more effective and help to develop a National Database Information System. CSGRC, Hosur has developed two software packages *viz.*, Mulberry Germplasm Information System (MGIS) and Silkworm Germplasm Information System (SGIS) to maintain the data with easy retrieval and survey base facility. MGIS and SGIS software packages can be utilised and required modifications be incorporated for uniformity of germplasm maintenance throughout the country vis-a-vis development of National Database. The uniformity in the maintenance can only be achieved through collaboration and co-operation between different units with meaningful exchange of material and information aimed at overall improvement of sericulture. For non-mulberry host plant and silkworms also similar information system may be developed.

GLOBAL DISTRIBUTION AND UTILISATION OF GENETIC RESOURCES

Documentary information on holding of host plant and silkworm genetic resources, their types and utilisation rate are either meagre or not available/ inaccessible in India and elsewhere. Globally, approximately 7000 mulberry and 4000 mulberry silkworm genetic resources are available. The pattern of distribution, emerged after reviewing various data published on sericulture so far, has been given in table 56 (Anonymous 2001, 1999c).

Japan and China are the richest countries from Seri biodiversity point of view. China maintains 37.51% mulberry and 15% silkworm genetic resources of the world. On the contrary, Japan harbours 19% mulberry and 39.38% silkworm genetic resources of the world. Hence, China is the richest country from mulberry genetic resources while Japan is richest from silkworm genetic resources point of view. India stands third from Seri biodiversity point of view as it holds 23.1% mulberry and 15% silkworm genetic resources of the world.

Apart from the above pattern of distribution, following facts are also noticed.

- The data on genetic resources may not be treated as complete.
- Duplicate accessions are substantial and maintained in one or the other form.
- Of the 2000 silkworm breeds, about half might have been maintained as duplicate.
- Percentage of total holding within and outside countries is not available.
- Actual number of genetic resources utilised under improvement programme is unknown.
- The rate of utilisation of genetic resources seems to be very low.
- Exchange of germplasm between countries is difficult and restricted.
- Only few molecular technologies are available for identification of germplasm.
- The estimates are based on pedigree informations which involve high costs and are rather subjective.

- Approximately 50 countries are involved in various sericultural activities.
- 46 institutes are presently involved in genetic resource maintenance.
- Only seven institutes maintain more than 200 accessions.
- Conserved genetic resources were not evaluated according to internationally agreed standards and conventions.
- Most of the important useful genetic resources have been kept out of exchange channel.
- Presently no international institute/organisation is established for maintenance and exchange of germplasm.

7. QUARANTINE AND PHYTOSANITATION

Phytosanitary and quarantine system has been introduced only recently in sericulture in our country. The unscientific import of genetic resources of plants as well as animals has brought many pests, diseases and weeds of dreaded nature in agriculture from outside the country along with biological materials. In sericulture also, the transport of genetic resources within the country has caused spread of many pests and diseases in new areas and import of silkworm eggs from other countries without strictly following the quarantine procedures has also caused introduction of some new diseases not occurring earlier in India. Tigerband and virulent pebrine diseases were recently introduced in India from China along with BY1 and B6 oak tasar silkworm races of *Antheraea pernyi* to Regional Tasar Research Stations at Imphal and Bhimtal during 1998. Hence, for controlling such menace there is an urgent need to have a quarantine and phytosanitary system in sericulture also. Further, advancement in genetic engineering and biotechnological research have also opened up new hazards of Biologically Modified Organisms (BMO), Genetically Modified Organisms (GMO) and transgenic organisms. Hence, enough caution needs to be exercised to prevent any unwanted GMO, BMO and transgenic organisms also. The need of effective and efficient quarantine and phytosanitary system in sericulture is all the more relevant in the face of new initiatives the world over for the protection of the sovereign rights of the countries in terms of the International Plant Protection Convention, Sanitary and Phytosanitary (SPS) agreement of World Trade Organisation and convention on biodiversity and at the national level the Destructive Insects and Pests (DIP) Act, 1914, Plants, Fruits and Seeds (PFS) Order 1989 revised as Plant Quarantine (Regulations of Import into India) Order, 1998, Indian Environment Protection Act, 1986 and Recombinant DNA Safety Guidelines, 1990 which ensure the safe import and export of all living organisms.

8. CONSTRAINTS AND MEASURES TO PROMOTE UTILISATION OF GENETIC RESOURCES

The constraints in utilisation of sericultural germplasm resources have been identified as presented here (Anonymous 1999c).

Table 51: Performance of CSR hybrids at CSR&TI, Mysore during 1993-99

Hybrid	Pupation (%)	Yield/ 1000 larvae by Wt. (Kg)	Cocoon weight (g)	Shell weight (g)	Shell ratio (%)	Raw silk (%)	Filament length (m)	Filament size (d)	Reelability (%)	Neatness (p)	Renditta
1. CSR2 x CSR4	92.1	18.54	2.024	46.8	23.4	19.2	1176	3.21	76.0	94.8	5.2
2. CSR2 x CSR5	89.8	17.84	1.992	46.6	23.6	19.4	1215	2.95	81.0	94.3	5.2
3. CSR3 x CSR6	90.5	18.44	2.022	49.3	24.4	19.5	1328	2.94	79.0	94.8	5.1
4. CSR13 x CSR5	89.7	19.24	2.104	50.4	24.2	19.4	1280	3.14	79.0	94.5	5.1
5. CSR12 x CSR5	94.7	20.46	2.118	51.0	24.1	19.9	1277	2.85	77.0	93.3	5.0
6. CSR16 x CSR17	93.4	20.13	2.134	49.2	23.0	18.0	1189	3.44	80.0	94.1	5.6
7. CSR20 x CSR19	92.7	18.56	2.058	47.7	23.1	19.1	1188	3.02	79.0	93.7	5.2
8. KA x NB4D2	89.9	19.40	2.111	42.9	20.4	15.4	1047	3.29	78.0	93.3	6.5

Source: Datta 2000c

Table 52: Performance of new authorised bivoltine hybrids

Hybrids	Cocoon colour	Pupation (%)	Cocoon wt (g)	Shell ratio (%)	Raw silk (%)	Denier	Reelability (%)
CSR2 x CSR4*	White	93.0	2.00	23.5	19.2	3.2	80
CSR2 x CSR5*	Creamish white	90.0	1.95	23.6	19.5	2.9	85
CSR12 x CSR6	White	94.0	2.10	24.1	20.0	2.8	77
CSR18 x CSR19	Cream	95.0	1.85	21.5	17.5	2.7	85
CSR16 X CSR17	White	93.5	2.13	23.0	18.0	3.4	80
CSR3 x CSR6	White	90.5	2.00	24.4	19.5	2.9	79
Kishu x Showa	White	95.8	1.96	24.4	20.7	2.8	83
Future race	White	97.0	1.90	24.5	20.5	2.3	85

* Now under popularisation in the field.
Source: Yamaguchi 2000

Table 53: Performance of the seven crosses of eri silkworm strains

Crosses	Fecundity (No.)	Hatching (%)	Larval Wt. (g)	ERR (%)	Cocoon Wt. (g)	Shell Wt. (g)	Shell Ratio (%)	Absolute silk yield	Rank*
YP x GBZ	384	96.6	5.89	92	3.33	0.48	13.91	163.71	7
YZ x GBS	386	95.8	6.31	95	3.64	0.54	14.89	189.80	1
GBZ x YS	374	94.9	6.30	91.3	3.59	0.52	14.85	170.88	5
YS x GBS	396	96.5	6.21	90.6	3.47	0.51	14.99	179.27	3
YZ x YS	430	96.1	6.14	85.6	3.41	0.52	15.35	184.64	4
GBZ x YP	391	95.6	6.07	90.0	3.40	0.51	14.41	173.32	6
GBS x GBZ	461	96.4	6.34	88.3	3.46	0.52	15.22	204.77	2

Absolute Silk Yield (ASY) = No.of cocoons harvested/dfls X avg. shell wt.
* Based on evaluation index value
Source: Debraj *et al*. 2001

Table 54: Rearing performance of two elite crosses of eri silkworm at farmers level

Particulars	ES1 (YZ x GBS)	ES2 (GBS x GBZ)	Control	Gain over control (%)	
				ES1	ES2
Fecundity (No.)	473.55	468.99	447.44	5.84	4.82
Hatching (%)	92.49	90.27	89.77	3.03	0.55
Cocoon weight (g)	3.52	3.42	3.30	6.66	3.64
Shell weight (g)	0.51	0.47	0.45	13.33	4.44
SR (%)	14.44	13.99	13.83	4.41	1.15
ERR (%)	90.17	87.13	86.04	4.80	2.10
Yield/100 dfls	39,540	37,213	31,658	14.10	7.37

Source: Debraj *et al.* 2001

Table 55: Germplasm evaluation in India

Units	No. of accns.	% of holding	No. of traits	% of total descriptors (33)	No. of accns.	% of holding	No. of traits	% of total descriptors (33)
CSRTI Mysore	380	23.75	15	45.45	207	26.70	15	55.55
CSRTI Berhampore	331	20.69	18	54.51	66	8.50	12	52.17
CSRTI Pampore	65	4.06	12	36.36	145	18.70	11	47.83
KSSRDI Bangalore	375	23.44	10	30.30	105	13.50	12	52.17
Others	449	28.06	08	24.24	252	32.50	09	39.13
Total	1600	100.00			775	100.00		

Source: Anonymous 1999a

Table 56: Current status of mulberry seri-biodiversity in the world

Sl. No.	Country	Mulberry Accessions.	Mulberry. Acc. available for exchange	Silkworm Accessions			Available for exchange
				Multivoltine	Bivoltine	Total	
1	China	2600	900	20	580	600	377
2	Japan	1312	700	30	1542	1572	943
3	India	1600	17	150	450	600	31
4	France	70	NA	—	53	53	35
5	Italy	50	NA	NA	NA	123	81
6	South Korea	300	NA	6	300	306	NA
7	North Korea	NA	NA	5	281	286	262
8	Brazil	NA	NA	10	65	75	NA
9	Others	1000	NA	NA	NA	NA	NA
	Total	6932	1617	221	3771	3992	1729

Source: Anonymous 1999c, 2001

Table 57: Mulberry silkworm hybrids authorised by Central Silk Board for different states and seasons

States/Regions/Zones	Combinations	Hybrids	Season	Date of authorisation
Andhra Pradesh	Multi x Biv	P2D1 x N	Winter	31-07-1995
West Bengal	Multi x Biv	MY1 x NB18	Spring/Autumn	31-07-1995
	Biv x Biv	P2D1 x NB18	Spring	
		N x (NB18 x P5)	Autumn	
		SH6 x KA	Spring/Autumn/Winter	
		CA2 x NB4D2	Spring	
Assam/Bihar/Orissa/ Madhya Pradesh	Multi x Biv	MY1 x NB18	Spring/Autumn	31-07-1995
		P2D1 x NB18	Spring	
		N x (NB18 x P5)	Summer/Autumn	
		PM x NB18	Summer	
	Biv x Biv	SH6 x KA	Spring/Autumn/Winter	
		CA2 x NB4D2	Spring	
		NB18 x P5	Winter	
Jammu & Kashmir	Biv x Biv	YS3 x SF19	Spring	31-07-1995
		SH6 x KA	Spring	
		SH6 x NB4D2	Spring	
		CA2 x NB4D2	Spring/Autumn/E. Winter	
		PAM 101 x NB4D2	Autumn/Early Winter	
		CC1 x NB4D2	Autumn/Early Winter	
		PAM 111 x SF19	Autumn/Early Winter	
		SKUAST-1 x SKUAST-6	Spring	19-02-1996
Uttar Pradesh/Uttaranchal	Multi x Biv	P2D1 x NB18	Summer/Early Autumn	31-07-1995
		RD1 x NB18	Summer/Early Autumn	
	Biv x Biv	YS3 x SF19	Spring	
		SH6 x KA	Spring	
		SH6 x NB4D2	Spring/Autmn/E. Winter	
		CA2 x NB4D2	Spring	
		PAM101 x NB4D2	Autumn/Early Winter	
		CC1 x NB4D2	Autumn/Early Winter	
		PAM111 x SF19	Autumn/Early Winter	
Rainfed Areas/ Irrigated areas	Multi x Biv	BL23 x NB4D2	Spring/Autumn	08-05-1997
		BL24 x NB4D2	Spring/Autumn	
South India				
Temperate/ Sub Tropical zones	Biv x Biv	CSR2 x CSR4	Spring/Autumn	
		CSR2 x CSR5	Spring/Autumn	
Temperate/Tropical zone	Biv x Biv	CSR12 x CSR6	Spring/Autumn	04-05-1999
		CSR18 x CSR19	Autumn	
		CSR16 x CSR17	Spring/Autumn	
		CSR3 x CSR6	Spring/Autumn	
		KSO1 x SP2	Spring	

Total authorised hybrids : 23 (7 Multivoltine x Biv and 16 Biv x Biv)

Source: Surendra Nath *et al.* 2000

Table 58: National level overall average performance of most popular authorised mulberry silkworm hybrids

Name of the hybrid	Hatching (%)	Total larval period (d.h)	Pupation rate (%)	Cocoon yield 10,000 larvae (kg)	Cocoon wt. (g)	Shell wt. (g)	Shell ratio (%)	Filament Length (m)	Wt. (cg)	Size (d)	Reelability (%)	Raw silk (%)	Boil off loss (%)	Neatness (%)
BL23 x NB4D2*	94.39	25.00	90.61	14.280	1.620	30.00	18.52	708	20.00	2.54	83.00	31.24	20.45	88.25
BL24 x NB4D2*	95.17	25.00	90.35	13.880	1.620	30.00	18.52	683	20.00	2.63	83.00	30.91	20.21	85.57
CSR2 x CSR4*	94.80	24.00	92.00	16.240	1.790	40.00	22.35	920	28.00	2.75	80.00	33.77	21.39	91.68
CSR2 x CSR5*	94.66	24.00	90.00	15.360	1.770	40.00	22.60	941	30.00	2.83	83.00	35.20	22.29	88.68
CSR3 x CSR6*	90.52	23.00	80.84	15.772	1.866	44.80	24.01	995	28.84	2.60	74.00	33.27	21.22	91.68
CSR12 x CSR6*	90.07	23.00	91.69	17.856	1.908	45.00	23.58	988	30.78	2.80	83.00	34.96	20.46	91.06
CSR16 x CSR17*	93.50	23.00	82.00	15.500	1.819	41.30	22.70	972	30.28	2.79	80.00	37.67	21.36	91.47
CSR18 x CSR19*#	90.00	23.00	91.10	16.352	1.725	36.90	21.39	823	24.79	2.71	83.00	33.61	20.63	94.88
KSO1 x SP22**	95.31	23.00	94.48	17.188	1.825	37.00	20.27	810	25.65	2.84	83.00	32.54	21.32	82.00

* Evolved by CSR&TI, Mysore.   ** Evolved by KSSRDI, Thalaghattapura, Bangalore   # Temperate tolerant

Source: Surendra Nath et al. 2000

Table 59: Characteristics and cocoon yield for most popular authorised mulberry silkworm hybrids

Name of the hybrid	Season	Condition	Characteristics and Yield
1. BL23 x NB4D2*	Spring & Autumn	Rainfed	More shell weight, longer filament length, average yield 26-28 kg/100 dfls at field level.
2. BL24 x NB4D2*	- do -	Irrigated	Higher productivity, longer filament length, average yield 44 kg/100 dfls at field level.
3. CSR2 x CSR4*	- do -	—	High shell ratio, longer filament length, average yield 64 kg/100 dfls at field level.
4. CSR2 x CSR5*	- do -	—	High shell ratio, longer filament length, average yield of 66 kg/100 dfls at field level.
5. CSR3 x CSR6*	- do -	—	High shell ratio, longer filament length
6. CSR12 x CSR6*	- do -	—	High shell ratio, longer filament length
7. CSR16 x CSR17*	- do -	—	High shell ratio, longer filament length
8. CSR18 x CSR19**	Summer & Autumn	—	Tolerant to high temperature upto 35°C, high shell weight, high shell ratio, longer filament length, average yield 43 kg/100 dfls at field level.
9. KSO1 x SP2**	Spring	—	Shorter larval duration, high disease resistance, high cocoon yield, high silk content, low renditta, average yield 40 kg/100 dfls at field level.

* Evolved by CSR&TI, Mysore   ** Evolved by KSSRDI, Thalaghattapura, Bangalore   # Temperate tolerant.
Source: Surendra Nath *et al.* 2000

Table 60: Country-wise/State-wise silkworm germplasm resources collected at CSGRC, Hosur

Country/State	I Phase (1993-97)		II Phase (1997-2000)		III Phase (2000-2003)		Total		Grand Total
	BV	MV	BV	MV	BV	MV	BV	MV	
1. Brazil	—	—	2	—	—	—	2	—	2
2. Bangladesh	—	3	—	—	—	—	—	3	3
3. China	16	4	25	—	—	—	41	4	45
4. France	4	—	6	—	—	—	10	—	10
5. Iran	—	—	3	—	—	—	3	—	3
6. Japan	10	3	29	—	—	—	39	3	42
7. Poland	—	—	3	—	—	—	3	—	3
8. Russia	17	—	1	—	2	—	20	—	20
9. South Korea	—	—	1	—	—	—	1	—	1
10. Thailand	—	—	2	—	—	—	2	—	2
11. Ukraine	2	—	—	—	—	—	2	—	2
12. Vietnam	—	—	3	—	—	—	3	—	3
Total (A)	**49**	**10**	**75**	**—**	**2**	**—**	**126**	**10**	**136**
1. Assam	2	2	—	—	—	—	2	2	4
2. Jammu & Kashmir	80	—	—	—	1	—	81	—	81
3. Karnataka	15	21	20	—	4	5	39	26	65
4. Maharashtra	—	2	—	—	—	—	—	2	2
5. Tamil Nadu	—	1	8	—	6	2	14	3	17
6. Uttaranchal	10	—	—	—	7	—	17	—	17
7. West Bengal	13	20	—	—	—	—	13	20	33
Total (B)	**120**	**46**	**28**	**—**	**18**	**7**	**166**	**53**	**219**
Grand Total (A+B)	**169**	**56**	**103**	**—**	**20**	**7**	**292**	**63**	**355**

Source: Anonymous 2001, Hosur; Sinha *et al.* 2002

Table 61: Major centres of mulberry germplasm conservation (other than CSGRC, Hosur) in India

Sl. No.	Centre	Exotic	Indigenous	Other	Total
1	Central Sericultural Research and Training Institute Mysore (Karnataka)	138	188	101	427
2	Central Sericultural Research and Training Institute Berhampore (W.B)	91	210	—	301
3	Central Sericultural Research and Training Institute Pampore (J&K)	33	23	9	65
4	Karnataka State Sericulture Research and Development Institute, Bangalore	181	170	—	351
5	Sher-E-Kashmir University, Srinagar (J&K)	26	31	18	75
6	E.R.R.C., Kerala	—	71	—	71
7	Bangalore University	21	21	10	52
8	Regional Sericultural Research Station, Kalimpong	12	31	—	43
9	Regional Sericultural Research Station, Dehra Dun	6	33	—	39
10	Regional Sericultural Research Station, Jorhat	1	10	13	24
11	Regional Sericultural Research Station, Coonoor	3	18	—	21
12	Padmavathi Mahila University, Tirupathi	2	26	—	28
13	Agricultural University, Mirgund, J&K	—	15	—	15
14	University of Mysore, Mysore	5	5	—	10
15	Andhra Pradesh State Sericulture Research and Development Institute, Hindupur	—	5	—	5
16	Other centres	31	45	1	77
	Total	**550**	**902**	**152**	**1604**

Source: Anonymous 2001

Table 62: Major centres of silkworm (*Bombyx mori* L.) germplasm conservation in India

Sl. No.	Name of the Institutions	Multi-voltine	Bivoltine	Total
1	CSGRC, Hosur	56	272	328
2	CSR&TI, Mysore	65	152	217
3	CSR&TI, Berhampore	61	20	81
4	CSR&TI, Pampore	—	145	145
5	KSSRDI, Bangalore	15	17	32
6	University of Mysore, Mysore	9	10	19
7	Bangalore University, Bangalore	9	4	13
8	University of Agricultural Sciences, Bangalore	9	4	13
9	Sher-E-Kashmir University, Srinagar	0	10	10
10	Sri Krishnadevaraya University, Anantapur	2	4	6
11	RSRS, Coonoor	0	11	11
12	SW Seed Technology Laboratory, Kodathi, Bangalore	5	8	13
13	Seribiotech Research Laboratory, Kodathi, Bangalore	6	7	13
14	RSRS, Jammu	0	35	35
15	RSRS, Jorhat	4	4	8
16	RSRS, Sahaspur, Dehra Dun	0	38	38
17	RSRS, Kalimpong	0	52	52
18	RSRS. Titabar	5	5	10
19	Govt. Silk Farm, Dehra Dun	0	8	8
20	Govt. Silk Farm, Dhar (M.P)	0	6	6
21	Dept.of Sericulture, Govt. of Jammu	0	20	20
22	Udaipur University, Rajasthan	5	5	10

Source: Anonymous 2001; Sinha *et al*. 2002

Table 63: Directory of Sericultural Active Germplasm Conservation Sites (AGCS) and sub-centres (SC) in India

Germplasm type	Active Germplasm Conservation Sites (AGCS) and sub-centres (SC) in India	No. of accessions maintained	
		Silkworm	Host plant
I Mulberry	1. CSGRC, Hosur (AGCS)	361	842
	2. CSR&TI, Mysore (AGCS)	204	427
	3. CSR&TI, Berhampore (AGCS)	81	301
	4. CSR&TI, Pampore (AGCS)	139	65
	5. RSRS, Jammu (SC)	35	—
	6. RSRS, Coonoor (SC)	11	21
	7. RSRS, Sahaspur (SC)	44	39
	8. RSRS, Kalimpong (SC)	49	43
II Tropical Tasar	1. CTR&TI, Ranchi (AGCS)	16	134
	2. RTRS, Dumka (SC)	4	—
	3. RTRS, Jagdalpur (SC)	5	—
	4. RTRS, Baripada (SC)	5	—
	5. RTRS, Chinoor (SC)	2	—
III Oak Tasar	1. RTRS, Imphal (AGCS)	12	6
	2. RTRS, Bhimtal (SC)	5	3
	3. RTRS, Batote (SC)	5	4
IV Muga and Eri	1. CMER&TI, Lahdoigarh (AGCS)	8	25
	2. RMRS, Boko (CS)	4	75
	3. RERS, Mendipather (CS)	14	39

Source: Anonymous 2001

1. Lack of proper coordination between the scientists of germplasm, breeders and the users.
2. Inadequacy of useful information on the conserved germplasm.
3. Absence of long term breeding programme.
4. Lack of proper National Sericultural Research Strategy.
5. Various problems affecting the spread of both traditional and new varieties.

Increased use of genetic variability is a pre-requisite to meet challenges of development in sericulture. For evolving genotypes well adapted to extreme and highly variable agroclimates of low productivity areas, utilisation of genetic variability should be enhanced efficiently. There is mounting pressure to reduce the use of agrochemicals, water and nutrient resources which should be used more efficiently, thereby placing even greater reliance on the utilisation of genetic variability in highly productive traditional areas. Therefore, following measures are necessary to promote utilisation of sericultural germplasm resources.

Table 64: Mass seedling banks or archives of T.arjuna and T.tomentosa under CTR&TI, Ranchi

Sl. No.	Name of the station/unit	State	Area under natural T. tomentosa plantation (ha.)	Area under T. arjuna plantation (ha.)	Total area (ha.)
A	**Central Tasar Research & Training Institute,**				
1	Piska Nagri, Ranchi.	Jharkhand	10.7	14.3	25
B	**Regional Tasar Research Stations**				
2	Dumka	Bihar	07	—	07
3	Jagadalpur	Chattisgarh	30	—	30
C	**Basic Seed Multiplication & Training Centres**				
4	Kathikund	Bihar	24.5	12.5	37
5	Kharswan	Jharkhand	60	20	80.0
6	Madhupur	Bihar	70	39	109.0
7	Noammdi	Jharkhand	—	2.5	2.5
8	Balaghat	Madhya Pradesh	80	—	80
9	Bastar	Chattisgarh	93	07	100.0
10	Bilaspur	Chattisgarh	67.5	5.5	73.0
11	Boirdadar	Chattisgarh	97.0	20.0	117.0
12	Pali	Chattisgarh	42.0	30.0	72.0
13	Nowrangpur	Orissa	6.0	31.0	37.0
14	Pallahar	Orissa	58.0	05.5	63.0
15	Sundargarh	Orissa	155.0	25.0	180.0
16	Chinoor	Andhra Pradesh	60.0	30.0	90.0
17	Rampachodavaram	Andhra Pradesh	36.5	16.5	53.0
18	Vikarabad	Andhra Pradesh	40.5	9.5	50.0
19	Bhandara	Maharashtra	88.0	25.0	113.0
20	Chandrapur	Maharashtra	139.0	11.0	150.0
21	Patelnagar	West Bengal	32.0	8.0	40.0
22	Dudhi	Uttar Pradesh	—	20.0	20.0
D	**Research Extension Centres**				
23	Bangriposi	Orissa	26.0	07.0	33.0
24	Hanamkonda	Andhra Pradesh	17.0	13.0	30.0
25	Hatgamharia	Jharkhand	—	5.0	5.0
26	Katghora	Chattisgarh	3.0	30.0	33.0
27	Sultanpur	Uttar Pradesh	7.5	6.5	14.0

Source: Priya Ranjan *et al.* 1994

1. Evaluation of available elite host plant and silkworms in varied environments.

2. Development of national network to disseminate information and documentation.

3. Formulation of national level host plant and silkworm breeding programmes for meeting the aims and objectives of National Sericultural Research Strategy with an effective monitoring system.

4. Strengthening capacities and imparting training in breeding at regional and national level for widening the scope of breeding.

5. Promoting the use of well adapted local varieties in breeding programme and also breeding for wide and specific breeding.

6. Introducing pre-breeding with stress on base broadening programmes through conventional as well as biotechnological methods.

7. Setting of guidelines to remove restrictions and facilitate greater use of genetic resources.

8. Encouraging breeders for Registration of host plant and silkworm germplasm.

9. Constitution of "Host Plant Variety Authorisation Committee" in the line of "Silkworm Authorisation Committee".

10. Adoption of breeding approaches to help maintain and enhance genetic variability and reduce genetic vulnerability in the farmers field.

11. Introduction of a paper in the sericultural courses *viz.*, Post Graduate Diploma in Sericulture (PGDS) and M.Sc., (Seri) on germplasm maintenance, protection and utilisation for both mulberry and non-mulberry sericulture.

9. REGISTRATION OF GERMPLASM

Germplasm is a national heritage which should be systematically conserved for posterity. In order to safeguard this national property and claim the national sovereign right on our own material, a system of registration, authorisation and patenting is a must. Indeed registration is a process of enumerating and documenting all the genetic resources available in the country for which it has the sovereign right. In the light of GATT agreement and patent policy it is very important. Registration will also protect the national wealth of genetic resources against clandestine movement and exploitation of developing countries by developed countries. Registration will also serve as documentary evidence for claiming ownership/right of the material as all the morphological and trait details are furnished along with material to be registered. Further, in the case of silkworms, the commercially authorised material is a hybrid of 2 to 4 parents and hence, pure races going for the commercial hybrid combinations need to be registered to prevent the unauthorised hybrid combinations and also to protect the propriety of concerned breeder/institution (Ramani and Singh 2000). Accordingly, all the original biodiversities, land races, traditional cultivars, evolved varieties/races, naturally available genetic stocks, mutants, polyploids etc. available in the country have to be identified along with their locations and enumerated with proper numbering by an appropriate agency followed by preparing a passport data. Therefore, all the biodiversity both natural and created in host plants and silkworms should be documented and registered urgently on priority basis.

Registration of germplasm has been recognised under National and International Agricultural Research Systems. Indian Council of Agricultural Research (ICAR) recently introduced "Registration of Plant Germplasm" under a set of guidelines. Central Silk Board has also introduced such system under National Sericultural Research System" for registration of the germplasm. For releasing a variety for commercial utilisation there is a "Race Authorisation System" for silkworms. Scientists associated with the improved germplasm and genetic stock (new source of resistance, varied types of mutants/cytogenetic stocks etc.) have no mechanism for recognition unlike the developers of released cultivars/breeds. Lack of formal recognition of such useful materials and role of scientists in development of these material discourages them from sparing the valuable material to other researchers. Consequently, most of such valuable materials remain under utilised or get lost. Registration of germplasm will help in according due recognition to scientists who develop, improved silkworms and their host plants in accordance with the release procedure laid down for the purpose. With recent development concerning "Breeders Rights" towards variety protection linked with "Intellectual Property Rights" (TRIPS) made effective by signing of General Agreement on Tariff and Trade (GATT) and overseen by World Trade Organisation (WTO) and other related issues, due recognition of these material has become all the more important.

National Bureau of Plant Genetic Resources (NBPGR) under ICAR system has been identified as the nodal agency for implementation of the registration procedure. Accordingly, CSGRC, Hosur has been recognised as nodal agency for the purpose and assisted by a committee constituted by CSB. The procedure for identification and registration of silkworm and host plant germplasm have been drawn by CSGRC, Hosur (Anonymous 1999d).

10. SILKWORM RACE AND HOST PLANT AUTHORISATION

There is a need to develop standard procedure for evaluating the performance of silkworm hybrids evolved by different agencies prior to their introduction and commercialisation in the field. Hence, it was contemplated to develop a systematic mechanism for authorisation and release of new mulberry silkworm hybrids to the field as followed in Japan. As such, Central Silk Board recently introduced a regular "Mulberry Silkworm Race Authorisation System" in consultation with Japan International Co-operation Agency (JICA) experts to identify the high yielding mulberry silkworm hybrids evolved by the breeders with reference to region and season specific requirements. The concept of authorisation is a phenomenon to ensure proper introduction of silkworm hybrids in the field, keeping in view the consistent performance of the hybrids. Various norms suited to Indian requirements and guidelines for assessing and evaluating various characters were also framed.

CSB issues a notification every year in Indian Silk and invite entries from breeders who have evolved silkworm hybrids for commercial exploitation. The sub-committee of Race Authorisation scrutinises, short lists the entries that have values above the fixed floor values and sends recommendations to CSB, which arranges for testing the consistency in rearing performance of selected hybrids in all the seri-zones of the country according to set norms in 25 Authorisation Test Centres. The data on important characters of primary and secondary importance are entered in prescribed formats compiled, scrutinised, statistically analysed and final evaluation report is placed before the Main Race Authorisation Committee for approval for commercial exploitation (Diag. 3). List of approved authorised silkworm

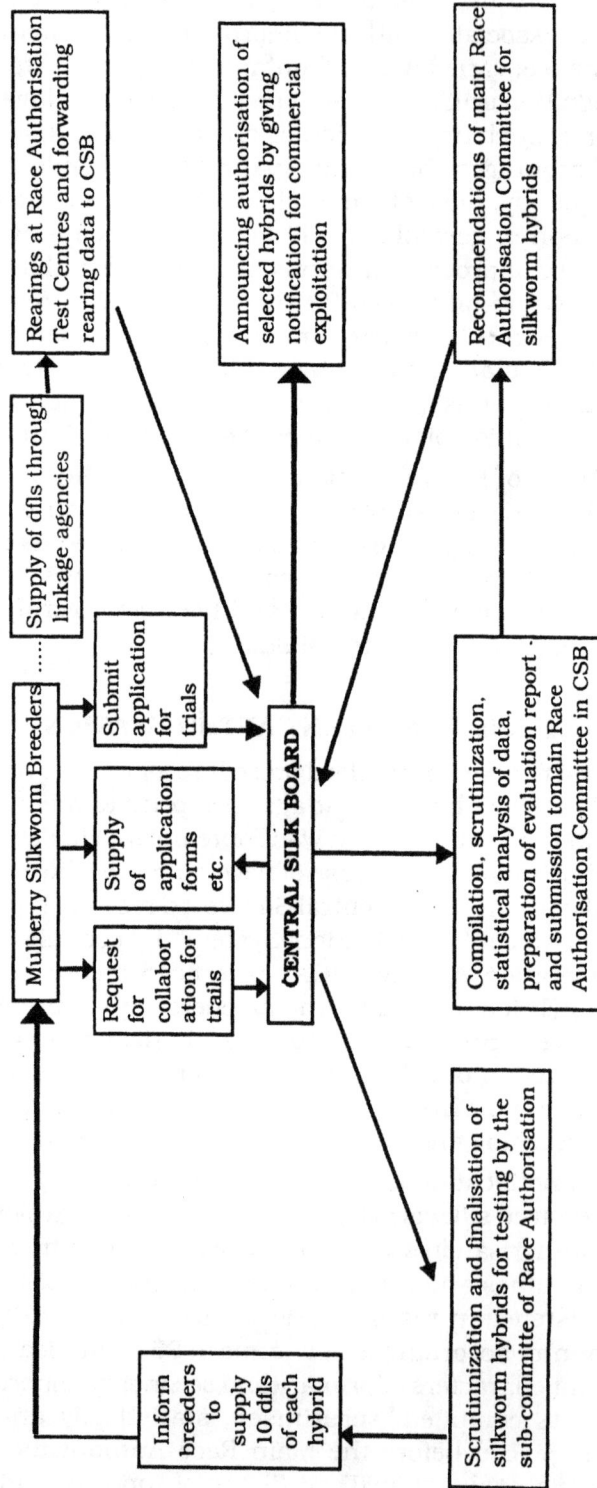

Diag. 3: Flow chart of (*B. mori*) authorisation system followed in India.

Source: Surendra Nath et al. 2000

hybrids is notified in Indian Silk journal. Only the authorised silkworm hybrids are allowed for multiplication and supply to the farmers for commercial rearings. A list of silkworm hybrids authorised so far by CSB is given in table 57. While table 58 and 59 depict the national level overall average performance data of the important breed characteristics of most popular authorised mulberry silkworm hybrids (Surendra Nath *et al.* 2000).

Central Silk Board has also constituted a "Mulberry Variety Authorisation Committee" and a Sub-committee of "Mulberry Variety Authorisation" during 1999 and initiated an "All India Co-ordinated Experimental Trials for Mulberry (AICEM) under which approximately 8 mulberry varieties along with two controls are being tested from July 2000 onwards for four years in different sericultural zones of the country.

11. DNA FINGER PRINTING AND BIOTECHNOLOGY

The DNA finger printing resolves problems of duplicates and disputes of Intellectual Property Rights to a great extent. Hence, development of a centre as National Facility for DNA finger printing is envisaged in future. DNA finger printing of released/registered varieties can also be used for establishing unique identity of genotypes/varieties. For studying genetic diversity and integrity of preserved accessions, RAPD, SSR, ISSR, EST, AFLP and RFLP analysis are essential.

Molecular characterization of 56 multivoltine and 80 bivoltine silkworm germplasm was made following ISSR, PCR method using two anchored primers. Besides, few races of multivoltine and bivoltine accessions were also characterized for isozyme pattern (a and b esterase, alkaline and acid phosphatase) and storage protein (Sinha *et al.* 2002)

The RAPD, Inter SSR-PCR assay and cluster analysis of 15 distinct mulberry species resulted into separation of wild and cultivated mulberry species into different groups. Distance matrix calculation based on RAPD data using 5 oligonucleotide primers revealed the dissimilarity among 44 mulberry verities ranging from 0.027 to 0.586. However, clustering pattern did not show correlation with regard to regional/geographical distribution of these verities. RAPD analysis using 13 primers were completed in 58 mulberry verities and 29 *M. laevigata* accessions. Primer OPY-10 showed cultivar specific bands in 58 mulberry accessions, which indicates that specific DNA marker can be used for identification. The RAPD analysis of 29 accessions of *M. laevigata* from six zones revealed that accessions from North Eastern region have association with those from Eastern India, Central India and South India.

Isozyme banding pattern in PAGE with peroxidase enzyme in 13 *Morus* species were studied to assess the allelic variation in isozyme loci to ascertain the genetic diversity (Tikader *et al.* 2002).

The conservation of genetic resources is very expensive and time consuming and field gene banks are prone to natural calamities and disorders apart from the ravages of pests and diseases. Hence, cryopreservation/*in vitro* conservation technology is essential to reduce the crop cycle in silkworm germplasm resources besides working out long term preservation methods for the safety back up both for silkworms as well as for host plants.

Table 65: Explorations required for collection and conservation of seribiodiversity in future

Sl. No.	Name of the species of host plants/silkworms	Variety/race/trait	Area to be explored
A	**MULBERRY & SILKWORM**		
1	Mulberry (*Morus* species)	Saline, drought and frost tolerant, stem borer, white fly and Tukra disease tolerant/resistant.	Andaman island coastal areas, North-eastern region, North-western sub-Himalayan regions.
2	Silkworm (*Bombyx mori*)	Shorter larval duration, high survivality, resistance to grasserie (BmNPV), early emerging male, high silk content ratio, longer filament length. All indigenous races *viz.*, Borpolu, Borpat, Leimaren Hawlok, Nistari & PM for high adaptation.	Assam, Manipur, West Bengal and Karnataka.
3	Wild relatives of *Bombyx mori*.	*Bombyx mandarina*, *Theophila* and *Ocinara* spp.	Sub-Himalayan regions and North-eastern India including Sikkim and Andaman & Nicobar islands.
B	**NON-MULBERRY HOST PLANTS**		
I	**Tropical Tasar**		
1	*Terminalia* spp.	High yielding, alkaline soil tolerant, drought resistant, gall resistant, leaf curl resistant. *T.crenulata, T.coricecea*, Kahvi, hybrids between both Arjuna and Asan, *T.alata* var. *alata* and *T.alata* var. *nepalensis*	Jharkhand, Orissa, Chandigarh, West Bengal, Maharashtra, Andhra Pradesh, Karnataka, Rajasthan, Uttaranchal, Jammu, Himachal Pradesh, Dadar and Nagar Haveli and North-eastern states.
2	*Shorea robusta*	High yielding, palatable to silkworms.	Jharkhand, Uttaranchal, Orissa, Chandigarh.
II	**Temperate Tasar**		
1	*Q. serrata*	High yielding, early sprouting, late maturing	North-eastern states/regions
2	*Q.grifithii*	High yielding, late maturing	- do -
3	*Lithocarpus dealbata*	- do -	- do -
4	*Q.semecarpifolia*	- do -	North-western region and Arunachal Pradesh
5	*Q.leucotricophora*	- do -	- do -
6	*Q. himalayana*	- do -	North-western region
7	Other *Quercus* sp.	Palatable to silkworms	North-eastern and North-western regions.
III	**Muga**		
1	Machilus bombycina (Som)	High yielding and palatable to silkworms	Assam, Manipur, Meghalaya, Tripura
2	*Litsaea polyantha* (Soalu)	- do -	Arunachal Pradesh
3	*L.salicifolia* (Dighleti)	- do -	Nagaland, Manipur
4	*L. citrata* (Mejankari)	- do -	Sikkim, Mizoram, Himachal Pradesh
5	*L.nitida* (Kothalva)	- do -	Uttar Pradesh, Uttaranchal
6	Other secondary food plants	- do -	Gujarat and Pondicherry

Sl. No.	Name of the species of host plants/silkworms	Variety/race/trait	Area to be explored
IV	**ERI**		
1	*Ricinus communis* (castor)	High yielding and palatable to silkworms	Assam, Meghalaya, Andhra Pradesh,
2	Heteropanax fragrans (Kesseru)		Nagaland, Manipur, Gujarat, West Bengal, North-western Himalayan
3	*Manihot utilitissima* (Tapioca)		region, Bihar, Jharkhand, Uttar Pradesh.
4	Other secondary host plants		
C	**NON-MULBERRY SILKWORMS**		
1	**Tropical Tasar**		
	Antheraea mylitta D.	Amnpv resistant, high SR%, short larval duration, high fecundity.	Jharkhand, Orissa, Chandigarah, Maharashtra, Andhra Pradesh, Assam, Nagaland, Rajasthan, Uttar Pradesh, Jammu, Himachal Pradesh, Dadar & Nagar Haveli.
2	**Temperate Tasar**		
	A. roylei *A. frithii* *A. compta* *A. pernyi* *A.proylei* *A.kmyvettii*	Apnpv resistant, true uni and bivoltine breeds suited to autumn/summer crops.	Jammu & Kashmir, Himachal Pradesh, Uttaranchal, Assam, Nagaland, Manipur, Meghalaya, Mizoram, Arunachal Pradesh, Sikkim.
3	**Muga**		
	A.assama	A anpv resistant, high fecundity and SR%, hibernating stock.	Assam, Uttaranchal, Nagaland, Manipur, Meghalaya, Gujarat, Pondicherry.
4	**Eri**		
	Samia cynthia ricini/ Philosamia ricini, Samia cynthia/Philosamia cynthia/ Attacus ailanthus and allied species	Prnpv resistant, high fecundity, short larval duration, hibernating stock.	Assam, Nagaland, Tripura, Arunachal Pradesh, West Bengal, Bihar and Orissa.

Source: Anonymous 2001, Hosur and Srivastav *et al.* 1996,1998a, 2000a,b, Srivastav and Thangavelu 1997,2000, Thangavelu *et al.* 2000

12. CONSERVATION SCENARIO OF SERI-BIODIVERSITY

For biodiversity conservation, Ministry of Environment and Forests, Govt. of India, the nodal agency on conservation has drawn a nation-wide action plan involving a network of other national organisations *viz.*, BSI, ZSI, FSI, ICAR, CSIR, ICFRE, NBPGR, NBAGR, Wild Life Institute of India, State and Central Pollution Control Boards, lead NGOs, WWF, MSSRF etc. The National Biodiversity Strategy Action Plan (NBSAP) has been prepared and finalised. The main activities envisaged under NBSAP are:

I. Identification of critical components of bio-diversity, *i.e.*, communities, ecosystem, species, genotypes etc.

II.   Enumeration of important generas, species, etc., which are endangered and needs protection on priority basis.

III.   Development of nation-wide data base.

IV.   Identification of methods of *in situ/ex situ* conservation support systems through a protected area, field gene bank, National Park, wild life sanctuaries concept.

V.   Development of strategies for documentation of traditional knowledge, innovations etc., and suitable reward.

VI.   Sharing of the benefits deriving out of use of local biodiversity, traditional knowledge, innovation etc. equitably.

The articles 8 and 9 of Convention on Biodiversity (CBD) regarding conservation, *in situ* conservation through establishment of Protected Areas (PA), envisages selection, establishment, management, access to biological resources etc. should be done by every nation. There are altogether 80 National Parks and 441 sanctuaries in our country and their number is likely to increase fast along with identification of biosphere reserves and plant communities *i.e.*, wet lands, mangroves, coral, reefs etc.

For the conservation of Seri biodiversity similar exercise is also called for. The entire Himalaya, North eastern states and Central India abodes a wide diversity of sericegenous fauna and flora. While mulberry silkworm is reared indoor the tropical and temperate tasar as well as muga are still wild and cultured outdoor. As a result of large scale deforestation, the wild mulberry, and food plants of tasar and muga are threatened. The sericegenous flora and fauna do not serve our clothing requirements alone but also fulfil the requirements of raw materials in pharmaceutical, timber, tannin, horticulture, match, cosmetic and food industries etc. to make our lives comfortable, besides they possess aesthetic and cultural values. Hence, conservation of Seri biodiversity is utmost need of the time.

A. CONSERVATION OF MULBERRY AND ERI SERI-BIODIVERSITY

Mulberry and eri silkworms are reared indoor and selected food plants are cultivated systematically. However, the food plants of both the silkworm species are also found in wild conditions and occur naturally in forests.

(a) *In situ* conservation

In situ conservation (on the site) in natural stands include national parks, wild areas, scientific reservations, natural areas or others of like nature. Seed stands/seed production areas and plus tree selection also come under *in situ* conservation. So far no practical approach seems to have been adopted for *in situ* conservation of mulberry and eri silkworms and their food plants in India. Hence, there is an urgent need to implement *in situ* conservation of these natural resources through ecosystem, biosphere and protected area approach to facilitate the protection of the important/endemic areas of food plants as well as dependent wild insect fauna *viz.*, *Theophila* and *Bombax mandarina* in Kalimpong, North-east region and *Ocinara* species in Andaman and Nicobar islands. The oldest known mulberry tree known at Joshimath (Tikader *et al.* 1999) and other wild mulberry trees in Himalayan belt as well as perennial Castor/*Heteropanax* plants scattered in their natural abode need to be identified and protected *in situ.* Similarly, wild eri silkworm *Samia cynthia ricini* Jr /*Attacus ailanthus* also need to be conserved *in situ* along with their wild host plants in their natural stand.

(b) *Ex situ* conservation

The *ex situ* conservation of 908 mulberry germplasm resources is being carried out at CSGRC, Hosur by maintaining them in base collection plot as dwarf trees. The accessions were collected from 26 countries and include 647 collections from different states of India (Table 43). Through periodical exploration trips covering 58 districts and 5 zones the CSGRC, Hosur has collected a total of 367 mulberry germplasm from 21 states. Maximum collection was made from North-west India (124) followed by South India (110) and North-east India (69). The maximum collection belongs to *M. indica* (151), folllowed by *M. laevigata* (12), *M. serrata* (34) and *M. alba* (20). Approximately, 50 collections are yet to be identified. (Tikder *et al.* 2002).The explorations revealed availability of wild relatives of *Bombax* sp. in Uttaranchal, *Theolphila* spp. in Sikkim and North-east states and *Ocinara* spp. in Mount Harriet range of Andaman islands. Besides, C.S.G.R.C., Hosur is also maintaining 355 (63 multivoltine and 292 bivoltine) silkworm races collected from different institutes and silk farms from 7 states of India and 12 other countries (Sinha *et al.* 2002, Table 60). Apart from CSGRC, Hosur, other research institutes also maintain mulberry germplasm for breeding programme, particularly the elite stocks called active germplasm. (Table 61) which include 550 exotic and 902 indigenous collections. Besides CSGRC, Hosur, 21 more major silkworm germplasm conservation centres are also operating in the country and they have been maintaining 195 multivoltine and 565 bivoltine silkworm races *in situ* (Table 62). CSGRC, Hosur, CSR&TI, Mysore, CSR&TI, Berhampore and CSR&TI, Pampore have been regarded as Active Germplasm Conservation Sites (AGCS) while RSRS, Jammu, RSRS, Coonoor, RSRS, Sahaspur and RSRS, Kalimpong have been recognised as sub-centres for conservation of mulberry and silkworm germplasm in the country (Table 63). These centres act as back-up centres and the collection maintained in the back-up centres serves as "safety back-ups". For muga and eri food plants as well as these wild silkworms, CMER&TI, Lahdoigarh has been proposed as AGCS while RMRS, Boko and RERS, Mendipathar have been recognised as sub-centres of germplasm conservation. In Regional Eri Research Station, Mendipathar 41 accessions of *Ricinus communis* and 12 accessions of *Manihot utilitissima* are being maintained.

Attempts may be made to establish clonal seed orchards of all the primary food plants of non-mulberry silkworms in hot spots with maximum variability and broad genetic base by representing all the morphotypes available in nature. Different Regional Tasar/Muga/Eri Research Stations should collect all relevant genotypes of respective silkworms and food plants and maintain, characterise, evaluate and document them properly. As such, like *Terminalia* descriptor, descriptors of other non-mulberry food plants and silkworms should be developed.

Exchange of inter-regional genotypes between all the Regional Stations in India should be effectively materialised in order to enrich the germplasm and also facilitate the expression of latent genes under diversified agro climatic conditions.

Introduction of non-mulberry food plants and silkworms from different parts of the world should be taken up under strict quarantine measures for enrichment of germplasm collection for future utilisation.

Molecular characterisation of all the available host plant and silkworm genotypes should be done to avoid duplication as well as conflicts related to "Intellectual Property Rights".

Long term cryopreservation technologies should be developed for all the important host plants to save time, energy, expenditures and space required during management/maintenance/conservation of genetic resources.

A list of endangered/threatened sericegenous flora and fauna should be prepared immediately and attempts should be made to conserve them *in situ* as well as *ex situ*. Kahvi and Ulta Saja in *Terminalia* species in Chattisgarh and various ecoraces of *Antheraea mylitta* D. should be given top priority in conservation measures.

Test plantings/archives of plus trees should be raised with respect to all the primary host plants either through half-sib seedlings or clonally to evaluate their performance as well conserve progenies *ex situ*.

Vast exploration of biodiversity in *Q. acutissima* Carr. syn. *Q. serrata* auct. non Thunb. and allied species complex in North-eastern region should be carried out in the pattern of *Pentaptera* section of *Terminalia* in Central India. This will also enable us to identify early sprouting and later maturing varieties for first and second crops respectively.

Collaboration between sister units of Central Silk Board, Indian Council of Forestry Research and Education, Botanical Survey of India, Zoolgical Survey of India, National Bureau of Plant Genetic Resources, National Bureau of Animal Genetic Resources etc. should be enhanced/established in conservation programmes.

Impact of non-mulberry host plants on protection of local environment should be published to avoid heavy deforestation for example, oak trees hold soil more strongly than pine forests and similarly *Terminalia tomentosa* contributes more humus to soil than any other species. This can check the conversion of oak forests into pine forests, so that land sliding in geologically fragile Himalayan region can be checked, and also check the conversion of Central Indian forests into non-fertile agricultural land.

Potentiality of *T.arjuna* for reclamation of alkaline/saline soils, water logged/peaty soils and ash dykes from thermal plants and *T. tomentosa* for reclamation of barren rocky lands should be publicised for inclusion under Social Forestry Programmes by Railways, coal industries (Western Coalfield Ltd., South Eastern Coalfield Ltd., Central Coalfields Ltd. etc.), National Thermal Power Corporation (NTPC), private industries and non-governmental organisations (NGOs).

Religious sanctity of *Arjuna* in relation to *Mahabharata* period about the birth of legendary hero *Arjuna* under *Arjuna* tree may be publicised among religious rural people to check felling of *Arjuna* trees. Similar efforts may also be made for other sericigenous flora of religious sanctity.

Efforts may be made for declaration of all the sericigenous primary food plants as National Trees like *Shorea robusta* for environmental protection as well as sustainable production of minor forest produce of non-timber value.

Instead of paying importance to silk oriented research and developmental activities alone, CSB institutes should also participate in R&D activities related to other minor forest produces *viz.*, indigenous drugs, oils, tannins etc. derived from sericigenous flora and fauna in the line of Indian Council of Agricultural Research (ICAR) which has also taken up animal husbandry, fisheries, horticulture, medicinal plants etc. apart from agriculture alone.

Biopiracy of *T. arjuna, T. bellerica, T. chebula* and *Ricinus communis* is being attempted in some countries which should urgently be taken care of to avoid national loss (PTI, Feb. 15, 1998).

We must realise that sericigenous genetic resources are our National Treasures and ancestral heritage which should be conserved for posterity by every body at all costs. Hence, a team of dedicated scientists should be properly trained and allowed to work continuously by providing incentives and infrastructural facilities.

B. CONSERVATION OF TASAR AND MUGA SERI-BIODIVERSITY

Tasar and muga culture are forest oriented industries. Hence, conservation of their genetic resources is being practiced in accordance with the proposals made for most of the forest trees and fauna.

(a) *In situ* conservation

i. Natural preservation

In India *in situ* conservation of wild species of flora and fauna are managed in the national parks, wild life sanctuaries, biospheres, botanic gardens etc. The management objective of these conservation processes vary considerably, hence they are not suitable for gene conservation of *Terminalia, Quercus, Perseaea* (*Machilus*) and *Litsaea*. However, many of the national parks like Kanha in Madhya Pradesh; British Arboretum, Manali (H.P), Langol, Nongmaijing, Heigang reserve forests in Manipur and Kaziranga in Assam have various genotypes of these tree species but they serve as recreation areas chiefly, hence they may be useful in conserving genes of only long lived species already growing there.

ii. Seed stands/seed production areas

Standard seed stands are forests of natural origin with neither plus nor minus quality. Their standard size is 5-6 ha. and their central core is actual seed collection area. Seed lots collected from standard stands can serve as control or check lots with which progeny tests can be compared.

Standard seed production area consists of a natural stand (or a plantation of known origin) heavily thinned to leave the best phenotypes to minimise pollen pollution from inferior trees in the vicinity. Seed production areas are seldom larger than 3-4 ha. In India, seed production areas of *Terminalia myriocarpa* (1.0 ha) and *T.arjuna* (30 ha.) are already established in Arunachal Pradesh and Uttar Pradesh (Emanual *et al.* 1990). There is an urgent need to develop the seed production areas of other tasar and muga host plants in collaboration with state forest department.

iii. Plus/elite trees

Superior (Plus/elite) trees used in tree improvement programme may be preserved in their natural stands. In *T. myriocarpa* 28 plus trees have been selected by forest department in Arunachal Pradesh while in *T. arjuna* and *T. tomentosa* more than 130 plus trees have been selected in Madhya Pradesh/ Chattisgarh, Jharkhand and Orissa by CTR&TI, Ranchi. These plus trees may be protected in collaboration with forest department and local people (Table 44). In oak tasar and muga host plants through morphotypes have been identified superior trees are still not been identified.

The *in situ* conservation of eco races of *Antheraea mylitta* D. *viz.*, Raily, Laria, Modal, Bogai, Andhra Local, Barharwa, Sarihan, Bhandara, Tira etc. are being carried out by units of CTR&TI, Ranchi in their natural abodes *viz.*, Regional Tasar Research Station at Jagdalpur, Baripada, Dumka, Chinoor and Bhandara.

(b) *Ex situ* conservation

i. Seed orchards

These orchards are not designed to conserve germplasm *per se*, but they may be used for such purposes temporarily under selective conservation. Ordinarily all good clones/progenies derived from them will go into a breeding orchard which may be utilised as a valuable bank or living collection of germplasm material for multiplication at will by vegetative means. In Arunachal Pradesh seed orchard of *T.myriocarpa* has already been established in 2 ha. by forest department. Seed orchards of other tasar and muga host plants needs to be developed by forest departments/Central Silk Board.

ii. Seedling banks/archives

These are collections similar to clonal banks but of seedling origin and serve as germplasm reserves. Mass plantations of *T. arjuna* and *T. tomentosa* have been raised in different states under field units of CTR&TI, Ranchi for tasar culture through mass seedling progenies which may also serve as germplasm reserves of both species since they inhabitate broad genetic base representing enormous genetic variability existing within and between both the species (Table 64). Similarly, mass seedling banks of muga and oak tasar food plants have been established by Regional Tasar Research Stations at Imphal (Manipur) and Bhimtal (Uttaranchal) and Regional Muga Research Station, Boko (Kamrup)/Central Muga Eri Research & Training Iinstitute, Lahdoigarh, Jorhat (Assam) at several places.

iii. Test Plantings

Test plantings may be made of provenances (or seed sources), small populations (or stands) within a provenance and individual trees (progeny testing) for individual species. In CTR&TI, Ranchi test plantings of 24 genotypes (22 plus trees + 2 checks/control genotypes) of *T. arjuna*, spontaneous hybrids and *T. tomentosa*, progenies of anomalous seedlings *viz.*, monocots, tricots, tetracots, twins, trins and progenies obtained from 4-12 winged fruits of above 24 genotypes were raised in 1987. Subsequently during 1993 progenies of 10 more plus trees were raised. These progenies are serving as good gene bank with broad genetic base for tasar food plants. The test plantings have not been raised/reported so far in oak tasar and muga host plants.

The *ex situ* conservation of all the eco-races of *A. mylitta* (tasar) and colour polymorphism in *A. mylitta, A. proylei* (oak tasar) and *A.assama* (muga) silkworms is being done by CTR&TI, Ranchi, RTRS at Imphal/Bhimtal and RMRS, Boko/CMER & TI, Lahdoigarh. RTRS, Imphal is also conserving PRP2, PRP5, PRP3, PRP12, RPP4 breeds of *A. proylei* and *A. pernyi, A. roylei* and *A. frithii* species. Besides, it is also maintaining BY1 and B6 breeds of *A. pernyi* received from China during 1998.

13. PROPOSALS FOR CONSERVATION OF SERI-BIODIVERSITY

The eco-races/geographical races of silk insects have been identified on the basis of availability of particular food plant species and local agro-climate in the region. Hence, these geographical races/ecoraces can only be conserved by conserving their food plants through a protected area approach *in situ. In situ* conservation is very important as it allows evolutionary process to continue in nature. Hence *in situ* conservation of seri-biodiversity needs to be given more importance than *ex situ* conservation. The *ex situ* conservation should only be regarded as a back up measure only. The localised/concentrated area of distribution of seribiodiversity should be identified and natural habitat, extent of area, prevailing realities etc., should be studied and follow up action for restoration be taken up immediately. Local NGOs and R&D organisations who are principal stock holders should be involved and local people/ communities/tribals should be educated and informed about the available seribiodiversity, its habitat and the need for conservation. The feeling of ownness must be developed among them by educating them about ethical/ethno biological/ environmental importance of the seri-biodiversity.

```
┌─────────────────────────────────────┐
│         CENTRAL SILK BOARD           │
└─────────────────────────────────────┘
                    │
┌─────────────────────────────────────┐
│            CSGRC, HOSUR              │
│   Co-ordination & Base Collection    │
└─────────────────────────────────────┘
                    │
┌─────────────────────────────────────┐
│ Active Collection & Conservation Centres │
└─────────────────────────────────────┘
```

Tropical	Sub-Tropical	Temperate

Humid
CSRTI
B'Pore

Arid
CSRTI
Mysore

Hilly Area
RSRS
Coonoor

Hilly Area
RSRS
Kalimpong

RSRS
Jorhat

RSRS
Sahaspur

RSRS
Jammu

CSR&TI,
Pampore

Himachal
Pradesh

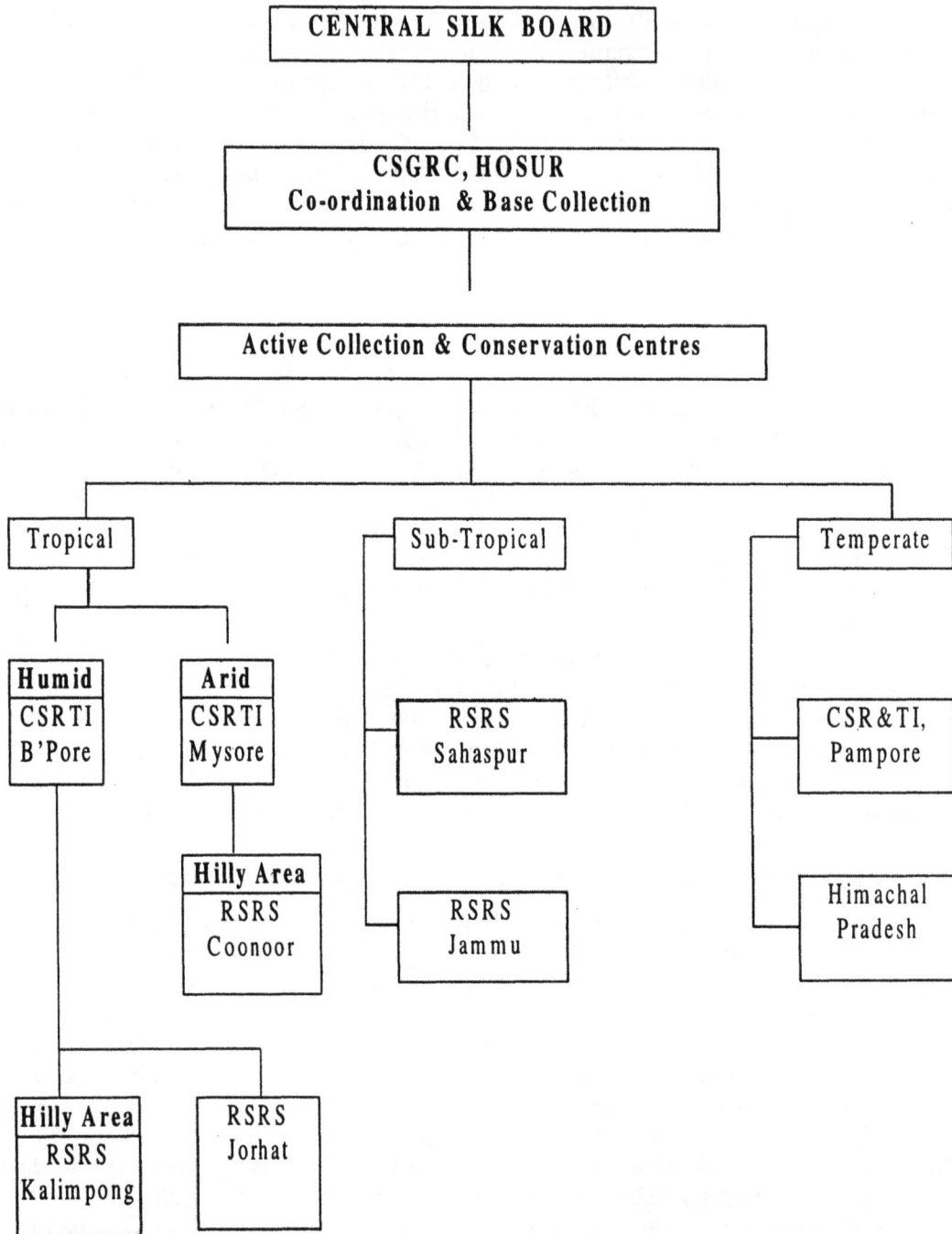

Diag. 4: Networking of mulberry (*Morus* spp.) and silkworm (*Bombyx mori* L.) germplasm conservation.

Source: Anonymous 2001

```
                    ┌─────────────────────────────┐
                    │     CENTRAL  SILK  BOARD    │
                    └─────────────────────────────┘
                                  │
                    ┌─────────────────────────────┐
                    │       CSGRC, HOSUR          │
                    │  Co-ordination & National database │
                    └─────────────────────────────┘
```

Co-ordination & base collection for Tasar	Co-ordination and base collection for Eri & Muga

CTR&TI, Ranchi	CMER&TI, Lahdoigarh

Active Germplasm Conservation Sites	Active Germplasm Conservation Sites

Tropical Tasar CTRTI Ranchi	**Oak Tasar** RTRS Imphal	**Muga** RMRS Boko	**Eri** RERS, Mendipathar

Sub Centres	Sub Centres
RTRS Bhandara	RTRS Bhimtal
RTRS Jagadalpur	RTRS Batote
RTRS Baripada	
RTRS Chinoor	
RTRS Dumka	

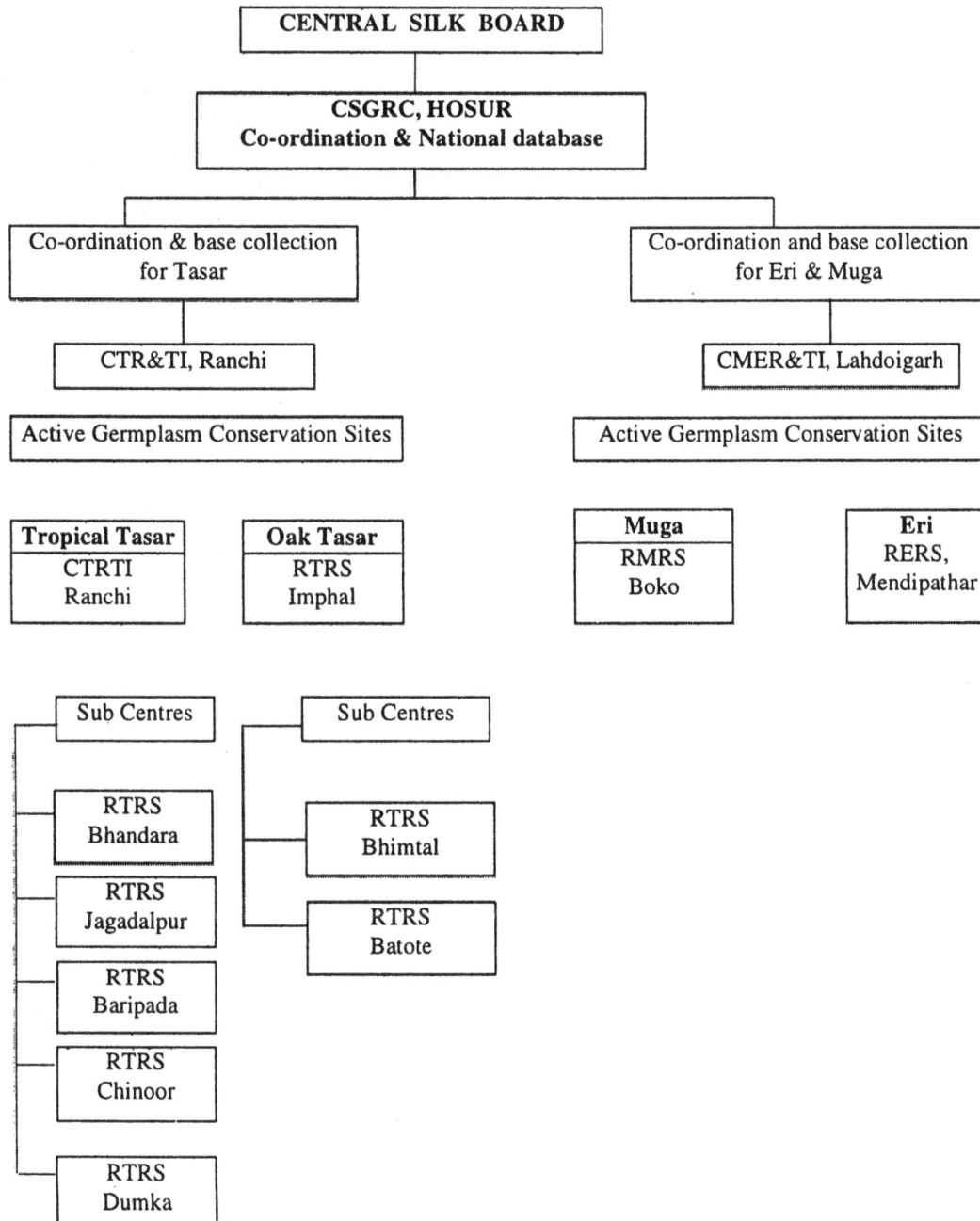

Diag. 5: Proposed network of Non-mulberry conservation.

Source: Anonymous 2001.

The mulberry (except *M.laevigata*) and non-mulberry food plants (excepting *Shorea robusta* and *Tectona grandis*) have low wood value, hence they have low priority for forest departments for initiating *in situ* conservation programme. Hence, efforts should be made to collect information on the availability of sericigenous flora in "Declared Protected Area Network of India" *viz.*, Biosphere Reserves, National Parks, Sanctuaries etc., and plus trees should be selected and wild silkworms be allowed to multiply there for restoration of deteriorated/disturbed ecological niche.

A sound data base for all the qualitative as well as quantitative traits of non-mulberry silkworms as well as their host plants should be developed through bioprospecting *viz.*, systematic survey and exploration, collection, characterisation and evaluation in the pattern of "All India Co-ordinated Mulberry and Silkworm Germplasm Evaluation Programme" in coordination with the main institutes and their Regional Research Stations.

The field gene bank for all types of host plants and silkworms may be established for conservation purposes. The identified ecoraces/geographical races should be reared in respective areas only for back up purposes. While doing so, it will be of utmost need to conserve the wild relatives of all food plants and silkworms in *in situ* as they are the source of dominant genes which may be utilised in future.

Uniform maintenance of germplasm in different agroclimatic regions can only be achieved through collaboration and cooperation between different units with meaningful exchange of material and information. Also, germplasm maintained at a place may be lost due to natural calamities, diseases and unavoidable reasons. Hence, a germplasm station should have its out stations in different ecosystems to serve as safety back-up units which will also facilitate the genetic materials to express their full genetic potential. Accordingly, a network on the conservation of mulberry as well as non-mulberry germplasm should start operation as early as possible (Diags. 4, 5). CSGRC, Hosur being the nodal agency is needed to coordinate various activities. The all India coordinated evaluation programme should be taken up in collaboration with different institutes of CSB located in different agro-climatic zones supported with numerous regional stations. However, it would be more pertinent to establish a separate non-mulberry Sericultural Germplasm Resources Centre or a separate wing should be opened in CSGRC, Hosur to coordinate the non-mulberry sector exclusively.

Under explored/un-explored non-mulberry silkworms and host plants, wild relatives and endangered species of sericigenous insect should be given priority for conservation through extensive explorations covering areas suggested/recommended by co-ordinated workshops/symposia in collaboration with CSB Institutes located in the different agro-climatic regions of the country and also with other institutes having similar nature of work *viz.*, NBPGR, BSI, ZSI, NBAGR, ICFRE etc. (Table 65).

As far as possible the *ex situ* conservation should be resorted through clonal means only, so that true to type representatives of the· food plant varieties are conserved. Hence, clonal propagation technology, macro-propagation as well as micro-propagation, needs to be developed on priority basis. Priority should be given for developing propagation technology for *T. tomentosa, Shorea robusta, Quercus acutissima, Quercus griffthii, Q. leucotricophora, Q. semecarpifolia* and *Lithocarpus dealbata* in which such technologies are lacking.

6

Sericulture and Allied Industries

The explosion of human population and the ever increasing need for food, clothing, shelter and medicine have forced human beings to find out ways and means for increasing economic returns from an unit area of land. Consequently, land has become very scarce and forest land is encroached for agriculture and other purposes; and a series of changes take place in the biosphere. The present trend of silk production indicate a declining trend in mulberry acreage due to water scarcity, fluctuating trends in rainfall, non-availability of skilled labour, high cost of inputs and comparatively low cost of cocoons in the market. Hence, blending of sericulture with other agro based product development has become urgent need of the hour. Integration of sericulture with agriculture, horticulture, pisciculture, silviculture and dairy not only increase the income of the farmers but also supplement and facilitate each other in eco-friendly manners. Further, a lot of scope exists for bye-product utilisation and commercial exploitation of new products from sericulture in cosmetic, furniture, tannin, leather, pharmaceutical, food, soap and other industries which may enhance the earning of farmers compared to traditional cocoon production alone. The non-traditional, non-textile sericulture products of South Korea have great demand all over the world. Hence, India has to explore in these directions also besides going into bivoltine sericulture in a big way, which will add new dimensions to Indian sericulture in the new millennium. Infact, mulberry and *Terminalia* trees have been regarded as Kalpavruksha and wonder trees due to their multipurpose uses (Dandin and Ramesh 1987, Singhvi *et al.* 1991, Srivastav 1991). Hence, our country should harness maximum benefits through their sustainable and judicious utilization.

A. INTEGRATION OF SERICULTURE WITH AGRICULTURE AND HORTICULTURE

For integrating sericulture with agriculture and horticulture, multicropping and intercropping are the best ways. In **mixed cropping system** more than one crop are grown together but all the crops are harvested during specific periods. On the contrary, in **intercropping system**, a crop is grown as the main crop whereas other crops, usually of shorter duration than the main crop, are cultivated in the spaces lying between the plants or between the rows. In intercropping, while the main crop continues to grow, the inter crops are harvested from time to time leaving inter-plant or inter-row spaces for future intercropping. The inter-cropping may be resorted either during the establishment or after the establishment of the main crop. In fact, intercropping in mulberry is already practiced by some farmers especially during its establishment. The low canopy of mulberry or non-mulberry plants during the establishment period and wider spacing between the rows may successfully be utilised for cultivation of short duration cash crops to fetch additional income from the same area of land, because proper establishment of mulberry from planting to first leaf harvest for silkworm rearing takes either 6-8 or 3-5 months in tropics in the gardens planted with cuttings and saplings respectively, whereas establishment of tree species *viz.*, Arjuna/Asan, Som/Soalu, Uyung/Banj, Castor/*Heteropanax* etc. takes at least three years.

In tropical regions the irrigated mulberry is cultivated under either row system with spacings ranging from 2' x 1/2' to 3' x 3' or under pit system with spacing ranging from 2' x 2' to 4'x4'. On the contrary, rainfed mulberry is cultivated only under pit system with wider plant spacing of 3'x3' to 6'x6'. In case of tropical tasar culture 4'x4', 6'x6', 8'x8'; muga culture 6'x6', 9'x9', 12'x12'; oak tasar culture 4'x4', 4'x6', 6'x6' and eri culture 3'x3', 6'x6', 9'x9', 12'x12' spacings are followed. In fact 4'x4' spacing for *Terminalia arjuna* and 6'x6' spacing for *T.tomentosa* have been recommended for tropical tasar culture, while 4'x4' spacing for *Q. serrata/Q. leucotrichophora/Q. grifithii/Q. semecarpifolia* for oak tasar culture, 6'x6' and 12'x12' spacing for *Machilus bombycina* for muga culture for individuals and village grazing reserves respectively, 3'x3' for *Ricinus communis* and 6'x6' for *Heteropanax fragrans* have been recommended for eri culture. All these spacings are indeed appropriate for inter cropping during the initial establishment phase in case of non-mulberry food plants, however after establishment 3'x3' and 4'x4' spacings do not allow penetration of sunlight, hence, if intercropping is to be practiced all the non-mulberry food plants should be raised at more than 6'x6' spacings. Hence, 8'x8' spacing for non-mulberry food plants will be most suitable for intercropping even after establishment also since their pruning is resorted at above four feet as against the bottom pruning resorted in case of mulberry. The bottom pruning of mulberry allows uniform sunlight to intercropping while pruning of non-mulberry plants at above four feet causes shadows in the field if spacing is below 6'x6' which is not much suitable for intercropping in both pruned as well as unpruned plants.

Since majority of sericulture farmers have very small land holdings and depend mainly upon family labour and simple tools, they neither have the capacity to take risk nor have enough land to diversify the cropping system. Thus, by growing other crops of short duration along with mulberry, the farmers can get additional benefit from intercrop besides the mulberry leaf yield thereby ensuring higher income per unit land area. Koul *et al.* (1996) have opined that intercropping of vegetables during winter in mulberry plantation can be taken without any harm to plant and the silkworm rearing. A detailed technology for intercropping has been discussed by Sinha *et al.* (1987). However, additional dose of fertilizer and FYM should be added to intercrop in addition to the recommended fertilizer applicable to mulberry plants (Anonymous 1995).

(a) Choice for intercropping

The intercrops fulfilling the characteristics of short duration, low canopy, shade loving, low input requirement and high remuneration should be selected for cultivation along with mulberry and non-mulberry food plants. The intercrops showing higher competition with sericegenous plants/flora with high canopy adversely affects their growth while heliophytic, low canopy intercrops do not grow under the shade of sericegenous flora. Not only this, besides increasing the net return per unit area, the right intercrop also improves soil fertility and facilitates better control of weeds, pests and diseases. Short duration legume crops like urd (70 days), soya beans (90 days) and black grams (90 days) are good intercrops for mulberry as well as non-mulberry food plants during their establishment. Intercropping in mulberry appears to be more suitable system when the dynamic properties like constancy in yield from year to year (stability), ability to survive in stress periods (resilience) and extent of system for operating for indefinite period (durability) is existing (Shukla *et al.* 1989).

Mane *et al.* (2000) found that the mulberry leaf yield in mono crop was at par with mulberry intercropped with Palak and French bean whereas mulberry leaf yield was low with Amaranthus intercropping. The highest intercrop yield was recorded in Amaranthus followed by Palak , French bean, Mentae and Coriander while no yield was obtained when Onion, Okra and Cluster bean were intercropped with mulberry which may be due to the fact that the duration of these crops are comparably more as compared to other crops used during their studies. Hence, intercropping of Palak can be taken up in the mulberry garden without affecting mulberry leaf yield and silkworm rearing.

Owing to the prevalence of high temperature upto 47°C and low humidity as low as 25% during summer in Purvanchal Region of Uttar Pradesh, silkworm (N x NB4D2 or PM x NB4D2) rearing should commence from 29th March to 2nd April, so that the rearing is completed by 25th April before the temperature and humidity become hostile for mulberry silkworm growth. Under these circumstances, farmers have been advised to undertake one intercrop of urd [*Vigna mungo* (Lin.) Hepper] or moong [*Vigna radiata* (Lin.) Wilczek syn. *P.radiatus* Linn.] in 3'x3' or 4'x2' spacing during mid April to first week of July when summer rearing is in progress. The germination and growth of urd and moong are not affected as mulberry leaves are used for rearing of silkworms. During and after silkworm rearing fortnightly irrigation is recommended which also saves mulberry plants during summer. Weeding and other cultural operations are not required as moong and urd do not allow growth of weeds. Pruning of mulberry and harvesting of these intercrops are undertaken during first week of July at the onset of monsoon. After harvesting from upper portion containing pods remaining plant material is incorporated into the soil by digging which provides organic matter to the mulberry plantation. Besides, urd and moong also add additional nitrogen in soil by fixing it from atmosphere. A total of 125 kg. of moong or urd may be harvested from such intercropping. Hence, besides income from silkworm rearing during April, the summer intercrop will provide an additional income of Rs. 2000/- from an acre of mulberry garden to the farmers apart from enhancing soil fertility and improving the soil texture (Gargi *et al.* 1997).

In Uttaranchal region intercropping of Lentil and Moong at (8'+2)x2'gives maximum returns of Rs. 33815/- from the cost of cocoons harvested by utilizing mulberry leaves and the cost of intercrops thereby facilitating net gain of Rs. 22958/- from one acre of mulberry plantation during spring and autumn seasons (Singh *et al.* 2003).

In Jharkhand region, Coriander can successfully be grown in the nursery beds without affecting the total growth of mulberry and as much as Rs. 12338/- per acre may be earned. Since there is no return from the land during the gestation period of mulberry, such intercropping would be beneficial for the farmers to generate additional income in Jharkhand (Dhingra *et al.* 2002).

In Tamil Nadu the preferential choice of farmers are onion, grams and coriander (Tables 66, 68). Onion, Coriander, Chillies and Maize are grown on the ridges usually in between the two rows of mulberry farming a ridge. Tomato and brinjals are cultivated at the edge of the ridges to provide them sufficient space for spreading and protecting their roots from excess of water flowing in the furrows during irrigation. On the contrary, pulses like urd bean, soyabeen, black gram and horse gram are grown in the furrows. Despite high profits from certain crops like Chillies (Rs. 5,000/- per acre) they should not be cultivated in sericulture farms due to their ill effects on the silkworms. Most of the intercrops are sown after south-west monsoon (June-July) and north east monsoon (Oct.-Nov.) when the weather conditions favour these crops in Tamil Nadu. Onion, tomato, brinjal, lady's finger and maize provided an additional income of Rs. 3000-4000/- per acre whereas pulses provided lesser income of Rs. 2500-2000/- per acre in Tamil Nadu before 1990, yet actual return varied from farmer to farmer due to variation in spacing of mulberry plants, density of intercrops and also the techniques of the farmers (Gangwar and Thangavelu, 1991).

The soil in the hilly region of West Bengal, is mostly sandy loam, with high levels of organic matter, humus and pH ranging from 4.1-6.7. The farming in the area is mostly rainfed or irrigated only through spring and small rivulets locally known as "Jhora" which also depend on rain. The prevailing agro climatic conditions are suitable for mulberry as well as turmeric. Turmeric grows well under warm and moist climate at higher altitude in loamy or alluvial soil. It cannot withstand water logging, gravely, stony or heavy clayey conditions, which are unsuitable for the development of rhizomes. However, it can tolerate pH range from 5-7.5. The farmers have been taking mulberry and turmeric as monocrops and not getting good yields due to various factors. The farmers harvest mulberry leave three times during April, June and September for silkworm rearing whereas turmeric is sown during April and harvested during November. Monocropping and intercropping of turmeric in between the rows of mulberry plants (2'x2' spacing) with the same doses of fertilizers and FYM application resorted to sole crop of mulberry revealed that leaf yield (30.68 kg/plot) decreases by 8.47% in mulberry but the yield of turmeric rhizome (7.68 kg/plot) is increased by 10.98% as against their sole monocrops (33.52 and 6.92 kg/plot respectively). When economics of intercropping with mulberry was calculated it was found that net profit from sole crops of mulberry and turmeric was Rs. 30,710/- and Rs. 14,400/- respectively. While intercropping of turmeric with mulberry yielded a net profit of Rs. 33,340/- during 1990-91. Hence, the net profit was increased by 26.3% from the same piece of land without any additional investment. Further, as the turmeric needs heavy manuring, cultivation of legumes may improve physical condition and fertility of soil through nitrogen fixation by root nodules (Tikader 1991).

Table 66: Major intercrops grown in Tamil Nadu and gross income from the inter crops.

Sl. No.	Intercrop	Season	Crop duration (days)	Gross income (Rs. per acre)
1	Soyabean	June - July	90	2,000-00
2	Urd bean / black gram	All season	65 - 70	2,000-00
3	Green gram	All season	70 - 75	2,000-00
4	Gram	October	110	2,500-00
5	Onion	June-July & Jan-Feb	90	3,000-00
6	Tomato	All season	120	3,000-00
7	Brinjal	All season	150	4,000-00
8	Lady's finger	June - July	100 -110	4,000-00
9	Coriander	October	90	2,500-00
10	Chilli	August	180	5,000-00
11	Maize	Sept.-Oct.	100 -110	3,000-00

Source: Shri R.Rajendra, Associate Professor of Agronomy, TNAU, Coimbatore(Pers.comm)
Adopted from: Gangawar and Thangavelu 1991

Table 67: Package of practices for the fruit trees and mulberry

Practice	Guava*	Sapota*	Mulberry
Climate	Moderate	Moderate	Moderate
Soil	Sandy, loamy	Sandy, loamy	Sandy, loamy
Planting system	Pits	Pits	Pits or rows
Spacing	6m x 6m	10-13 m x 10-13 m	0.15 m x 0.60 m or 0.6-0.9m X0.6-0.9m
No. of plants/ha	250-275	65-100	12,350 -1,11,000
Planting material	Grafts	Grafts	Saplings
FYM per year	25 kg/plant (Approx. 6500 kg/ha)	50 kg/plant (Approx. 3500-5000 kg/ha.)	20,000 kg/ha
Fertilizer (N:P:K) 1-3 years	50:25:75 g/pt.	50:20:75 g/pt.	300:120:120 kg/ha (2nd year onwards)
4-6 years	100:40:75 g/pt.	100:40:150 g/pt.	
7-10 years	200:80:150 g/pt.	200:80:300 g/pt.	
11 years onwards	300:120:150 g/pt.	400:160:450 g/pt.	
First crop	After 3-4 years	After 4 years	After 6 months
No. of crops/year	Three	Three	Five-six

* Extension information from Dept. of Horticulture, Bangalore
NB: Guava and Sapota yield 1000-1500 fruits/tree while mulberry yields about 30000 kg.leaf/ha. annually.
Source: Bongale 1990

Table 68: Additional income from intercrops in mulberry in Salem Dist. of Tamil Nadu during 1987-89

Sl. No.	Name of the farmer	Village	Mulberry acreage	Spacing	Intercrops	Extra input (Rs.)		Yield/ crop (Kg.)	Return from intercrop actual (Rs.)	Additional profit from intercrop (Rs./acre)
						Seed	Labour			
1	Velappa Gr.	Kolakattupudur	0.25	2' x 2'	Blackgram	18.00	15.00	40	320.00	1148.00
2	Ramachandran	- do -	0.40	2' x 1'	Onion	100.00	30.00	400	600.00	1175.00
3	P. Marappan	Pudupatti	0.40	2'6"x2'6"	Onion	225.00	30.00	800	960.00	1762.50
4	P. Muthu Gr.	- do -	0.40	2'6"x2'6"	Onion	225.00	30.00	600	720.00	1162.50
5	Arunachalam	N.K.P. Road	0.35	2' x 1'	Onion	50.00	15.00	210	315.00	714.30
					Blackgram	13.00	—	57	456.00	1265.70
6	Arunachalam	P.M.Malayam	0.40	2' x 1'	Coriander	10.00	—	40 Ltr.	200.00	425.00
7	Subramani	Pudupati	0.20	2' x 1'	Onion	40.00	15.00	200	300.00	1225.00
8	R. Vijayakumar	- do -	0.35	2' x 2'	Blackgram	25.00	15.00	50	400.00	1028.60
9	Subramani	- do -	0.50	2' x 1'	Coriander	15.00	—	8	240.00	450.00
10	M. Subramani	- do -	0.50	2'6" x 1'	Onion	140.00	30.00	640	960.00	1580.00
11	K. Selvam	- do -	0.40	2'6"x2'	Maize	20.00	15.00	400	600.00	1412.50
12	Ganesan	- do -	0.60	2'6"x2'	Onion	250.00	30.00	800	1200.00	1533.30

Source: Gangwar and Thangavelu 1991

Studies have revealed that a farmer who opts for intercropping of mulberry with Saffron in Kashmir can get a good quantity of mulberry leaf from the same field he used to get Saffron alone . Therefore, silkworm rearing can be taken up in the lean period when there are no operations related to Saffron cultivation. It would not only generate work but also add a good deal of returns to farmers in Kashmir (Kaur *et al.* 2002).

(b) Choice for mixed cropping

Cultivation of mulberry in other perennial plantations as a mixed crop is not a common practice; yet in Mandya, Bangalore and Tumkur districts of Karnataka coconut is popular associate of sericulture. Besides, in Dharwad districts (Karnataka) some farmers have planted mulberry in existing guava plantations while some other farmers have simultaneously raised the two plantations together due to following reasons:

1. Mixed cropping generates subsistence income from sericulture during gestation period till the main crop comes to bearing. Guava, sapota and mango etc. take 3-4 years while coconut takes 6-8 years for bearing after plantation.

2. Mixed cropping generates higher income in the long run from the same piece of land through horticultural as well as sericultural produce.

3. Mixed cropping increases the efficiency of solar energy harvesting, soil nutrients and other agricultural inputs due to multi-tier crop pattern. While the fruit trees require proportionately higher quantity of potash, mulberry requires more of nitrogen. Such differential requirement for the three major (NPK) plant nutrients reduces the competition between the plant species to some extent.

4. Mixed cropping provides a protective covering to the exposed soil Exposure of land directly to sunlight, wind, rainfall etc., raises the soil temperature, accelerates decomposition of organic matter, reduces percolation of rain water, enhances rate of evaporation and reduces/prevents entry of air into the soil.

Mulberry a deep rooted perennial plant with rapid vegetative growth comes to bearing within 6-8 months after planting and under bush system of cultivation it grows luxuriantly as a bush. There is no soil exposure after its establishment and repeated tillage is not required. At the same time it also reduces the intensity of shade considerably due to repeated harvesting of leaves at an interval of 2-3 months. Hence, it can serve as a most suitable mixed crops for guava, sapota, citrus, lemon, pomegranate and papaya and similar other fruit plantations due to their moderate crown size and measurements (Table 67). According to Bongale (1990) following points help in establishing mulberry as a mixed crop in plantation of fruit crops.

i. An area of about five feet radius be left uncultivated around each tree and mulberry should be planted in rest of the area.

ii.   A trench of about two feet depth be dug around each tree to prevent the roots of mulberry competing with those of fruit trees.

iii. Mulberry be planted preferably at a close spacing by row system between the fruit trees so as to restrict the root growth of mulberry for reducing competition between the two crops.

iv.   Nutritional requirements of mulberry and fruit trees be met separately as per respective recommendations.

v.   Spraying of plant protection chemicals be carried out with sufficient safe period for silkworm rearing.

vi.   Mulberry plantation be renewed every ten years by uprooting old ones and planting new ones for preventing severe competition between the two crops and facilitate optimum harvest of the fruits.

The subject area for innovation of more efficient agriculture system for cultivation of mulberry with various fruit trees and other horticultural/agricultural crops is widely open.

(c) Sericegenous plants as food/Ornamentals

i. Mulberry as a food for human

While *Morus alba* var. *multicaulis* has been recommended for leaves, *M.alba* var. *atropurpurea* is cultivated widely as tree crop for its large, cylindrical, dark purple and succulent fruits known as sorosis in addition to its leaves. *M.nigra* is also cultivated as small tree for fruits which are ovoid to oblong, 2.0-2.5 cm long, purple to black in colour, while fruits of *M.serrata* are purplish and 2.5-5.0 cm long and 8.0 cm long in *M.macroura*. Normally, fruit setting is observed during Nov.-Dec. and June-July but in *M.nigra* it is during May-June in South India. The fruits of *M.laevigata* are sweet and insipid. Mulberry fruits have indeed been considered as an ideal food for our health (Phillip 1988).

In M-5 mulberry variety at Bangalore in 3'x3' spacing majority of the plants bear more than 100 fruits (69-165). The average girth of fruits is 0.876 cm (ranging from 0.827 to 0.934 cm) while average length is 1.5 cm (ranging from 1.429 to 1.639 cm). The average number of fruits/plant, weight of single and 100 fruits have been found to be 111.69, 0.804 gm and 59.9 gm respectively, thereby yielding 6.6 qtls of fruits/acre/year (Venkatesh *et al.* 1993). *M.nigra* has been found to yield 20-30 lb.fruits/tree/year. However, yield is correlated to yield governing factors *viz.*, species, cultivation practices, season and geographical conditions.

Mulberry fruits are eaten directly as fresh fruits. Besides, they are also used for preparing pickles, jam, jelly, sherbut and wine because of their high nutritional and medicinal value. Mulberry fruits are also consumed directly by avies fauna (crows, minibirds), insects (pentatomid bugs and ants) and canines (dogs) (Venkatesh *et al.* 1993).

Mulberry curry

The growing mulberry leaves contain 6-6.8% protein, Vitamin A, B, C, D, 4% soluble carbohydrates, 0.6% crude fat and several compounds of calcium, phosphorus, silicon, manganese, magnesium, iron, copper and zinc etc. and hence they satisfy all the dietary requirements of human body. Rural people in south have found mulberry leaves equally good for consumption like Amaranthus, pea, cucumber and drumstick leaves. Mulberry curry prepared from tender leaves as hereunder can be a good table delicacy (Sreekumar *et al.* 1994).

Ingredients

1.	Chawki mulberry leaves	20 nos.
2.	Coconut	1/2 (2-3 cm long 1/2 cm wide)
3.	Edible oil	3 table spoons
4.	Mustard	1/2 tea spoons
5.	Green chilly	5 Nos.
6.	Hens egg	1 No.
7.	Curry leaf	One small branch
8.	Salt	According to taste
9.	Onion	5 nos.

Ingredients may be altered/changed/added according to individual taste.

Preparation

Chop the mulberry leaves and cook in a pressure/idli cooker for 3 min. Heat the oil and put mustard seeds, onion and chilly pieces and roast. Add curry leaves and subsequently put steam cooked mulberry leaves and mix. Add hens egg directly and stir to mix it thoroughly. Then mix grated coconut and add salt according to taste. Serve this delicacy to 5-6 members. Mulberry leaves are also used for preparing a decoction known as mulberry tea. By adding the roots and stem of mulberry a broth is also prepared. The alternate host plant of mulberry silkworm lettuce (*Lactuca sativa*) is also eaten and used as salad.

Chitin, a derivative of cellulose is derived from cuticle of silkworms and exo-skeleton of pupae. It is used as an additive to increase the loaf volume of wheat flour bread from about 4.9 to 6.0 cm^3/gm since it has more water binding capacity than micro-cellulose (Knorr 1982). The degreen silkworm faeces yields about 10% pectin which can be used as a thickener in candy, jelly, jam, concentrated fruit juice and as stabilizer in ice cream. It can reduce blood triglyceride and cholesterol and has bacteriostasis and detoxification activity (Majumder 1992).

One of the Russian mulberry variety with drooping branches, when grafted at a height on suitable stock, forms the best weeping lawn tree popular in America known as "teas" weeping mulberry. The variety Venulosa with attractive lobed leaves is said to be one of the best ornamental mulberries (Singhal *et al.* 2000).

ii. Non-mulberry silkworm food plants and their food value for human

(a) Tropical tasar secondary food plants

The leaf and flower buds of Kachnar (*Bauhinia variegata*), a secondary host plant of tropical tasar silkworm are popular vegetables. Besides, it is also planted as an avenue tree for presenting heavenly sight of beautiful pink, white and purple varieties (Randhawa 1983). Likewise, buds of *Bombax ceiba* a secondary host plant of tasar silkworm are also used as vegetables (Shukla and Mishra 1979). The seeds *of Buchanania latifolia* (chironji) are used as substitute for almond and are used in confectionery. They also yield "chironji oil" upto 51.8% (Jain *et al.* 1990, Shukla and Mishra 1979). The fruits of *Carissa* carandas (karonda) are eaten either raw or made as pickles while berry of *Diospyros melanoxylon* (Tendu) are eaten in small quantity. The fruit of Embelic myrobalan (*Helianthus embelica*) are eaten and pickled due to rich source of Vitamin C. The fruit of kel (*Ficus tjakela*) are eaten by tribals in Kerala (Nair and Jayakumar 1998). The fruits of Kharpat/Kaikar (*Garuga pinnata*) are eaten raw and pickled (Brandis 1906). The ripe fruits of Mahawa (*Madhuka indica*) are eaten and the outer cover of the fruits and seeds are used as vegetables. The Mahawa as well as sal (*Shorea robusta*) seeds yield edible oil/ghee. Jamun (*Syzyzium cumini*) and Gulab Jamun (*Syzyzium jambos*) yield delicious astringent fruit rich in iodine. The kernels of *Terminalia catappa* (Desi Badam) are eaten. The fruits of *Zizyphus jujuba* (Beri) and *Zizyphus mauritiana* are eaten fresh, dried or preserved.

(b) Eri silkworm food plants

Fruits of Papaya (*Carica papaya*) are very delicious and Tapioca (*Manihot utilitissima*) roots/tubers are eaten as cassava/sakarkand. Bajramani (*Zaxthoxylum rhesta*) fruits are used as spice (Pandey 1992).

(c) Muga silkworm food plants

Leaves of *Cinnamomum obtusifolium* (Tejpata) are used as spice and *Michelia champca* which is grown in gardens for large sweet smelling flowers yields an oil used in perfumery. Flowers of Champa also yield a dye (Shukla and Mishra 1979). Golainchi/Chameli (*Plumeria acutifolia*) are popularly known as temple tree and their flowers are used in wreath making and offered to deities (Randhawa 1983).

(d) Oak tasar food plants

The fruits of Sareng (*Castanopsis indica*) are edible (Negi and Naithani 1995). The nuts of Chaukhu (*Castanopsis purpurella*), *Lithocarpus xylocarpa* and *L.dealbata* are eaten in very limited quantity.

Thus, only secondary host plants serve as either vegetables and/or fruit crops. Since secondary host plants cannot be raised as economic plantations due to their secondary nature their role in integration with sericulture is of limited scope.

B. SILKWORMS AS FOOD

In Hong Kong, China, Korea and Japan, the healthy silkworm larvae are sterilized, vacuum dried and sold as commercial food. In powder form, it is used as common source of animal protein and sauce preparation specially for cardiac and diabetic patients due to its low cholesterol content (Singhal *et al.* 2000).

Silkworm pupae is a nutritious food for human diet as it contains a balanced amount of moisture, chitin, water soluble proteins, carbohydrates, amino acids and Vitamin C, besides crude protein as a major constituents (Majumder 1992, Majumder *et al.* 1994). In addition, it contains fairly large amount of nutritionally important minerals like Na, K, Ca and P, Vitamins like B_2, nicotinic acid, folic acid and Vitamin B_1 and essential as well as non-essential amino acids (Bose and Majumder 1990, Bose *et al* 1989).

Yokoyama (1962) opined that nutritive value of silkworm pupae is even superior than that of fish meal and not inferior to beef. In Assam eri pupae are fried in oil, spices are added and eaten as delicious dish. The ancient Greek and Romans used the pupae powder in the preparation of cakes and bread; in China, children prefer pupae as food to chocolate. Native American Indians used the dry pupae in flour and chick lets (Bose 1992) and Vietnamese eat silkworm larvae as it is (Singhal *et al.* 2000).

Medicinal wine, popularly called as "moth wine" has been prepared by Shaanxi Sericultural Technology Station, China to treat impotence, abnormal menstruation and menopausal symptoms (Mandel 1991, Rajiv and Kumar 1996).

Pelade, the inner layer of silkworm cocoon shell is valuable ingredient of human food in China and Japan. One Kg of pelade is obtained from 10 Kg cocoon after reeling. Commercially known as "Shinki fibroin"in Japan, it is rich in bi—and tripeptides and easily digested and assimilated. It has the property to reduce the cholesterol and blood sugar and provides additional energy (Ramakanth and Anatharaman 1997). 1-5% of "Shinki fibroin" can be added to food material and generally 5 g per day is consumed (2g/cup) with milk or coffee (Singhal *et al.* 2000).

C. INTEGRATION OF SERICULTURE WITH SILVICULTURE/SPORTS/PAPER INDUSTRIES

i. Mulberry Sericulture

Growing teak trees along with mulberry has been found to be a suitable combination of agro forestry model, which can be adopted by farmers in India for enabling them to earn more from same unit of land without much investments. Under United Nations Development Programme (UNDP) assisted extension programme, five villages each in Karnataka and Andhra Pradesh were adopted by the Institute of Wood Science and Technology, Bangalore for free distribution of teak saplings. The progressive sericulture farmers came forward to grow teak trees in their mulberry gardens. Sericulturists who showed interest in teak growing on an average formed a major group comprising of 40% (5-80%) and 12% (5-18%) in the total farmer community in Karnataka and Andhra Pradesh respectively. Sericulturists planted saplings either in between the rows of mulberry plants or on the boundary lines. The programme started in 1992 and teak trees have been found to grow there without additional expenditure or adverse effect on the growth of the mulberry plants. Teak trees survive in the dry agro climatic conditions but grow best in the moist, warm and tropical climate in sandy to clayey loam soils with pH 6.5-7.5 by producing large deep root system. It thrives well in areas with 1000-1500 mm rainfall and demands profuse light. Hence, conditions suitable for moriculture are more or less suitable for growing teak along with mulberry. Teak does not create much shade because its side branches are usually trimmed off during early years of growth for formation of a clean bole. There is no need of separate irrigation for teak since it efficiently absorbs water leaked out of the surroundings, however, watering of teak seedlings are required during the early years for good growth. When teak trees are planted on boundaries, ploughing and other operations of soil for mulberry can be undertaken without any inconvenience and no difficulty is faced in harvesting of mulberry leaves also. The pests of teak *viz.*, *Hyblaea puera* and *Haplia machaeralis* do not attack mulberry and their control through chemical spraying may be resorted only when there is no rearing or leaf harvest; otherwise the mulberry leaves may be used after recommended safety period (Ramadevi and Khali 1996). Besides being economically very important as a timber tree, teak also serves as secondary food plant of tropical tasar silkworms. The mulberry chips (excess stem, barks and roots) may be converted into fibre boards through conditioning, deflaking, drying, chopping and pulping, shifting, glueing and mixing, mat forming, pressing, curing and polishing (Majumder 1997).

Likewise, besides serving as host plants of various silkworms many plants serve as highly valued economically very important timber trees also. Such plants may be left abandoned as and when required to grow undisturbed without utilising them for silkworm rearing for yielding timber only. Only their side branches should be trimmed to provide a clear bole. However, such plants should necessarily be located on the boundaries only so that they may not create much shade for other silkworm food plants. The wood of white mulberry (*Morus alba*) is used for house building, agricultural implements, sports goods and furniture etc. It is considered as medium grade fuel wood and their shoots can be used in paper industry when grown as bush and contain tannins for tanning and colouring purposes (Singhvi and Sinha 1991).

Mulberry wood is used in the manufacture of sports goods and fancy articles due to blend of strength, flexibility and elasticity. Being highly shock resistant and not liable to split, it is preferred in manufacture of sports article like hockey sticks, cricket bats and stumps and tennis rackets. Jalandhar (Punjab) and Meerut (U.P.), in India have approximately 80 small scale industries engaged in production of sports goods from mulberry wood which is easy to saw and peels well in turnery and hence, used in manufacture of bobbins, tool handles and toys. The furniture made up of mulberry wood has very good finishing (Singhal *et al.* 2000).

Chinnaswamy and Hariprasad (1998) found that the fuel value of mulberry is much superior over most of the agriculture residues. Hence, mulberry can solve the energy crisis of rural areas where mulberry can easily be grown but sericulture cannot be practised.

In China and Europe, the bark is stripped from waste and pruned branches are used in paper industries. Young shoots, which contain high cellulose are also used for manufacturing paper (Dandin and Ramesh 1987).

Table 69: Chemical composition of common forages grown in South India (% on dry basis)

Sl. No.	Name of fodder	Botanical name	Crude protein	Ether extract	Crude fibre	Total carbohydrate	Ash content
1	Maize	*Zea mays*	9.00	1.00	28.90	77.70	12.20
2	Jowar	*Andropogon sorghum*	4.60	1.50	38.80	85.30	9.10
3	Napier grass	*Pennisetum purpurium*	10.70	2.0	30.50	71.20	16.00
4	Rhodes grass	*Chloris gayana*	10.10	1.60	34.10	78.20	10.10
5	Mulberry	*Morus alba*	26.10	4.20	12.00	68.76	7.97

Source: NDRI, Bangalore (adopted from Radhakrishna *et al.* 2000)

ii. Tasar Culture

The timber of *Anogeissus latifolia*, food plant of tasar silkworms, is used for making agricultural implements, and construction purposes besides serving as fuel wood (Maheshwari and Singh 1991, Nair and Jayakumar 1998, Khanna *et al.* 1998). The woods of *Bauhinia varigata* (Kachnar) are used for agricultural implements and fuel purposes. The timber of *Bomax ceiba* (semul) is used for match wood, tea chests and rough plankings (Sagreiya 1967) while woods of *Careya arborea* (Kumbhi) are used for agricultural implements (Singh 1990). The timber of *Dalbergia sissoo* (Shisham) is used for building, furniture, carriage and carving while its wood pulp is used for preparation of writing and printing papers. (Sagreiya 1967). The wood of *Chloroxylon sweitenia*, Indian Satinwood is used for agricultural implements. The wood of Dikamali (*Gardenia lucida*) is used for comb preparation (Brandis 1906). The wood of Anjan (*Hardwickia binata*) is used for construction, agricultural implements and building purposes (Nair and Jayakumar 1998), while the wood of *Lagerstroemia parviflora* is moderately hard and used for building, furniture, boxes and is useful for preparation of agricultural implements (Singh 1990, Shukla and Mishra 1979). The wood of *L.speciosa* (Jarul) is valued next to the teak as a timber tree and is used for building, boats, carts etc. (Randhawa 1983). The wood of Bijasal (*Pterocarpus marsupium*) is very hard, close grained, heartwood yellowish-brown, durable and much used for building, agricultural implements, carts, wheelwork, boats etc. (Sagreiya 1967). The Sal (*Shorea robusta*) wood is used for piles, beams, planking and railing of bridges, doors and windows, furnitures, bodies of carts and above all for railway sleepers, boats and canoes (Pandey 1992). *Shorea talura* (Lac tree of south India) yields timber of moderate quality used for building and construction purposes and furnitures. The wood of *Syzyzium cumini* (Jamun) is white, close grained, durable and used for construction, agricultural implements, and building purposes (Randhawa 1983, Pandey 1992). The timber of *Terminalia arjuna* (Arjuna) is used for agricultural implements, building of boats, carts, tool handles and construction purposes (Singh *et al.* 1983, Randhawa 1983). The timber of *T.bellarica* (Bahera) keeps well under water and used for boats and agricultural implements (Jain 1979), temporary construction work, general purpose class II plywood, black boards, heavy packing cases and boxes (Najma *et al.* 1981). The wood from *T.catappa* (Desibadam) is used for construction work, especially suitable for posts, beams and rafters (Singh *et al.* 1983). The timber from *T.chebula* (Harre) is used for construction work as beams, scanthings and planks, carts, tools, mathematical engineering and drawing instruments (Singh *et al.* 1983, Najma *et al.* 1981). The wood of *T.paniculals* (kinjal) is used for beams, rafters, planks, tea chests, commercial grade plywoods, black boards, agricultural implements, boat building, railway sleepers and lorry boards (Singh et.al 1983, Najma *et al.* 1981, Agarwal 1986). The wood from *T.tomentosa* is used for plywood chests, ceiling boards, domestic purposes, cheap utility plywood, match boxes, splints, rice pounders, ship, boat, furniture, cabinet, parcel, electric transmission poles, paper pulp, fire proof buildings, fuel wood and charcoal (Pandey 1980).

iii. Muga Culture

The wood of *Michelia champaka* (Champa) is used in mill work and making of boxes, musical instruments, boats, toys and drums etc. (Shukla and Mishra 1979).

iv. Ericulture

The wood of Maharukh (*Ailanthus excelsa*) is suitable for making match splints (Segreiya 1967). The wood of Gumhar/Gamhar (*Gmelina arborea*) is used as mortar in Kerala and for making plough parts and drums in Uttar Pradesh. Bhitera (*Jatropha multifida*) is cultivated as a beautiful ornnamental tree/shrub. (Rao 1991). Vilayati shisham (*Sapium sebiferum*) serves as a beautiful ornamental tree (Randhawa 1983). While walking sticks are made from the wood of *Zanthoxylum alatum* (Tejbal), the wood of *Z.rhesta* (Bajramani) are used for making shuttles (Pandey 1992).

v. Oak Tasar Culture

The wood of sarang (*Castanopsis indica*) is used for shingles and construction work whereas the wood of *Castanopsis lancaefolia* (Bucklai) is used for house building. The wood of chawkhu (*C.purpurella*) is used for making shingles, cart shafts, axles, naves, poles, yokes and ploughs and after preservation railway sleepers of fairly good quantity may be made from its timers. The wood of *Lithocarpus dealbata* (Sahi) and *L.fenestrata* are used for agricultural implements while that of *Q.acutssima* is used for building purposes. The wood of *Q.baloot* (Ber-chur) is used for agricultural implements and tool handles while the wood of Moru (*Q.floribunda*) is used for agricultural implements, constructional purposes, tool handles, sledge runners, walking sticks and carrying poles. It is found highly suitable for making strong and heavy wheel spokes, cask and barrel making, firewood and for charcoal (Pearson and Brown 1932, Gamble 1922). The wood of *Q.glauca* (Banni) is also used in the same fashion as that of *Q.floribunda* especially for soles for sledges in Tehri Garhwal area in Uttaranchal (Gamble 1922). According to Shukla and Kukreti (1983), the wood of *Q.grifithii* (chaka-Uyung) can be used for construction work, poles, ballies, tool handles, anvil blocks, mallet heads and agricultural implements besides being used as fuel. The timber of Bajrant/Buk (*Q.lamellosa*) is used for beams, doors, windows, frames, posts, rafters, bridge construction, shingles (aplit), plough, cart, axles, naves, spokes and felloes, fuel and charcoal. The timber of Banj (*Q.leucotricophora*) is used for wall construction, agricultural implements, tool handles, building works, fuel, charcoal making and its wood pulp may be used for high grade hard board with water resistance and good strength. Kharsu (*Q.semecarpifolia*) yields fine timber locally used for building purposes as beams, rafters, door frames, wall plates, agricultural implements, bed steeds, mule saddles, bent wood for furniture, cart wheel rims, barrel staves, whisky vats, roof sticks of lorry bodies, handsome furnitures, firewood and charcoal (Rahman *et al.* 1956, Pearson and Brown 1932, Gamble 1922). The timber of Sehop (*Q.semiserata*) is easy to saw and work to a smooth surface. It is used for pins with which cart wheels are joined in Myanmar (Pearson and Brown 1932).

D. INTEGRATION OF SERICULTURE WITH PISCICULTURE

In many countries, pupae are fed to carps and other fishes for better yields. The people of Pearl river delta region in China have developed the "mulberry dike-fish pond complex" an ecosystem which makes full use of available land and water resources. The peasants of pearl river delta fulfil the food requirements of fish raised in a mu of fish pond by the amount of silkworm excreta and crystalises produced by a mu (1 hac=15.15 mu) of mulberry trees. Thus, the ratio of mulberry to fish is 1:1. Hence, "mulberry dike-fish pond" is an inter-related ecosystem which promotes development of different branches of agriculture by utilising full production potential of human beings and their environment. In this system mulberry is fed to the silkworms who serve as primary consumers and whose excreta and crystalises are inturn fed to the fishes who serve as secondary consumers. The fish excreta and algae are decomposed by aquatic organisms who serve as reduction agents and break down the organic matter in the pond to produce nitrogen, phosphorus and potassium. This organic pond mud is returned to the mulberry as fertilizer. Mulberry trees require ample sunlight and fertile soil otherwise the quantity as well as the quality of mulberry production is reduced. The temperature must be above 10°C and below 40°C otherwise growth of mulberry is affected. The growth of silkworms also ceases when the temperature is below 15°C and above 32°C and when humidity is less than 70 or above 85 percent. For fish cultivation the temperature of water should be between 5°C and 35°C while 5.0 mg per litre of oxygen is ideal for their survival. When oxygen content reaches 1 mg/litre the fishes come up to the surface and when it is less than 0.2 mg/litre or temperature is below 3°C they die. Thus mulberry trees, silkworms and fishes have different environmental needs yet they are easily manipulated with due care under "mulberry dike fish pond ecosystem".

It is estimated that 20 kg of grass can be transformed into 0.5 kg of grass carp plus 0.25 kg of big head carp or silver carp. Mulberry and sugarcane leaves, elephant grass, peanut and corn form staple food for fishes. Silkworm excreta contains 2.2-3.5% nitrogen, 2.0-2.5% phosphorus and 1.5-2.0% potassium which is higher than those found in excreta of pig, sheep or cattle. It is estimated that 100 kg mulberry leaves will yield 60 kg of silkworm excreta which inturn can generate 8 kg of fishes. These fishes can serve as important source of additional income and protein for sericulturists. In addition the pond supplies fertile mud two-three times a year. Chinese peasants have estimated that it contains 5% organic matter needed by plants. Pond mud also act as weed killer, retards water evaporation and maintains soil fertility for a longer period (Bari, 1990).

In mulberry dike fish pond ecosystem different species of fishes may be cultivated. The bighead and silver carp thrive in the upper part of the pond, the grass carp in the middle zone and black carp live in the lower zone. In fact, grass carp accounts for about 50% of the fish population and feeds on silkworm excreta and phytoplanktons. Phytoplanktons serve as the major food for silver carp and also consumed by zooplankton which are inturn consumed by bighead carps. The mud carp and the blackcarp eat the unconsumed silkworm excreta and plankton which sink to the bottom (Bari, 1990).

Mulberry and some non-mulberry host plants cannot withstand waterlogging hence sericulture could not be introduced in places which remain water logged for most parts of a year. A large amount of waste land in India belong to low lying water logged areas. These areas are utterly unsuitable for any agricultural use and hence kept fallow. Besides, areas which face periodical floods also are not suitable for sericulture. Such areas may be suitably utilised by following the Chinese mulberry dike fish pond ecosystem or precolombian cultivation technique used on the low lying plains of the peruvian Andes around lake Titicaca which involved raising rectangular platforms of earth alternating with canals across the area, which ranged from 13-22 feet in width, suitably long and about 3-4 feet in height. The platforms were constructed of earth including top soil that was removed to make canals. On these raised platforms bumper crops were produced without the use of fertilizer or modern tools. Sediments in canals, nitrogen rich algae, plant and animal remains provided fertilizer to the roots through seepage and capillary action and layer of rich top soil also contributed for success of the crop. In winter, the water in canals absorbed maximum heat from sun which was radiated back by night to increase the night temperature.

Under the Intensive Sericulture Development Programme (ISDP) of Central Silk Board, Dhubulia, West Bengal a location which remains water logged for more than 9 months in a year, eight raised beds were prepared to cover 16,600 ft² or nearly 0.38 acres of totally unusable submerged wasteland. The eight raised beds were of 70'x14' (x 4" depth). The platforms were made on raised beds in June-July 1989 under 2'x2' spacing with saplings of S1 variety. The mulberry plants established completely by the end of Dec. 1989 when they were pruned 9" above the ground level. The raised plant beds were weeded but no fertiliser was applied. Only canal residue which contained green algae and plant and animal residue was shelved onto the platform during January,1990 when the canals mostly dried. During the third week of June when monsoon appeared in full swing, plants were pruned. On an average each bed produced 158 kgs. of leaf or nearly 17,354 kgs of leaf/ha. Even considering the amount of land which was wasted for preparation of canals, the total of 0.38 acres of plantation consisting of eight raised beds and nine canals produced 8,219 kg of leaf/ha. which may be termed a bumper production (Sarkar and Prasad 1991). If the canals are also used for fish cultivation like the people of Pearl river delta region of China, the income of the peasant can be further enhanced from such regions of India, which have been completely left fallow due to low lying nature or suffering from frequent or periodical floods.

E. INTEGRATION OF SERICULTURE WITH DAIRY AND ANIMAL HUSBANDRY

White mulberry has been considered as a very important fodder tree (Dwivedi 1988). A study undertaken to know the extent of participation of rural youth in different activities of dairy and sericulture and their socio economic characteristic revealed that most of the farm youth participate in dairy activities by assisting the family members and only 10-15% have taken up dairying themselves exclusively. On the contrary in sericulture, majority of the farm youths (75%) participation is through self doing rather than assisting their family members (Lakshmi Raju and Nataraju 1998). Animal which can feed on silkworm litter are cow, sheep, pigs, buffaloes and poultry (Majumder 1997).

Currently, an acre of irrigated mulberry garden with improved mulberry variety, high yielding silkworm hybrids and adoption of new technology can generate as much as one lakh rupee of income through the transaction of cocoons per year and at the same time it provides employment to 5 persons throughout the year in Karnataka. This profit margin can substantially be increased from its wastes such as fuel wood, feed for cattle/poultry/fish and pupal oil etc. It is presumed that around 30% of mulberry waste will be available after silkworm feeding which comprises of leaf waste and small twigs and this waste is being used as a cattle feed (Radhakrishna *et al.* 2000)..**"Silk and Milk"** concept depends on the symbiotic relationship between dairying and sericulture. **"Silk and Milk"** has been an established practice in some parts of Karnataka. In fact integration of dairying with sericulture rescued some progressive farmers from the set back they suffered. Shri Mathukutty Vaidhyan undertook dairy running as one of the major ventures but gradually he found that dairy was no longer a paying venture and became a drain on his resources, so he was likely to close down the same. However, he started mulberry plantation in 3 acres and took up sericulture also which rescued his dairy from the set back he suffered. Sericulture and dairying not only strengthened each other but together became stronger than they were singly under his management. While the milch cows fed on the mulberry leaves left over after the silkworm feeding yielded 10% more quantity of milk, the excreta of the cows proved to be a good manure for the mulberry plants, as they yielded juicy and thick leaves of dark green colour as compared to other unmanured mulberry plants. The mulberry farm of Shri Vaidhyan is crises crossed with underground pipes carrying fresh liquid manure to the mulberry.and other plants from the dairy refuse. The urine from the cow sheds and other liquid refuse is first stored in an open pit and diluted with water before being sent through the pipes to the garden. Today he has 15 acres of land under mulberry and in a month, harvests the cocoon three times. During the last three years (1988-1990) he harvested cocoon crop worth over Rs. 12.0 lakh besides additional profits from dairying. While he defends his mulberry plantation of 15 acres, he claims that six acres, if managed well, should comparatively give more (Editor, Indian Silk, Oct. 1991).

It is estimated that one acre of mulberry garden can produce at least 10-12000 kg. of mulberry leaf/year from which 500-600 kg. of cocoons are produced and about 20% (350-400 kg/acre/crop or 2000-2400 kg./acre/year) of leaf and/or twigs considered as bed refusal (uneaten leaves and tender stumps) is sufficient and suitable as fodder for three cows. Apart from this, approximately 5% of the mulberry either soiled or unfit due to over maturity can also be used as fodder to the cattle. Rearing of three cattles along with rearing of silkworms will inturn meet the FYM requirement of one acre of mulberry garden and pupae may be used for extraction of oil/feed for poultry etc. (Radhakrishna *et al.* 2000).

' The studies carried out at National Dairy Research Institute (NDRI), Bangalore revealed that mulberry leaves consist of highest protein and ether extracts and lowest crude fibre and ash contents as compared to other alternatives feeds for cattles *viz.*, Maize, Jowar, Napier grass, Rhodes grass etc. (Table-69). Although mulberry leaf stalk/stump forms the low cost/no cost alternative fodder for cattle, yet bed refusals contaminated with bed disinfectants should be avoided. Only bed refusal of the succeeding days where the bed disinfectants are not used can be effectively used. The mulberry leaf/twigs are nowhere inferior in providing the body growth and production/energy requirements to cattles and dairy animals have good appetite and eat mulberry leaf due to good palatability.

Research findings have shown that feeding a cow (weighing 300-400 kg.) with 5 kg. of mulberry bed refusal/day during lactation period increases the milk yields by 10%. Yet according to NDRI, Bangalore excessive feeding of mulberry leaf causes misconceiving/mis-carriage and repetitive breeding further leads to extended period of lactation with reduced milk production. Hence, feeding of mulberry leaves should be avoided during the first half of the pregnancy period and generally only 20-25% of the total feed requirement of cattles should be replaced by mulberry waste as a fodder. In rural areas of Karnataka especially in Kolar and Bangalore generally a pair of cow is reared with land holding of an acre of mulberry and "**Milk & Silk**" goes hand in hand and are the major squares of income in this area. However, according to Radhakrishna *et al.* (2000), the usage of bed refusal directly as a cattle feed may be restrained in the absence of proper awareness in silkworm disease management as bed refusal will be carrying maximum pathogen load in the event of out break of diseases which further spreads the diseases and contaminates the premises.

In branch feeding/shoot feeding method bed cleaning is totally avoided during fourth and fifth instars hence bed refusal is not available for cattle. However, farmers are using bed refusal of last three days (upper portion) as fodder for cattle which amounts to 150-170 kg. in row system and 225-250 kg. in pit system. Bed refusal of last three days is dried by some farmers and used as fodder for goat/ sheep. Substances extracted from mulberry shoots are effective in improving hair growth in animals like rabbits and sheep. Rabbits have great liking for mulberry leaves. Body weight and growth of fur are said to be better when they are fed on mulberry leaves. The sheep are maintained well and yield more milk, if they are fed with mulberry leaves. By the end of rearing season, if mulberry leaves are still left in the garden they can be utilised as cattle feed efficiently (Majumder 1997). When fat extracted from pupae (30% of dry weight) is mixed with poultry feed, improve their egg laying capacity and colour of the egg yolk is affected. When litter is increased to 20% in poultry feed, the xanthophyll content of yolk increased to 0.017 mg/egg (Majumder 1997). The excess mulberry shoot along with shoot rearing refusal can also be converted as nutritionally rich silege. The research organisations may collaborate for effective utilisation of sericultural waste and sericulture and dairy and animal husbandry can be made as a package for economic upliftment of rural mass to improve their living standards.

F. INTEGRATION WITH PHARMACEUTICAL INDUSTRIES

The food plants and the silkworms which provided much for our silken dress have lot of scope to identify, screen and evolve various medicines for our ailments. Various kinds of crude extracts from flowers, fruits, seeds, leaves, barks and roots of many silkworm food plants have tremendous medicinal value in pharmaceutical industries and constitute major portion of the traditional pre-scientific as well as modern pharmacopoeia. Indian sericegenous medicinal flora is one of the richest and cosmopolitan flora with high therapeutic values/potentialities as described below briefly.

(a) MULBERRY

Mulberry has lot of pharmaceutical values (Chinnaswamy *et al.* 1993). Root extracts of *M.nigra* which grows profusely in Jammu and Kashmir and Darjeeling has the innate power to suppress rise in the blood sugar if taken with meal. Moranoline (after Morus), more widely known as Deoxynojirimycin (DNJ) is a sugar analogue. It has been found that the glycoprotein 120 found at the very surface of the Human Immunodeficiency Virus (HIV) carrying an unusually large number of sugar side chains becomes inert and looses its ability to invade new cells in the presence of the alkaloid DNJ. It was also found that DNJ can also control intestinal alphaglucosidoses. The drug and its various forms when tested against diabetes led to identification of a new drug, homonojirimycin, which can control diabetes in an effective manner (Ray 1990). Mulberry leaves also have anti-oxidation property and their regular use as tea may result as anti-aging. Besides, mulberry leaves are very effective in controlling high blood pressure, lowering cholesterol level, anti-cancer effect, heavy metal absorption in the drink, relief of arterio-sclerosis and prevention of hyperlipemia. Young mulberry leaves collected in spring season are more effective. Washed, dried and powdered mulberry leaves are consumed as a drink in cold water. Boiling of leaves is not advisable (Sharma and Sinha 2000). In Japan, tea prepared from mulberry leaves is consumed as traditional healthy diet (Nomura 1988). Tsushida *et al.* (1987) and Machi (1990) measured gamma-amino butyric acid (GABA) in tea and mulberry leaves and developed "gaboron tea" using N_2 gas. This tea has property to control high blood pressure (Omori *et al.* 1987). Mulberry leaf derived tea "Kuwacha-(R)" has also been commercialised by Honshina Co. Ltd., Saitama, Japan. Shimizu *et al.* (1992) reported that DOPA content which occurs in very low concentration in mulberry leaves may not have an anti-feed ant effects to insects. As caffeic acid is very active for anti-bacterial activity (Iizuka *et al.* 1976), Shimuzu *et al.* (1982) opined that presence of L-DOPA in mulberry leaves may act as precursor to caffeic acid in mammals. Since, biogenic amines and their precursors occur in trace amounts "Kuwacha tea" from mulberry leaves can be taken regularly without any pharmacological or poisonous effects, yet, its overdose should be avoided due to the presence of gamma- GABA (Singhal *et al.* 2000). The fruit juice from mulberry is used for curing sore throat, fever, dyspepsia and melancholia while the alcoholic extract from roots is effective in lowering blood pressure. A chemical called solanasol present in mulberry leaves and silkworm faeces has pharmaceutical significance. Chlorophyll extracted from the silkworm litter by using a suitable solvent and product is used in medicine in China and Japan (Majumdar 1997). Singhal *et al.* (2000) have reviewed importance of mulberry in pharmaceutical industries.

Chinese scientists have found that anti-bacterial peptides present in mulberry silkworms work against diseases in human beings (Iyengar 1998). Hongulu *et al.* (1987) has found that anti-bacterial peptides from *Bombax mori* have killing actions against K_{562}, *Myelogenous leukemia* cells. Zheijiang Academy of Traditional Chinese Medicine has developed a medicine "ganxuebao" from silkworm excreta which has an efficiency rate of 95.6% for treatment of cancer patients (Mandel 1991). Japanese scientists from Sericulture Experiment Station, Miyagi Prefecture and Iwate University have prepared anti-fungal fraction from haemolymph of the silkworm which inhibits the mycelial growth of *Pyricularia oryzae, Fusarium oxysporum, F.cucurnerinum, Glomerella cingulata* and *Alternaria pori* (Iyengar 2000).

The use of chitin preparations for the postoperative treatment of the internasal operations *viz.* conchotomy, deviatomy, polypectomy etc., has shown that it is easier to use and it also hastens healing of wounds (Sagar *et al.* 1991).

Chitosan derived from chitin shows potent anti-microbial activity against various bacteria and fungi (Machida 1992). It performs functions of immuno adjuvant or an anti viral agent and also stimulates lymphokine and works as anti viral and anti-tumour agent. As an anti sordes agent it prevents carcinogenic bacteria from teeth (Katti *et al.* 1996). Due to low toxicity, the regenerated biocompatible chitosan membrane checks bleeding in major surgeries and has an excellent wound curing effects (Sagar *et al.* 1991, Katti *et al.* 1996). Both chitin and chitosan have been used for wound dressing, contact lenses and enhancement of dissolution properties of poorly soluble drugs such as griseofulvin, phenytoin, phenoborbitol, prednisolone, flufenamic acid indomethacin. Hence, chitin and chitosans derived from pupae have very wide applications in pharmaceutical industries accordingly.

Silk is used to make sutures and can be added to the medical textiles as fibres, tendons, artificial ligament, blood vessels and skin grafts due to its excellent mechanical properties and strength. It can also be used in contact lenses or drug rehabilitation system involving tissue regeneration or even burn victims (Singhal *et al.* 2000). Silks are also being developed as scaffolds for tissue engineering, biodegradable films and gels for filtration and for drug delivery system (Personal communication from Dr. Catherine Graig, Museum of Comparative Zoology, Centre for Biotechnology, Horward University, Cambridge, U.S.A. in 2003).

Some organic compounds like paste chlorophyll, carotene, phytol and tricontanol may be extracted through industrial processing. The paste chlorophyll is utilised in pharmaceutical industry for treatment of gastric diseases like ulcer, liver hepatitis and some blood diseases. Copper-chlorophyll sodium can be used for the treatment of stomach diseases, hepatitis, acute pancreatitis and chronic nephritis. In U.S.A., there are more than 30 kinds of medicines which contain Copper-chlorophyll-sodium. Paste chlorophyll has been extracted from faeces of bivoltine KA races of silkworm in India (Revanasiddaiah and Yashoda Bai 1999). The carotene is used for manufacturing medicines for treatment of alimentary canal and respiratory organ diseases. Phytol is the medical raw material for Vitamin E and K while tricontanol is extensively used as a wonderful plant growth regulator (Singhal *et al.* 2000).

Glucose lowering effect of powdery silkworm was discovered at Dept. of Sericulture and Entomology, National Institute of Agricultural Science and Technology, Suwon, South Korea. Consequently, silkworm rearing increased there from 31,627 boxes in 1995 to 54000 during 1997 and 1998 as powdery silkworm found a large role as **function food** favoured by many diabetes. Silkworm powder has proved as a powerful blood glucose lowering agent due to high amount of deoxynojiri-mycin, a nitrogenous compound, which is more effective than insulin (Kang Sun Ryu *et al.* 2002). As sericin, the non-digestible protein is produced in silk gland just after 3rd day of the 5th instar, the silkworm is dried after this stage to get silkworm powder preferably by cold and dry technique rather than mature heat and dry technique. Not only this, the fibroin silk protein is in great demand for its use as artificial skin in plastic surgery. The fibrous scleroprotein is used in making surgical thread, which is easily digested inside the skin (Sharma and Sinha 2000).

The scientists of Natural Product Research Institute, Seoul National University and National Sericulture and Entomology Research Institute, South Korea developed the methodology to grow **cordyceps**, the rare miraculous herb on the body of silkworms. It occurs naturally in the highlands of China, Tibet and Nepal at an altitude above 10000 feet ASL. This smaller than finger size fungi is priced more than its weight in gold as it exerts medicinal effects such as fatigue recovery, anti aging and anti cancer. Also, the amino acids obtained by biochemical resolution terminate hang over and activates liver function. The **cordyceps** production in South Korea reached 6 MT in 1997 and increased to 30 MT in 1999 and farmers are earning 5 times more for producing silkworm powder and 6.5 times more for producing cordyceps as compared to traditional cocoon production (Sharma and Sinha 2000). The alternate host plant *Lactuca sativa* is given in bronchitis and asthma. Its stem and root are used as poultice in burns and painful ulcers (Sinha 1996). The silkworm is useful as a bioreactor and has several pharmaceutical and industrial application (Thangavelu *et al.* 2003).

(b) NON-MULBERRY SILKWORM FOOD PLANTS

Non-mulberry host plants, besides providing quality foliage to wild and domesticated silkworm possess abundant medicinal properties of immense value.

I. Tropical Tasar Food Plants

1. *Terminalia arjuna*: The leaf juice of arjuna is used in earache and fruit and bark are used in native medicine as a tonic. It is antidysenteric and reduces fever. Powered bark with milk is used in hypertension and also as a cardiotonic. It helps in secretion of urine and is given in cirrhosis. Bark extract is also used to wash ulcers. Ayurvedic medicines prepared from Arjuna tree are available in the market also. (Ramakrishna Naik *et al.* 2000).

2. *T.bellerica*: Fruit is a laxative and an ingredient of 'Triphala' and other famous preparations in Ayurveda. It is an effective medicine for bronchitis, asthma, sore throat, thirst, cold pits, joint pain and also used to prevent external bleeding, discharge and reduces fever. In English medicine it is considered as a good brain tonic. It is also used in the preparation of tonic given for piles, diarrhoea, ingestion and headache. Its extract is used to cure leprosy (Ramakrishna Naik *et al.* 2000).

3. *T.chebula*: Fruit is used to prevent external bleeding, discharge, wounds, scalds, and chronic ulcer. Fruit extract is used as gargle in toothache, inflammation of mucous, membrane of mouth, bleeding gum, wash for watering eyes and asthma. Bark is diuretic and cardiotonic. The fruit is also used as laxative and digestive and have some effect on blood pressure as cardiotonic. The powder of the fruit is used as a dentifrice for strengthening the gums. The fruit is an ingredient of well-known preparation "Triphala". It is also a good medicine for leprosy and used for skin disorders, itching and jaundice (Ramakrishna Naik *et al.* 2000).

4. *Zizyphus jujuba*: The fruits of this wild variety are palatable, acidic and astringent. The dry fruit is emoluent, aperient and relieves cough, cold, bronchitis and fever. Bark is astringent and used as a simple remedy for diarrhoea while alkaloidal extracts of leaves significantly reduces blood sugar level (Ramakrishna Naik *et al.* 2000).

5. *Zizyphus mauritiana*: Leaves are used with Catechue in typhoid. Bark stops external bleeding and discharge, decoction of bark prevents dysentery and seeds are used in diarrhoea. Fruits are cooling, relieve pain and used as tonic while kernels are sedative and relieve abdominal pain during pregnancy. It is also used as poultice in wounds and given as an antidote for acute poisoning (Ramakrishna Naik *et al.* 2000).

6. *Syzizium cumini*: Bark is used for blood purification, as a gargle and an astringent. It is also used in sore throat, bronchitis, asthma, ulcers and dysentery. Fresh juice of bark is given in diarrhoea with goat milk. Bark, stem, leaf and flowers are used in diabetes and known to possess antibiotic properties. Edible fruit reduces external bleeding and discharge. Its vinegar is a good appetiser and is prepared from juice extracts from unripe fruits (Ramakrishna Naik *et al.* 2000).

7. *Anogeissus latifolia*: The crushed bark is used for curing cough and stomach pains (Singh 1990).

8. *Bauhinia variegata*: The infusion of leaves is laxative and cures diarrhoea, dysentery and piles (Kurian 1995). The decoction of root prevents obesity and used as a carminative, expels gases, flatulence and gripping pains from stomach and bowels. The decoction of bark is a good tonic, anthelmintic and used against tuberculoses, skin ailments, ulcers and leprosy (Kurian 1995). The decoction of bark mixed with honey is given in leucorrhoea. flower and floral buds are laxative and dried buds are used for diarrhoea, dysentery and piles (Sinha 1996).

9. *Bombax ceiba*: Flowers are good for colitis and cause sterility hence used in birth control. Fruits are used as stimulant expectorant and antidote in snakebite. Besides fruits are also used for curing colon infection and ulceration of bladder and kidney. Gum is demulcent, haemostatic and astringent and used in diarrhoea, dysentery and menorrhagia. Bark paste is used against dysentery, while in female diseases bark extract is used along with honey (Singh 1990, Sinha 1996). In sexual impotency, roots (musali) are used as a good medicine since they are stimulant, tonic and aphrodisiac (Maheshwari and Singh 1991).

10. *Buchanania latifolia*: Gum is used for diarrhoea and fruit is used for relieving burning of body and fever (Singh 1990).

11. *Careya arborea*: Juice of the bark along with honey is given for cough and cold while bark is used as a astringent (Singh 1990, Ansari 1997).

12. *Carrisa carandas*: Decoction of roots is used as laxative, anthelmintic and tonic. Unripe fruit cures biliousness, stimulates appetite, reduces fever and prevents/cures scurvy while ripe fruits check bleeding of internal organs. The decoction of leaves reduces fever (Kurian 1995, Sinha 1996).

13. *Celastus paniculatus*: Decoction of leaves promotes menstrual flow while decoction of seeds promotes removal of mucous secretions from bronchial tubes and used for cough, asthma, Leprosy, headache and leukoderma. The decoction of bark is used for abortion. The oil from seeds is nerve stimulant, brain tonic and checks rheumatic pain (Kurian 1995).

14. *Chloroxylon swietenia*: Crushed leaves are used to cure wounds (Singh 1990).

15. *Dalbergia sissoo*: Decoction of leaves is used against gonorrhoea, root is astringent, wood is useful in leprosy, boils, eruptions and to allay vomiting (Sinha 1996).

16. *Diospyros melanoxylon*: The ash of bark mixed with coconut oil and ash of *Calotropis procera* is used on boils and blisters. The bark paste is used in wound healing (Maheshwari and Singh 1991). The bark is also used for diarrhoea and dysentery (Ansari 1997).

17. *Dodonea viscosa*: Leaf extract is applied in fresh injuries. Wood is carminative, slightly purgative, applied on wounds and considered as useful in rheumatism (Sinha 1996).

18. *Embellica officinalis*: The fruits are astringent, diuretic, laxative, brain and nerve tonic, richest in Vitamin C and used in diarrhoea, amoebic dysentery, haemorrhage, anemia, gastritis, theist, flatulence, gripping pain from stomach and bowels, cough and jaundice etc. Flowers are refrigerant. Seeds are used in asthma, leucorrhoea and bronchitis. The fruits are also useful in constipation, pyorrhoea, piles, haemoptysis. The boiled extract of leaves is known to control blood sugar. The fruits also reduce fever, vomiting, congestion of liver, leprosy, painful urination, cold in nose and nasal haemorrhages (Sinha 1996, Kurian 1995). Fruits are important ingredients of "Triphala" and "Chayavanprash".

19. *Gardenia lucida*: The resin is used as a stimulant and given in dyspepsia.

20. *Ficus benjamina*: Decoction of leaves mixed with oil is applied against ulcers (Sinha 1996).

21. *Ficus hispida*: Bark, fruit and seeds are used as purgative. They also promote secretion of milk. Leaves are used to rid off ringworms, latex on boils and seeds in dysentery (Sinha 1996).

22. *Ficus religiosa*: Bark is astringent, leaves and young shoots are purgative, fruit is laxative and seeds are laxative as well as cooling. Bark infusion is useful in scabies, boils gonorrhoea and rheumatism (Sinha 1996, Ansari 1997).

23. *Ficus retusa*: Bark extract is used in billious disorders and powdered leaves are applied in headache. Powdered adventitious roots mixed with salt are applied to decayed or aching tooth. Root, bark and leaves boiled in oil are applied to wounds and bruishes (Sinha 1996).

24. *Lagerstroemia indica*: Bark is stimulant and febrifuge. Leaves and flowers are purgative and roots are astringent and used as gorgle while seeds are narcotic (Sinha 1996).

25. *Lagerstroemia speciosa*: Decoction of leaves and bark is diuretic and controls blood sugar. Bark infusion checks fever and diarrhoea. Fruit cures small ulcers in mouth. Bark, leaves and flowers are laxative and purgative (Kurian 1995, Randhawa 1983, Sinha 1996).

26. *Madhuca indica*: Decoction of bark cures ulcers, leprosy, fractures, and biliousness. Decoction of flowers help in curing heart diseases, ear complaints, soothing of alimentary tract, relieving inflammation, bronchitis and cough. Decoction of fruits helps in treatment of blood diseases, heart diseases, bronchitis, tuberculosis and ear troubles. Oral consumption of flowers fried in ghee cures piles and skin diseases. Milk sap from stem, branches/branchlets is used for rheumatism (Randhawa 1983, Kurian 1995).

27. *Melastoma malabathricum*: The bark and leaves are used against skin troubles. The root and bark extract are used as goggle and antiseptic. The plant is very useful in diarrhoea, dysentery and leucorrhoea (Sinha 1996).

28. *Pterocarpus marsupium*: Paste of leaves cure skin diseases and pain when applied externally. Bark and wood decoction used for treatment of diabetes (Singh 1990).

29. *Shorea robusta*: The resin is astringent, detergent, aphrodisiac and given in dysentery, indigestion and gonorrhoea (Sinha 1996).

30. *Tectona grandis*: The decoction of root is given in anuria, infusion of flowers is given in biliousness, urinary discharges, bronchitis and congestion of liver and decoction of bark is useful in piles, leukoderma, dysentery, headache, burning pain over the region of the liver and cures inflamed eyelids, inflammatory swellings, itchiness of the skin. Flowers and seeds are diuretic while bark is laxative, anthelmintic, astringent and expectorant. Leaf extract is given in hysteria. The boiled extract of leaves along with small fishes (*Punctius phutino*) is used for normalising blood circulation (Sinha 1996, Kurian 1995).

31. *Terminalia catappa*: Decoction of leaves is used as a remedy for dysentery, diarrhoea, fevers, worms and inflammation of mucous membranes of urethra and vagina. The juice of young leaves is used as an ointment for rashes, itching, scabies and applied on forehead to relieve headache. The sap of young leaves boiled in the oil of kernel is used for curing leprosy (Kurian 1995).

32. *Terminalia paniculata*: The bark juice is diuretic and cardiotonic and applied with ghee and rock salt in parotitis. (Anonymous 1976).

33. *Terminalia alata*: Bark is bitter, styptic, diuretic and cardiotonic. Decoction of bark is used as a remedy for dyspepsia, atonic diarrhoea and applied locally to weak indolent ulcers. Gel like gum is purgative (Sinha 1996).

34. *Terminalia coriacea*: Bark is a potent cardiotonic and heart stimulant (Sinha 1996).

35. *Terminalia crenulata*: Decoction of bark checks blood pressure and used in Chattisgarh for painless delivery by tribals (Anonymous 1976 & Pers. Comm.)

36. *Quercus leucotricophora*: Acorns are diuretic and astringent and used for indigestion and diarrhoea in children and for gonorrhoea in adult (Negi and Naithani 1995).

II. Muga Food Plants

1. *Actinodaphne angustifolia*: Leaf infusion is given in diabetes and urinary disorders (Sinha 1996).

2. *Cinnamonum obtusifolium*: Bark is used in dyspepsia and liver complaints (Sinha 1996).

3. *Litsaea polyantha*: Leaves are applied in muscle pains of legs and arms. Bark is astringent and used in diarrhoea. Seeds yield a fat useful in rheumatism. Powdered bark is applied for relieving pains caused due to blows, bruises or hardwork and applied to fractured bones in animals (Sinha 1996).

4. *Litsaea citrata*: Fruits are carminative, used against dizziness, hysteria, paralysis and loss of memory (Sinha 1996).

5. *Michilus bombycina*: Extract of leaves is useful in mouth ulcers while bark extract is given in asthma.

6. *Michelia champaca*: Plant is aromatic, anthelmintic, diuretic and stimulates sweating. It helps in ejection of mucous and other secretions from upper respiratory tract. The bark is bitter, aromatic, antiperiodic and employed as an abortifacient for 2-3 months old pregnancy. As an oral contraceptive the root is made into paste with fruits of black pepper and given after menstruation for 3 days. Infusion of flowers is considered as a stimulant, tonic and carminative (Ramakrishna Naik *et al.* 2000).

7. *Plumeria acutifolia*: Decoction of leaves or bark brings down fever and cures asthma while decoction of root is a remedy against sores. Milky latex mixed with coconut oil relieves the rheumatic pains. Heated leaves are applied as a poultice over swellings (Kurain 1995, Randhawa 1983).

8. *Symplocas paniculata*: Bark is used in ophthalmia and for checking threatened abortion (Sinha 1995).

III. Eri Food Plants

1. *Ailanthus excelsa*: The bark as well as leaves are used as a tonic especially in debility after childbirth and also in dyspepsia. The bark is aromatic, febrifuge and anthelmintic. It is used against indigestion and helps in ejection of mucous from upper respiratory tracts hence given in chronic bronchitis and asthma. Preparations from bark prevents external bleeding, discharge and reduce fever. During diarrhoea and dysentery it is used to expel intestine worms (Ramakrishna Naik *et al.* 2000).

2. *Carica papaya*: Fruit is digestive, carminative and diuretic, rich in Vitamin A, used in night blindness. Green fruit is laxative and yield "Papain" a broad spectrum proteolytic enzyme and "Pectin", an important food adjunct. Papain was used as a bacteriophage during cholera epidemic. It is used for treatment of dyspepsia, other digestive disorders and tonsil infections. Seeds are vermifuge, rich in protein and a fatty acid, which has certain insecticidal properties. The leaves contain an alkaloid and are also rich in ascorbic acid (Ramakrishna Naik *et al.* 2000).

3. *Gmelina arborea*: Leaves are demulcent, used in gonorrhoea, cough and to remove foetid discharges and worms from ulcers. Flower is used in blood diseases and fruit decoction in fever and bilious affections. Plant is also used in snakebite and scorpion sting. Dasamularishta consists of its roots as one of the ingredients (Sinha 1996, Maheshwari and Singh 1991, Nair and Jayakumar 1998, Pandey 1992).

4. *Jatropha curcas*: Seed oil is anthelmintic, purgative, hair stimulant to abortifacient. Seed oil is taken in dropsy and applied externally in sciatica rheumatism and skin diseases. Leaf extract is used for increasing lactation and treatment of toothache. Bark extract in water is used against rheumatism and leprosy. Bed bugs are removed by fumigation of leaves. Latex is applied to boils, skin sores and rheumatic joints (Sinha 1996).

5. *Manihot utilitissima*: Tuber flour is given in fever and stomach disorders. While juice is aperient, the leaves are purgative and vermifuge and used for sores, ascites, eczema, coriasis, scabies and sycosis (Kurian 1995, Sinha 1996). "Bitter cassava" variety is extremely poisonous. The roots contain a proportion of hydrocyanic acid and root and stem have purgative properties (Ramakrishnan Naik *et al.* 2000).

6. *Ricinus communis*: Castor oil is purgative and also used in ointment as a soothing agent and an oil vehicle in eye drops. It is also used in making contraceptives, jellies and creams. A gel prepared from castor oil is useful in dermatosis and protective in occupational eczemas and dermatitis. Consumption of 1-2 tea spoon castor oil with milk at bed time for 2-3 days cures constipation (Ramakrishna Naik *et al.* 2000). Leaves are warmed, crushed and applied to anus for treatment of bleeding piles, leaf paste is used as a poultice on sores, gout, abdominal ailments and rheumatic swellings whereas decoction of roots is given for treating lumbago (Sinha 1996, Ansari 1997). Administration of the pulp of one or two castor seeds during menstrual period is stated to prevent fertilization for one or two years respectively (Dash 1993). The oil from the seed has several medicinal and industrial application of high value, which are well known.

7. *Zanthoxylum armatum*: Bark and fruits are carminative, stomachic, anthelmintic, aromatic, tonic and given in fever and dyspepsia. Fruit are used in dental problems and scabies. Bark is used for intoxicating fishes and repelling insects. Seed oil is considered as antiseptic and antibacterial (Sinha 1996).

8. *Zanthoxylum rhesta*: Unripe fruits are given in dyspepsia, asthma, bronchitis, heart troubles and rheumatism. Pericarp is astringent, stimulant and digestive. Seed tincture is given in cholera. Essential oil of fruits is locally applied on skin swellings (Sinha 1996).

G. INTEGRATION OF SERICULTURE WITH OIL INDUSTRIES

(a) Oil from Silkworm Pupae

During the biological development, pupae synthesize fatty acid which can be extracted as oil from dried pupae powder by two methods *viz.*, solvent extraction method and pressure application methods described by Bose and Majumder (1990). Hot extraction method gives more percentage of oil than cold extraction and chlorophyll is the best extracting solvent. The oil can be refined by steam distillation or by filtering through activated charcoal. The iodine number and saponification number of *Bombax mori* pupae oil are 110 and 210 respectively which are comparable to sunflower oil. The saponification value, acid value and ester values of *Antheraea assama* pupae oil are 190.2, 131.2 and 59.0 respectively. Italy is the second largest producer of pupae oil followed by Japan. Linseed oil (75%) mixed with pupae oil (25%) can be used for preparation of paints and varnishes. Sterol isolated from pupae oil is a very good hair tonic and can be used for manufacture of soap and candles (Bose and Majumder 1990). Pupae oil is also used for manufacturing of a compound "Gunmaile", which is used in liver and blood diseases (Sugun and Kumar 1996). Production of oyster mushroom *Pleuriotus sajrcaju* raised under the strata of eri silkworm waste gives higher net returns as compared to paddy straw strata (Baruah *et al.* 1999).

Lecithin is used in various food and pharmaceutical products. It is also used as an anti oxidant for vegetable oils and fish liver oil due to its high solubility in fats and oils with properties of edibility and absence of deleterious effects on products (Ghatak and Krishamurthy 1995, Calkin 1947). More than 200 tonnes of Lecithin is imported in India annually. Hence, large quantities of Lecithin found in pupal waste can be exploited as a natural indigenous resource (Majumder *et al.* 1996).

(b) Oil from Host Plants

Seeds of *Morus alba* yield 25-35% edible oil (Jain *et al.* 1990). Besides, seeds of some non-mulberry food plants also yield oil and fatty acids and can serve as raw materials for oil industries. A brief description of such raw materials and plants is given below:

I. TASAR FOOD PLANTS

1. *Bauhinia variegata*: Seeds yield oil. The qualitative and quantitative studies of such oil is required.

2. *Bombax ceiba*: The seeds yield 22.3% oil which may be exploited commercially for production of edible oil (Jain *et al.* 1990).

3. *Buchanania latifolia*: The seeds are used in confectionery as a substitute for almond and yield an aromatic oil known as "Chironji oil" upto 51.8% (Jain *et al.* 1990, Shukla and Mishra 1979).

4. *Celastrus paniculatus*: The seeds contain 35-50% oil content, which is edible and hence may be exploited commercially (Jain *et al.* 1990). The seed oil is also used in gout, paralysis and epilepsy (Ansari 1997).

5. *Embelica officinalis*: Seeds yield 16% oil (Jain *et al.* 1990).

6. *Mahuka indica*: The seeds yield 35-40% oil, a substitute for ghee, which is used in cooking, hence, it has been named as butter tree. Mahua oil is also used for illumination and for making margarine or soap.

7. *Shorea robusta*: The seeds are highly rich with 12.5-18% fatty oil and feed the raw material to edible vegetable oil industries.

8. *Shorea talura*: Seeds yield vegetable fatty oil, which is edible.

9. *Terminalia bellerica*: Kernels yield 25% oil (Jain *et al.* 1990). Oil from half ripe fruit is purgative, good tonic for hairs and substituted for ghee in Madhya Pradesh (Anonymous, 1976).

10. *Terminalia catappa*: The kernels contain myristic, palmitic, stearic, oleic and linoleic acids. The leaves and fruits contain corilagin, gallic acid, ellagic acid and brevifolin carboxylic acid whereas bark and wood contain ellagic acid, gallic acid, catechin, epicathechin and leucocyanidin (Anonymous, 1976).

11. *Terminalia chebula*: The kernel yield as much as 36.4% oil (Jain *et al.* 1990)

II. MUGA FOOD PLANTS

1. *Michelia champaca*: The oil from flowers is used in perfumery and useful in cephalalgia (Shukla and Mishra 1979). The seeds yield 32.2% edible oil, which may be used commercially (Jain *et al.* 1990).

III. ERI FOOD PLANTS

1. *Jatropha curcas*: Seed oil is used for manufacturing candles, soaps and as a lubricant and for illumination. The seed oil is medicinally very important.

2. *Ricinus communis*: Seeds yield 35-55% thick, colourless or greenish castor oil which is excellent lubricant and medicinal. The castor oil is also used for manufacturing varnishes, paints, typewriter inks, aromatic perfumes, transparent soaps, protective covering for insulation, food containers, guns, airplanes, leather and for coating fabrics (Pandey 1992, Kochar 1981).

H. INTEGRATION OF SERICULTURE WITH TANNIN AND LEATHER INDUSTRIES

Artificial leather can be made by processing bark of mulberry (Majumdar 1997). Also the precipitate of the hydrolysed silk protein when dried and powdered could be used as a coating over the foam to give a good touch of silk leather (Sharma and Sinha 2000).

I. TASAR FOOD PLANTS

1. *Hardwickia binata*: The pod, leaves and bark contain 6%, 7% and 9% tannins respectively (Prasad *et al.* 1974, Masilamani and Vadivelu 1997).

2. *Terminalia alata*: Tannins obtained from bark are promising material for soil stabilisation (Bhatia *et al.* 1987).

2. *Terminalia bellerica*: Fruits contain 5-17% tannins (Chopra *et al.* 1965).

3. *Terminalia chebula*: The fruits yield one of the most valuable tannins (Chopra *et al.* 1965).

4. *Terminalia paniculata*: Bark and fruit yield 14% tannins.

5. *Terminalia tomentosa*: Fruit and bark yield tannins upto 20% for the industry.

6. *Castanopsis indica*: Bark contains 6-12% and leaves contain 10% tannins (Negi and Naithani 1995).

7. *Castanopsis lancaefolia*: Bark, wood and leaves contain tannins.

8. *Castanopsis purpurella*: Bark contain 11-13 and leaves 12% tannins (Negi and Naithani 1995).

9. *Lithocarpus dealbata*: Bark contains 10-13% tannins (Negi and Naithani 1995).

10. *Lithocarpus fonestrata*: Bark contains 10-16% tannins and can be used to give leather of light colour (Negi and Naithani 1995).

11. *Quercas acutissima*: The tannin contents in leaves show 3.5 fold increase over a period of 45 days.

12. *Quercus baloot*: Bark is used for tanning.

13. *Quercus floribunda*: The leaves contain 9.85% and bark contain 4.99% tannin contents and used for tanning locally (Negi and Naithani 1995).

14. *Quercus glauca*: The bark contains 12% tannins (Negi and Naithani 1995).

15. *Quercus grifithii*: Bark contains 5-10% tannins (Gamble 1922, Kanjilal *et al.* 1940).

16. *Quercus kamroopi*: Dry leaves, twig, bark and mature bark contain 11.06%, 10.5% and 9.69% tannins respectively (Fraysmouth and Pilgrim 1918).

17. *Quercus lamellosa*: Bark contains 12.6% tannins (Negi and Naithani 1995) which is used for tanning.

18. *Quercus leucotricophora*: Bark is used for tanning due to 6-23% tannin contents (Negi and Naithani 1995).

19. *Quercus semecarpifolia*: Bark from branches yields 24% tannins while stem bark yields only 11.63% tannins (Troup 1921).

EPILOGUE

The mulberry silkworm larvae feed at the rate of 42 kg. leaves/1000 larvae of which about 22.5 kg are ingested and 19.5 kg. are wasted. An excreta of about 13 kg/1000 larvae is obtained as silkworm litter. Residual mulberry leaves/shoots as well as pupae are useful. The unutilized silkworm eggs, egg shells, diseased silkworms, exuviae (moults), moths, excrements can also supplement the income of sericulturists through their effective and efficient utilization (Majumder 1997).

From the phloem of mulberry shoots, fibres can be produced and used as supplementary. Such fibres may be used in spinning like jute fibers are produced for making ropes, nets etc. The alpha fiber content varies from 28-44% in mulberry.

The mulberry pulp is also a valuable material and the bark of the tree is used for manufacturing paper. Mulberry twigs are used in basket industry for which lot of demand already exists in civil construction works. On an average 2500-3000 medium sized basket may be fabricated from mulberry twigs available per year from one hectare of mulberry garden which can provide an additional income of about Rs. 15000.00 in Northern belt of India (Singhal *et al.* 2000).

Substances extracted from mulberry shoots are effective in improving hair growth in animals like rabbits and sheep. In cosmetic industry also, some of the isolates from mulberry shoots have applications. Possibly the polyphenol or its derivatives, polysaccharides including pectin and glucose form a viscose material, which may be responsible for making the skin and hair soft. Chlorophyll extracted from silkworm litter is used in cosmetics in China and Japan. Various types of pectines, polymer or plastic materials can also be prepared from silkworm litter (Majumder 1997). Generally, silkworm discharges about one kg of faeces after eating 10 kg of mulberry leaves. Such faeces can be exploited for manufacture of tooth paste, medicines, cosmetics, edibles and plant growth regulators etc. The pupal fat is a good raw material for manufacture of cosmetics, glycerines and soap. Sericine is also used for making cosmetic products, soap and silk wine. Hygienically processed products from pupae could be better used for human consumption. In fact, the protein of pupae is better than the protein of soya bean, fish or beef and used for preparing aminoacid, and flavoured products. In Japan and China silkworm pupal cakes and other edible items are prepared after removing pupae odour by a fermentation and deodorisation process.

Chitin and Chitosan are useful for enhancing dissolution of poorly soluble drugs. China has already marketed chitosan based products. In India also, Shriam Institute has extracted chitin from silkworm pupal skin obtained from grainages. The yield of chitin is 3-4% based on dry weight of pupa. Tricontanol, a plant growth promoter is produced from exuviae of insects as exocrine compounds. The highest yield of n-tricontanol is found in eri silkworm. Trehalose, artificial fibres, membranes and peptones may also be prepared from silkworm larvae and/or pupae (Majumder 1997). Thus high value chemicals/materials may be produced from wastes, which inturn will also lead to the reduction in the cost of finished goods. In fact, the list of bye-products from silkworm eggs, larvae, pupae and moths is enormous and silkworm may be utilised as an efficient tool for producing materials other than silk. A technology has also been developed for production of commercially important protease enzyme and lysine amino acid by utilising the bacterial culture isolated from decomposed pupal waste as inoculum in the fermenting media (Chinya *et al.* 2000, Ashok Kumar *et al.* 2000).

Utilization of cut and pierced cocoons is steadily increasing in handicraft industries and gaining popularity. Pierced cocoons are cut, cleaned and dyed to make beautiful garlands, flower vases, wreath, dolls, pen stands, jewellery, wall hangings, wall plates, craft clocks, greeting cards, bouquets etc. (Uma *et al.* 1993, Vathasala 1997). These handicrafts are generally sold in handicraft emporium at net profit of about 50%. Now, it has become a source of income for many rural families for their livelihood. The silk carpets manufactured in Kashmir, Iran and Turkey etc. are very popular throughout the world (Dhar 1997). Such handicraft industries should be promoted extensively.

The sericultural scientists in India are now required to hunt for non-traditional and high profit generating sericultural products. Availability of dry pupae in India is approximately equal to the raw silk production of about 14000 mt. (Sonwalkar 1998). Exploitation of pupae for any useful industrial products in India is now necessary to make silk reeling sector more viable and stable besides generating more employment opportunities. Now sericulture needs to be preferred more for food and other medicinal/industrial bye-products rather than dress alone.

The blending of sericulture with agriculture, horticulture, silviculture, dairy and animal husbandry, pisciculture, pharmaceutical, cosmetic, timber, tannin, leather, food and other industries will not only make India a prosperous and leading country in sericulture but shall also alleviate poverty from the country even from grass root levels through efficient management and proper distribution of profits in post GATT era if our R&D understands the need of the hour and identifies and develops technologies which can make full use of mulberry and silkworm refuse as a valuable bye-product in bio-resource life-science industries.

7

Sericulture and Environment

Man's association with nature, particularly with trees, dates back to time immemorial and his needs for good food, good environment, good health, bright future and security are all depending upon trees and forests. Trees render multifarious services to mankind after rising from womb of mother earth and growing towards the sky. Trees conserve and purify the abiotic factors of environment *viz.*, soil, water and air and provide sustenance to all living beings on the earth. Late Indira Gandhi believed that the survival of man is dependent on the survival of animal and plant life. Late Dr. Rajendra Prasad also once said "I wonder some times if there is any natural resource which gives so much and of which we know so little as the forests". Dr. E.F. Schumacher believed that there is no salvation of life for India except through trees which is in conformity with the saying of Nand Rishi "Food would last as long as the forests". Realizing the importance of trees and forests Padam Puran inspired all human beings to plant trees by saying "Man planting trees by the wayside will enjoy the bliss of heaven for as many years as there are fruits, flowers and leaves on that he planted." Gandhiji said "Earth provides enough to satisfy every man's need but not for a new man's greed", because he realized that man has ravaged trees and forests not only for food, clothing, shelter, wood, and fibre but also sacrificed them due to ever increasing demand of development in agriculture, urbanization and industrialization. Often it is said "Forests precedes civilization and deserts follow" since acute pressure on forests due to genuine needs as well as extravagant greed of mankind still continues unabated which is endangering sustenance of mankind by depleting forest resources rapidly. Overexploitation of forest wealth caused various ecological hazards such as floods, soil erosion, changes in climate and drought. Besides, other socio-economic problems like fuel-famine and unbridgeable gap between demand and supply position of forest produce have also become alarmingly rampant.

According to Forest Survey of India (State of Forests Report, 1997) during 1995, 385037, 249309, 4533 Km² were covered with dense forest, open forest and mangrove forests which increased/reduced to 367260, 261310 and 4827 Km² during 1997 thereby amounting to loss of 17777 Km² and increase of 12001 Km² and 294

Km² area respectively in respective forest covers. Thus, the total forest cover of 638879 Km² during 1995 has come down to 633397 Km² during 1997 thereby exhibiting loss of 15482 Km² of forest cover during the period of two years. This statistics reveals that 7741 Km² or 0.7741 million hectare of forest cover is lost every year and rate of loss of forest cover is 2121 hectare/day in India which is very alarming. Loss of total forest area in seven tropical tasar producing states during 1986 to 2000 may be visualized from Tables 71 and 72. Population explosion has indeed caused overexploitation of forest wealth. Besides, other fundamental problems of development, agricultural stagnation, grossly unsustainable land use and rising unemployment etc. also contributed to deforestation, ecological destruction and biodiversity extinction etc.

(A) SERICULTURE AND FOREST CONSERVATION

Deforestation creates adverse circumstances to life and gives rise to poverty which causes further deforestation. Hence, deforestation and poverty are two basic ingredients of two vicious circle. Deforestation causes wide spread unemployment, hunger and migration of tribals and village folks. The greed of few, together with necessity of poor persons destroys limited forest resources. Hence, sustainable management of forests for more benefits is the only remedy for deforestation as well as Jhum cultivation. Accordingly, concept of "Joint Forest Management" was evolved under National Forest Policy, 1988 in India. In "Joint Forest Management" government agencies and people share the benefits and burden (excluding fund) together while working for growth and sustenance of forests. Following needs were stressed under National Forest Policy, 1988.

(1) Maintenance of environmental stability through preservation and restoration of ecological balances.

(2) Conservation of remaining forests and genetic resources for the benefit of posterity.

(3) Meeting the basic needs of people, especially fuel wood, fodder and timber.

(4) Maintaining the intrinsic relationship between forests and poor people living in and around by protecting their customary rights and concessions on the forests.

Of the total raw silk production of 14035 m.t. (1997-98), the share of mulberry and non-mulberry silk was 77% and 23% respectively (Shetty and Samson 1998). In the light of National Forest Policy, 1988 both mulberry as well as non-mulberry sericulture are discussed as hereunder.

(1) Mulberry sericulture

Nearly 286496 hectares of land was covered under mulberry plantations in the country during 1995-96 which decreased to 280651, 282244, 270069 and 227151 hectares during 1996-97, 1997-98 and 1999-2000 respectively (Table 9). The rearing and grainage are conducted indoor and intensive agricultural practices including full use of fertilizers and irrigation are resorted outdoor to produce good quality foliage in high quantity for silkworm rearing . Although mulberry sericulture does not help directly in forest management yet it indirectly helps in sustenance of forest through employment generation among tribals and villagers on their own land holdings, thus reducing their dependence on forest for fuel, fodder and money (Singh and Sinha 2000). During 1995-96, mulberry sericulture provided employment to 59.50 lakh persons which increased to 59.67, 60.57, 62.14 and 66.11 lakh persons during 1996-97, 1997-98, 1998-99 and 1999-2000 respectively (Table 9). The total mulberry acreage of 227151 hectares helps in environmental protection also since mulberry plantation reduce soil erosion and contribute in oxygen and nitrogen cycles of the nature.

(2) Ericulture

Like mulberry sericulture, ericulture is also multivoltine and practised indoor. Ericulture accounts for 69.78% of total non-mulberry silk production and has close links with the culture and traditions of people of North-eastern region (Shetty and Samson 1998). While domestic demand of warm clothing is fulfilled by eri rearing, the edible pupae form the major source of protein for village folk of North-eastern region. Apart from North-east this culture is marginally practiced in West Bengal, Bihar, Orissa and Andhra Pradesh primarily for production of castor seed and its oil. Castor is grown over an area of 6,39,000 hectare with a production of 3,05,200 tonnes of seed (Anonymous 1987). Castor leaves can be harvested to the tune of 25% at mid maturity stage of capsules without affecting the seed yield and can be used for eri silkworm rearing. Thus, castor growers can earn additional income by ericulture which plays an important role in rural economy and employment (Devaiah *et al.* 1981). Castor plants directly contribute to environmental protection but ericulture indirectly helps in forest conservation and environmental protection through employment generation among tribals and village folks on their own land holdings or natural castor plantations, thus reducing pressure on forests for fuel, fodder, food and money.

The rearing of tasar and muga is in general being practised outdoors on suitable food plants and grainage is conducted in door. In fact, wild/semi-domesticated silkworms of tasar and muga are an integral part of forest ecosystem and their culture directly helps in forest management through employment generation as well as afforestation with tasar and muga host plants. Besides, obligatory dependence of tribals and village folks on host plants of tasar and muga silkworm encourages them to protect these flora in the forests for their sustainable commercial exploitation.

(3) Tropical tasar culture

Tasar culture accounts for 24.85% out of the total non-mulberry silk production approximately (Shetty and Samson 1998). Out of approximately 63.3397 million hectares of forests, 33.4845 million hectares of land fall within tropical tasar producing states. Tropical tasar is normally practised in Central and Southern plateau of Jharkhand (erstwhile Bihar), Chattisgarh (erstwhile Madhya Pradesh),

Orissa and West Bengal extending to fringes of Uttar Pradesh, Andhra Pradesh and Maharashtra in the humid and dense forest areas. The estimated 11.16 million hectares of tropical forest in tasar producing states are covered with 9.7 million hectares of Sal forests and 1.46 million hectares of Asan, Arjuna and other food plants (Tables 70&71). However, only a meager percentage of Sal forests and 0.56 million hectares of Asan, Arjuna and other food plants have been put to use for tasar culture while deep interior forests have remained unexploited (Singh and Sinha 2000, Sengupta 1986). Besides, Central Silk Board has implemented Inter State Tasar Project (ISTP) during 1981-86 with the financial assistance from Swiss Development Cooperation in eight tropical tasar producing states. Under ISTP, 7845-8000 hectares of *Terminalia arjuna* block plantations were developed for tasar culture which is beneficial to more than 1000 tribal families. Besides, mass seedling banks/archives of *Terminalia arjuna* and *T. tomentosa* have also been raised in Central Tasar Research & Training Institute, Ranchi and its eight Regional Tasar Research Stations, 22 Basic Seed Mulitiplication & Training Centres and 10 Research Extension Centres at various places of tropical tasar growing states. The total area covered by *T. tomentosa* and *T. arjuna* accounts to more than 1250.2 and 393.3 hectares respectively in these stations/centres. Besides, governments of all tasar growing states have also opened many tasar farms which contain huge plantations of both the *Terminalia* species (Table 64). It is estimated that around 1.40 lakh tribal families are engaged in tasar silkworm rearing (Rao *et al.* 1998) and they are directly contributing in Joint Forest Management and environmental protection by maintaining tasar host plants in and around the vicinity of forests.

(4) Temperate (oak) tasar culture

Oak tasar culture is practised in North-eastern and North-western sub-Himalayan belt of India where abundant oak flora is available. Out of 2,25,45,200 million hectares of forests, only 18,41,000 hectares of land is under oak tasar host plants in Arunachal Pradesh, Assam, Nagaland, Manipur, Mizoram, Meghalaya, Uttaranchal, Himachal Pradesh and Jammu & Kashmir (Tables 70, 72, Singh and Sinha 2000). But their utilization is very limited as traditional rearers are not there and oak tasar culture is of very recent origin unlike China where it has been practised since time immemorial. The 18,41,000 hectares of oak trees effectively and efficiently contribute towards environmental protection directly as well as through oak tasar culture since more or less all the oak species are suitable for oak tasar silkworm rearing.

Central Silk Board has established three Regional Tasar Research Stations at Imphal (Manipur), Bhimtal (Uttaranchal) and Batote (Jammu and Kashmir), five Research Extension Centres at Palampur (H.P.), Gopeshwar (Uttaranchal), Kikruma (Nagaland), Umragsho (Assam) and Yaikongpao (Manipur) in addition to one oak tasar grainage at Imphal (Manipur). Besides, there are 35 Seed Farms in State sectors of North-western region and 28 Oak Tasar Grainages and 40 Tasar Extension Centres in State sectors of North-eastern region. All these stations, centres and farms have their own orchard plantations of various species for oak tasar culture as well as environmental protection.

Sericulture and Seri-biodiversity

Table 70: Statewise forest area and its availability for sericulture

Sl. No.	State	Total forest (ha.)	Area under silkworm host plants		
			Tasar	Oak tasar	Muga
1	Bihar (including Jharkhand)	26,52,400	9,18,000	-	-
2	Madhya Pradesh (including Chattisgarh)	1,31,19,500	50,44,000	-	-
3	Orissa	46,94,100	20,24,000	-	-
4	West Bengal	8,34,900	3,53,000	-	-
5	Andhra Pradesh	43,29,000	13,02,000	-	-
6	Maharashtra	46,14,300	10,04,000	-	-
7	Karnataka	32,40,300	5,21,000	-	-
8	Arunachal Pradesh	68,60,200	-	12,25,000	-
9	Assam	23,82,400	-	24,000	1,41,469*
10	Nagaland	14,22,100	-	15,000	-
11	Manipur	17,41,800	-	40,000	-
12	Mizoram	18,77,500	-	15,000	-
13	Meghalaya	15,65,700	-	23,000	-
14	Uttar Pradesh (including Uttaranchal)	33,99,400	-	3,05,000	-
15	Himachal Pradesh	12,52,100	-	1,39,000	-
16	Jammu & Kashmir	20,44,000	-	55,000	-
	Grand Total	**560,29,700**	**1,11,66,000**	**18,41,000**	**1,41,469**

Source: Singh and Sinha 2000

Table 71: Availability of tropical tasar food plants in various states

S. No.	State	Total forest area (in lakh hectares)	Area under tasar food plants (in lakh ha.)
1	Jharkhand (including Bihar)	30.59	9.18
2	Chattisgarh (including Madhya Pradesh)	168.13	50.44
3	West Bengal	11.83	3.55
4	Orissa	67.46	20.24
5	Andhra Pradesh	65.18	13.02
6	Maharashtra	66.96	10.04
7	Kanataka	35.10	5.21
	Total	**445.25**	**111.68**

Source: Sengupta 1986

Table 72: Availability of oak species in India and their exploitation for tasar rearing

Sl. No.	State/Region	Area under oak flora (Hec.)	Exploitable area (Hec.)	Major oak species available	Exploitable oak species for tasar (1.2 m = 4 ft.) rearing (Altitudes)
A. North-East					
1.	Manipur	40000	20000	Q. serrata, Q. dealbata, Q. grifithii, Q. semiserrata	Q. serrata (600-1800m AMSL) Q. dealbata (2000-3000'AMSL) Q.grifithii, Q. semiserrata (intermixed).
2.	Nagaland	15000-20000	5000	Q. serrata, Q. grifithii, Q. lineata Q. lamellosa, Q. dealbata	Q. serrata (600-1800m AMSL), Q. dealbata (2000-3000 AMSL) Q. grifithii (Above 3000' AMSL)
3.	Assam	24000	2000	Q. serrata, Q. semiserrata, Q. dealbata Q. grifithii.	Q. serrata (600-1800m AMSL) Q. semiserrata (600-1800, AMSL)
4.	Arunachal Pradesh	1225000	5000	Q. serrata, Q. grifithii, Q. dealbata, Q. semecarpifolia, Q. lanuginosa, Q. lineata, Q. lamellosa, Q. kamroopi	Q. grifithii (Above 3000' AMSL).
5.	Mizoram	15000	5000	Q. serrata, Q. dealbata, Q. grifithii Q. semecarpifolia, Q. semiserrata	Q. serrata (600-1800m AMSL) Q. grifithii (Above 3000' AMSL)
6.	Meghalaya	23000	500	Q. serrata, Q. dealbata	Q. serrata (600-1800m AMSL), Q. dealbata (2000-3000' AMSL)
B. North-West					
7.	Jammu & Kashmir	55000	1000	Q. incana, Q. himalayana, Q. semecarpifolia, Q. glauca, Q. ilex (exotic) Q. rubur, Q. serrata	Q. incana (1200-2500m AMSL) Q. himalayana (5000-7000' AMSL) Q. semecarpifolia (7000-9500' AMSL)
8.	Uttaranchal	305000	5000	Q. incana, Q. semecarpifolia, Q. glauca, Q. ilex, Q. floribunda, Q. lanuginosa, Q. serrata (exotic)	Q. incana (1200-2500m AMSL), Q. himalayana (5000-7000' AMSL) Q. semecarpifolia (2500-3500m AMSL)
9.	Himachal Pradesh	139000	3000	Q. incana, Q. dealbata, Q. semecarpifolia, Q. glauca, Q. ilex (exotic), Q. rubur, Q. cerris, Q. palustris, Q. rubra, Q. serrata. (exotic)	Q. incana (3500-6500 AMSL). Q. himalayana (5000-7000' AMSL), Q. semecarpifolia (7000-9500' AMSL).
		1841000	46500	Total species : 15 species	Exploitable : 7 species.

* Q. serrata syn. Q. acutissima, Q. incana syn. Q.leucotricophora, Q.himalayana syn. Q.dilatata syn. Q. floribunda

Table 73: Muga food plantation under A.M.F.P. Scheme

Sl. No.	State	Year			Total land (ha.)
		1994-95	1995-96	1996-97	
1	Assam	85	265*	103	453
2	Meghalaya	14	31	06	51
3	Nagaland	-	15	-	15
4	West Bengal	-	10	02	12
	Total	**99**	**321**	**111**	**531**

* Including 50 hectare of village grazing reserves

Source: Siddiqui and Das 1999

(5) Muga culture

Muga silk accounts for 5.37% of the total non-mulberry silk production (Shetty and Samson 1998). Muga culture is an integral part of life and culture of Assam. Recently, it is also introduced in the neighbouring states of Nagaland, Meghalaya, Manipur and Arunachal Pradesh. It has also spread to Coochbehar district of West Bengal. The muga silkworm is polyphagous and semi-domesticated since it feeds on variety of trees and reared outdoor but spins the cocoon indoor. Hence, contribution of muga culture is enormous in environmental protection directly through host plants as well as indirectly through employment generation and reduction of pressure on forests for livelihood of tribals.

Earlier upper Assam was full of muga food plantations but an earthquake in 1950 and unabated deforestation have affected muga culture drastically. To arrest the declining trend of muga silk production, the Government of Assam has established 12 muga nurseries and raised 450 hectares of plantations in 53 Village Grazing Reserves. It has been estimated by Planning Commission, Govt. of India that a total plantation area (with higher spacing of 10 x 10 mtrs.) has declined to 2500 ha. from 5765 ha. (1980-81). Against a total requirement of 6420 ha., only 2500 ha. muga food plants are currently available. Hence, 3920 ha. of additional plantation of muga food plants is required at existing rate of utilization. A total of 531 ha. of muga food plantation has been already raised under Augmentation of Muga Food Plant (AMFP) scheme during 1994-95 to 1996-97 in Assam, Meghalaya, West Bengal and Nagaland (Table 73). Under AMFP scheme 1202 farmers have been benefited and an additional income of Rs. 17500-21000 per annum is likely to be generated as per existing rate to each farmer by rearing 500 dfls in two commercial crops to produce 35000-42000 muga cocoons. Out of 531 ha., 481 ha. lie in private land (2 x 2 mtrs.) and 50 ha. (4 x 4 mtrs.) lie in Village Grazing Reserves (Siddiqui and Das 1999). Recently, area under muga food plants has increased to 3500 ha. in North-eastern region including West Bengal (Rao *et al.* 1998). However, there exist 1,41,469 ha. of forests having muga food plants in the states of Assam and Meghalaya (Singh and Sinha 2000) which contributes towards environmental protection directly since they are not fully utilized for muga culture.

Central Silk Board has established one Central Muga and Eri Research and Training Institute (CMERTI) at Lohadoigarh, Jorhat (Assam) and a Regional Muga Research Station at Boko (Assam). CMERTI has three Research Extension Centres, seven Basic Seed Production Centres and two Commercial Muga Seed Production Centres (Shetty and Samson 1998). All these establishments have their own orchard plantations of various *Persea* (*Machilus* Jr. syn)/*Litsaea* species for muga culture as well as environmental protection.

Population explosion, growth of civilizations and ever increasing greed caused the degradation of frontier forests to dense forests, dense forests to open forests and open forests to scrub forests and scrub forests to waste lands. Currently, our country has 5721100 hectares of scrub forests and every year nearly 6000 Km² of dense forests is converted into open forests. The scrub forests may soon be converted into waste land if the situation is not tackled immediately. Hence, if sericulture is promoted in these waste lands it will not only rejuvenate the lost forests but also provide livelihood to many tribal families.

(B) SERICULTURE AND SOIL CONSERVATION

Sericulture has multi-faceted role in regenerating the forests. Loss of leaf due to silkworm rearing is only for short time as fresh leaves appear within 30-40 days. As silkworm rearing requires light pruning for tasar and muga culture and heavy pruning for mulberry sericulture, the new flush of leaves receive more sun light to trap more energy for photosynthesis. The leaf litter, excreta of tasar and muga silkworms and dead worms fall on the soil surface and enrich the soil with organic matter, minerals and water for passing on to second stratum of growing plants which help in recycling of these materials to increase the fertility of forest land and adjoining piece of land through their roots as well as shoots. The root system of second stratum of plants *viz*. herbs, shrubs and creepers transport the minerals to forest floor, loosen the soil for better retention of water and enhance the growth of root systems by penetrating to the different depths of soil. Besides increasing fertility, soil and moisture conservation are also facilitated by increased vegetation consistently. When the protective cover of vegetation on the soil is removed, the structurally unstable tropical soil are exposed to the rains. Losses due to erosion immediately after land clearing are normally alarmingly large. Hence, trees planted for sericulture reduce run off and erosion losses in natural forest communities by providing a multiplayer defence against the impact of rain drops. Further, the litter and the humid layers on the soil surface acts as a cushion against soil erosion. The barren lands around forest area having regular sericultural activities turned fertile and enhanced forest regeneration (Singh and Sinha 2000).

(C)   SERICULTURE AND SOIL RECLAMATION/FERTILITY

Several species of *Terminalia* are very useful in soil reclamation, conservation and pollution control and hence, play a very vital role in changing the land scape and in improving the degraded soil and waste land in the country. Species like *T. arjuna* can be raised in acidic (P^H 4.5) as well as alkaline soil (P^H 9.5) and also in water logged conditions. *T. arjuna* was also successfully raised on fly ash resulting from burning of large quantities of coal in thermal power plants in Korba district of Chattisgarh (Srivastav *et al.* 1998). *T. tomentosa* and *T. arjuna* are very useful in soil reclamation as they add every year large quantities of organic matter to the soil in the form of leaf and silkworm litter through tasar culture. Frequency of litter production by *T. tomentosa* as a dominant tree species in the forest has been reported to be 93% and that of *Tectona grandis* 100% which is very high. Other food plants of tasar silkworm *viz.*, *Anogeissus latifolia*, *Lagerstroema purviflora*, *Diospyros melanoxylon*, *Pterocarpus marsupium* and *Madhuca indica* have also been reported to have 20%, 47%, 67%, 13% and 6.5% frequency of litter production for retention of soil fertility in the forests of tropical region (Prasad and Mishra 1985).

Besides litter fall through dead and falling leaves, twigs, branches, fruits etc., about 20-25% of the total living biomass of the trees is in roots and there is constant addition of organic matter to the soil through dead and decaying roots of sericultural trees. Besides, fertility status of soil is also improved through additional amount of nitrogen added to the soil system by the tree legume component of silkworm host plants *viz.*, *Dalbergia sissoo*, *Bauhinia variegata*, *Hadwickia binata* and *Pterocarpus marsupium* etc. Improvement in the organic matter status can result in an increased favourable activity of the favourable microorganisms in the root zone which not only solubilizes phosphates but also promotes growth promoting substances and cause commensalistic effects on the growth of plant species.

(D) SERICULTURE FOR STABILIZATION OF ECOSYSTEM

Large scale conversion of oak forests, the natural habitat of oak tasar silkworms, into the Pine forests in geologically fragile Himalayan region is one of the reasons for occurrence of land slides. Oak trees roots hold the soil strongly due to higher water/moisture retaining capacity than Pine trees which release water in a single go. Prof. Ramakrishna of School of Environmental Science at Jawahar Lal Nehru University, New Delhi exclaimed this view on the occasion of heavy land slides occurred during August, 1998 in Ukhimath region in which more than 200 persons lost their lives and many more have been victimised during August, 1998 on pilgrimage to Kailash Mansarover. Oaks not only regulate water cycle efficiently but facilitate nitrogen cycle also. The large scale conversion of Oak forests in to Pine forests has severely affected nitrogen cycle of ecologically fragile Himalayan ecosystem (Singh *et al.* 1984, Srivastav and Singh 2002).

Like tasar host plants, muga and eri host plants also facilitate conservation of soil and moisture in their respective regions because of their shrubby/arborescent habit. The role of protective action of trees in imparting stability to whole ecosystem is thoroughly understood. The clearing of trees not only affects the farm lands in the immediate vicinity but also destroys the catchment areas causing flooding of rivers and rapid silting of dams. Hence, forests and the non-mulberry sericultural trees also influence the ecoclimate of their regions. The favourable effects of soil conservation, physical properties, hydrological balance, microclimate and shelter belt significantly indicate the beneficial effects of sericultural trees in conserving and stabilizing the ecosystem.

(E) SERICULTURE FOR BIO-DIVERSITY CONSERVATION

Food plants of all the silkworms help in biodiversity conservation also. Various insects, rodents, reptiles, birds and microorganisms find shelters on the food plants of silkworms. Many of these insects and microorganisms serve as their parasites while some serve as symbionts, for example, all pathogenic microorganisms, leaf defoliators and stem borers are parasites while vesicular arbuscular mycorrizae are symbionts on host plants of various silkworms. Many beautiful orchid species have been found to grow as epiphytes on various food plants of oak tasar as well as muga silkworms in North-eastern region of India. Likewise, many epiphytes have also been found to grow on various food plants of tropical tasar silkworms in all the tropical states of India.

(F) SERICULTURE FOR FOREST PROTECTION THROUGH TRIBALS

While Rs. 2000/-per person may be earned either by forest collection of 2000 tasar cocoons within 20-30 days or through traditional methods of rearing 100 disease free layings in forest within 50-60 days, Rs.3750/- may be earned from economic plantations raised in 4'x 4' spacings. On the contrary, Rs. 21937/- may be earned by rearing of 450 disease free layings of tasar silkworms if integrated package of rearing is followed. A farmer may get Rs. 37500/- from rearing of muga silkworms in 150 trees/hectare/year by conducting two rearings while Rs. 6000-8000/- may be earned by oak tasar rearing of 1000 dfls in one hectare of natural plantations. These incomes may further be doubled when these cocoons are converted into yarns (Singh and Sinha 2000).

According to 1991 census, the tribal population in India accounts for 8.01% of the total population. Approximately 30 million (77%) out of 38 million tribals, are located in tropical (25.67 million) and temperate (4.23 million) belts. Yet, only 13.2 and 3.2 million tribals are living in tropical and temperate tasar cultivating districts respectively. Of these, only 1.40 lakh tribal families are engaged in tropical as well as temperate tasar rearing and almost equal number is engaged in post cocoon processing. In addition, a good number of tribals are also engaged in rearing, reeling and weaving of muga and eri cocoons/silk. Approximately 8000 reelers, mostly women are engaged in muga reeling. The direct and indirect employment utilisation in non-mulberry sector is estimated as one man year for producing one Kg. of raw silk. While the tribal population engaged in tasar cultivation varies from 45% in Jharkhand (erstwhile Bihar), 66% in Chattisgarh (erstwhile Madhya Pradesh) to around 80% in Orissa, the overall employment generation from practising non-mulberry sericulture covering both on farm as well as off farm activities has been estimated to be around 10-12 lakh persons/year which is negligible (Rao *et al.* 1998, Singh and Sinha 2000).

About 13 million hectares of forest is available with sericegenous flora as dense, open and scrub forests and most of the tribals are below poverty line, hence, if properly managed the tribal poverty can be removed and vigour of the forest also could be restored in ten to fifteen years. Since tribals and forests depend on each other a mutual harmony may be established between both the resources for environmental protection.

(G) NON-MULBERRY SILKS: ECO-FRIENDLY TEXTILES

Germany and the Netherlands have banned certain chemicals and dyes, especially azo dyes which contain harmful amines in the textile production. Government of India have also banned manufacture of nearly 112 dyes containing toxic and hazardous substances, which cause irritation in the skin and irreparable damage to the environment.

The use of chemical manure, pesticides and hazardous chemicals is either absent or limited since tasar, eri and muga culture are practised in the tribal zone/ forest areas. Further, manufacturing of non-mulberry silk products involves use of soap or alkali for cooking or degumming and minimum amount of enzymes are involved in yarn production. Sizing of the yarn is done in a limited way by using starch and chemical processing is either barest minimum or negligible. Bleaching is almost not done and hydrogen peroxide or sodium hydrosulphite is used for negligible bleaching. Chemicals such as soap, starch, soda, enzymes, bleaching agents (peroxides) do not pose any environmental problem since they are used in very small quantity and removed from the textile material after their purpose is served.

Dyeing/printing involves some application of metal complexes. Since the quantum of dye used is limited in non-mulberry silk production, the materials are safe. Likewise, the use of eco-friendly vegetable dyes is found in some cases, which confirm the eco-friendliness of the non-mulberry silk products. Hence, due to the eco-friendly operations involved in host plant cultivation, silkworm rearing and post cocoon technology associated with ethnic touch by tribal operators, the products manufactured out of non-mulberry silk could be considered as eco-friendly (Nadiger 1998)) and sericulture as a whole may be regarded as a greatest tool for environmental protection in new millennium to save mother earth and human beings.

(H) SERICULTURE: A REMEDY FOR SHIFTING CULTIVATION

Shifting cultivation is a method of cyclic cultivation, chiefly in vogue in tropics, where cultivators cut the trees, burn and raise field crops for one or more years, subsequently move on to another site and repeat the same process. It is practised in Madhya Pradesh, Chattisgarh, Andhra Pradesh, Tamil Nadu, Kerala, Assam and all the North-eastern states. The total area utilized annually is more than 55 lakh hectares and nearly 26 lakh people live in this way. According to an estimate nearly 81 million hectares, *i.e.*, approximately 25% of the countries land is subjected to soil erosion due to continuous destruction of forests and shifting cultivation seems to be mainly responsible for such large scale soil erosion (Sagreiya 1967).

Non-mulberry sericulture in forest trees may serve as an effective alternative for shifting cultivation because when tribals and village folks will earn their livelihood from the forests containing non-mulberry silkworm host plants then instead of cutting these trees for their shifting cultivation they will not only protect these natural resources but also contribute effectively and efficiently in Joint Forest Management for their sustainable utilization.

Thus, sericulture is indeed a very important avenue of our country for conservation and sustainable utilization of forests, conservation of natural resources like soil and biodiversity and maintenance of ecosystem for posterity.

8

Silkworm Seed and Rearing Technology

A. SEED TECHNOLOGY

Silkworms are multiplied from the eggs produced by the mother moths. In common parlance the silkworm seed in India is called laying. One laying contains the eggs laid by one moth and one healthy female moth lays more than 400 eggs in *Bombyx mori*, 200 to 250 eggs in wild silkworms, and all the eggs laid by one moth is called one laying. It is absolutely essential that the eggs laid by the moth should be free from disease, because pebrine (protozoan) disease is seed borne. The most dreaded disease of silkworm namely pebrine caused by a microsporidian, *Nosema bombycis* Nageli is transovarially transmitted by the mother moths to the progeny through the eggs and hence the layings are strictly subjected to microscopic examination for disease. Only the disease free silkworm seeds are multiplied and supplied to the farmers to conduct silkworm rearing, such disease free silkworm seeds are generally called disease free layings (DFL). Silkworm seed production center is generally called grainage and the silkworm seed production activity is called grainage activity. The seed production details discussed in this chapter is applicable to mulberry silkworm (*Bombyx mori*) and seed production in wild silk moths is discussed separately.

TYPES OF SILKWORM SEED

There are three types of silkworm seeds: they are (a) Pure races (b) Cross breeds and (c) Hybrids. Generally, pure races are reared only in the seed area and the seeds of these races are produced only in the basic seed grainages. The pure races are used for production of cross breeds and hybrids. The other two types *viz.*, cross breeds and hybrids are produced for commercial farmers to produce reeling cocoons for silk industry. In India about 90% of the silk comes from cross breeds. In cross breeds, the multivoltine race female moth is coupled with bivoltine race male moth to obtain hybrid vigour. The multivoltine races are resistant to adverse climate, diseases and feed also on poor quality mulberry leaves, but they yield low and poor quality silk, whereas bivoltine races are susceptible to diseases and adverse climate, therefore, yield very less cocoons under low management levels but under improved management practices, yield high quality and more quantity of cocoons and silk. The cross breed of multivoltine female and bivoltine male gives more yield of cocoons with reasonably good quality and quantity of silk. Hence, majority of the farmers in India prefer cross breeds. In temperate climate zones like Jammu & Kashmir, Uttar Pradesh, Uttaranchal, Himachal Pradesh, Haryana and Punjab, the farmers prefer bivoltine hybrids. In such hybrids both female and male parents are different races of bivoltine, usually one parent of oval cocoon race and another parent of dumb-bell shaped cocoon race. The hybrids and cross breeds exhibit heterosis. The bivoltine hybrids give high quality, international grade silk. Bivoltine hybrid rearing is more risky as they are highly susceptible to diseases and these worms prefer 24-26° C temperature and 70-80% relative humidity and hence most of the farmers in tropical regions only prefer cross breed, consequently most of the grainage in the tropical India, Karnataka, Andhra Pradesh, Tamil Nadu and West Bengal produce cross breeds silkworm seed.

GRAINAGE

Grainage is the place where silkworm seeds are produced. A grainage should be kept utmost clean. "Cleanliness is next to Godliness", this epithet is more relevant in hospitals and grainages. The grainage premises and equipments used are cleaned by normal routine methods and then in addition, the place and the equipments are washed with 5% bleaching powder solution and later with 2% formaline solution. The bleaching powder, slaked lime and formaline are most commonly used for cleaning and hence called disinfectants. Recently, chlorine di-oxide and chlorofet are used as disinfectants, instead of formaline, since it is carcinogenic. The quality of these disinfectants are of utmost importance, otherwise one will not achieve satisfactory level of disinfection. If the grainage and the equipment are not properly disinfected, it is likely to cause contamination of pebrine spores from the previous grainage activity. Hence, after each seed production operation and before processing another set of activity, the grainage and the equipments should be thoroughly disinfected. There are two types of grainages: the basic seed grainage, which multiply basic seed stock of pure race and commercial grainages, which produce cross breed and hybrid silkworm seed on a commercial scale and supply the seed to the farmers. The basic seed grainages are maintained only by Government Departments in India. Whereas commercial grainages are under Government Department and also under private sector.

GRAINAGE EQUIPMENTS

The grainage activity *i.e.*, the silkworm seed production involves a chain of very delicate activities. For production of silkworm seed some very simple equipments are essential (Table 74). Except microscope, centrifuge and acid treatment bath, the other items are common and cheaper.

CHEMICALS USED IN THE GRAINAGE

Bleaching powder, slacked lime, formaldehyde, chlorine di-oxide, chlorofet, hydrochloric acid, potassium carbonate (K_2CO_3) or potassium hydroxide (KOH) are frequently used in the silkworm seed production process. Bleaching powder, slacked lime, chlorine di-oxide, chlorofet and formaldehyde are used for disinfection and keeping the premises clean and free of pathogens. Hydrochloric acid is used in the acid treatment of eggs particularly in the bivoltine silkworm seed production process. Generally the bivoltine races undergo diapause at the egg stage. In order to terminate the diapause and make the bivoltine eggs hatch according to the requirement, the hibernating eggs are treated with hydrochloric acid. Potassium carbonate is used in the microscopic examination of moth smear. Pebrine disease is transovarialy transmitted from mother moth to its progeny, through the eggs. Pebrine disease is the sorrow of sericulture. Therefore, every care should be taken to produce and supply silkworm seed, which are absolutely free from pebrine spores. Hence, the mother moths after oviposition undergo microscopic examination. In this process, the abdomen of the mother moth is crushed after oviposition, along with few drops of potassium carbonate or potassium hydroxide which facilitates disintegration of tissues and cells and release of *Nosema bombycis* spores for examination under microscope.

PREPARATION OF CHEMICALS AS PER REQUIREMENTS

Slacked lime Ca $(OH)_2$ is used as such. Good quality bleaching powder should contain 36-37% chlorine content. For general disinfection of the grainage 5% bleaching powder solution is useful. To prepare 5% bleaching powder solution 50 gm of good quality bleaching powder should be thoroughly mixed in one liter water. Commercial formaldehyde available contains 40% concentration. For grainage use 2% formaldehyde is sufficient. To obtain 2% formaldehyde solution one part of 40% of formaldehyde is mixed with 19 parts of water. Usually the formaldehyde dissolved in water is called formaline. Potassium carbonate (0.5% concentration) solution for moth smear preparation is obtained by dissolving 5 gm of Potassium carbonate in one liter of water. Hydrochloric acid is required in two concentration depending upon the type of acid treatment. There are two types of acid treatment viz., cold acid and hot acid treatment. For cold acid treatment hydrochloric acid should be of 1.1 specific gravity, whereas 1.075 specific gravity is required for hot acid treatment. Accordingly the hydrochloric acid is prepared. Commercial formulation of hydrochloric acid available in the market has specific gravity of 1.18 at 15°C. There is correlation between the temperature and specific gravity of hydrochloric acid. 555 ml of HCl is dissolved in 445 ml of water to get one litre of HCl solution having 1.1 specific gravity and 416 ml HCl is dissolved in 584 ml of water to get 1.075 specific gravity at 25°C. For hot acid treatment, the hydrochloric acid is not heated directly. Hot acid baths are available for this specific purpose. In these acid baths, the acid kept in a container is heated indirectly through heating the water batch. These acid treatment baths are double containers, the outer container contains water and the acid is kept in the inner container and these baths are electrically operated.

RAW MATERIALS FOR SILKWORM SEED PRODUCTION

Parental seed cocoons of pure races are the raw materials for silkworm seed production. Parental seed cocoons are multiplied in four tier multiplication system, generally denoted as P4, P3, P2, and P1. 'P' refers to parental stock and the numerals refer to the stage of multiplication. Generally the parental stock at P4, P3 and P2 levels are done only in Government farms called Basic Seed Farms (BSF). The parental stock multiplication is required in small quantities only. A Basic Seed Farm may multiply 50 to 150 DFLs at a time, but the basic seed stock should be multiplied with utmost care and at each level the quality of silkworm, the seed cocoon and the layings are examined to ensure high quality silkworm seed production at commercial level.

BASIC SEED FARMS AND BASIC SEED GRAINAGES

Presently, India requires nearly 600 million dfls to utilize the mulberry leaves produced by the farmers and produce reeling cocoons through silkworm rearing to manufacture 20,000 metric tons of silk. This 600 million dfls are produced in commercial grainages, for which the parental seed cocoons, the raw materials are generated in the basic seed farms. Primarily the basic seed farms maintain the parental seed stock of pure races following the standard norms of silkworm race maintenance as stipulated by the silkworm breeders, as they handle the breeders stock. The breeders stock should be handled properly to maintain the original race characters and maintain the vigour through proper selection at various levels, *viz.*, the eggs and the seed cocoons. The P4 farm always maintain the races as per silkworm breeders norms and makes one level multiplication and thereafter, the seeds are supplied to P3 farms. Similarly the P3 farms make one level multiplication and supply the basic seed to P2 farms. In turn the P2 farms multiply the seed one more level and produce large quantity of basic seed and supply to selected P1 farmers, who are seed cocoon producers. These P1 seed cocoon producers further multiply the silkworm seeds in large scale through silkworm rearing and produce the parental seed cocoons, which become the raw materials for commercial grainages. These seed cocoon crops are reared in the seed area by selected seed cocoon producers. (The chain goes like this: P4 farms à P3 farms à P2 farms à P1 seed cocoon producers who will produce and supply seed cocoons to commercial grainages for production of cross breed/hybrid seed.)

SEED AREA

The seed cocoons are produced by about 5,000 farmers in each of the states *i.e.*, Andhra Pradesh, Tamil Nadu and West Bengal and more than 20,000 farmers in the State of Karnataka. Similarly, in other states like Uttar Pradesh, Uttaranchal, Jammu & Kashmir and other states also there are seed farmers. Most of the states, where sericulture is a major crop there are legislations governing the sericulture industry, particularly the five traditional states *viz.*, Karnataka, Andhra Pradesh, Tamil Nadu, West Bengal and Jammu & Kashmir have enacted Silkworm Seed Act, which prohibits multiplication of silkworm seed cocoons in other than the prescribed zones. Thus, most of the parental seed cocoons are produced by the farmers in the "seed area" or selected seed farmers notified by the department of sericulture.

SEED COCOON CROP

The seed farmers receive basic seeds of silkworm through the extension network of the sericulture departments and conduct silkworm rearing adopting seed cocoon production techniques. These farmers regularly receive technical guidance from the extension staff of department working in Technical Service Centres (TSC). The extension staff periodically inspect the health of the silkworm and examine them for pebrine disease. If pebrine disease is noticed at young stage of the silkworm during first or second or third stage, such seed cocoon crops are destroyed to prevent spread and contamination of other seed crops. Such farmers get free supply of basic silkworm seed as compensation for the loss. Also such farmers who lost their cocoon crops due to pebrine receive special attention from the department officials who assist the farmers to disinfect their silkworm rearing house and equipments. If pebrine disease is noticed at later stages of larval development *i.e.*, fourth or fifth stage, then such cocoons are transacted separately and sent for silk reeling, so that diseased cocoons do not find their way into the grainage.

SEED COCOON MARKETS

The seed farmers obtain the basic seed from basic seed farms and produce seed cocoons. These seed cocoons are transacted only in the seed cocoon markets under the administrative control of sericulture departments. These seed cocoon markets are regulated through Silkworm Seed Act. The seed cocoons are sold by the farmers to commercial grainages and the sale rates are fixed by the Sericulture Department, the rates vary depending upon cocoon quality, cocoon availability and demand from commercial seed producers. The surplus seed cocoons and cocoons unfit due to pebrine or muscardine, flacherie, grasserie disease are sent for silk reeling. Good quality seed cocoons are always in demand and fetch high rates. These seed cocoons are the raw materials for the commercial grainages to produce commercial silkworm seed called Disease Free Layings (DFL).

COMMERCIAL GRAINAGES

To produce one lakh dfls every month, an ideal grainage requires 6,000 to 6,500 sq. ft built-in area, with atleast 7-8 well ventilated rooms. The grainage equipments are listed in Table 1. The quality as well as the cost of silkworm seed directly depends upon the seed cocoon quality and hence care should be taken to procure quality seed cocoons only from the regulated seed markets. At the time of purchase itself one can conduct pupal gut examination and determine if the seed cocoons are free from pebrine disease; only such seed cocoons which are disease free should be purchased and carefully transported to the grainages. Cocoon which are melted, cocoon crops affected by muscardine, grasserie and flacherie are also unfit for seed production. Silkworm seed production involves a series of delicate operations.

PROCESS OF SILKWORM SEED PRODUCTION

In the production of silkworm eggs, the seed production techniques should be followed systematically to ensure quality of silkworm eggs. Trained technicians and workers play an important role in commercial grainages to produce quality silkworm eggs in large scale. There are several techniques involved in egg production.

DISINFECTION

Prime importance should be given to this aspect, before the seed cocoons are received in the grainage. All the rooms of grainage should be cleaned by using the disinfectants like bleaching powder and formaline in order to keep the grainage free from pathogen. Similarly, all the equipments used in grainage for seed production are also properly disinfected before using them.

RECEIPT OF SEED COCOONS

Immediately after the arrival of seed cocoons in the grainages, they should be checked for quality. From each batch of seed cocoons 20 pupae are collected randomly and gut examination is conducted for pebrine disease. Even if one pupa shows few spores of pebrine, the entire lot is rejected and sent to market for silk reeling. In order to ascertain the cocoon quality, 10 male and 10 female cocoons are weighed and their cocoon weight, shell weight and shell ratio are recorded, to ensure quality of the raw material in order to produce high quality silkworm seed for commercial purpose. The multivoltine seed cocoon weight should not be less than 1.1 gm/cocoon and shell weight should not be less than 0.1 gm/cocoon. Similarly, the bivoltine cocoon weight and shell weight should not be less than 1.4 gm/cocoon and 0.3 gm/cocoon respectively.

SORTING OF COCOONS

After having confirmed the disease free nature of seed cocoons and their quantitative characters, they should be brought into seed cocoon preservation chambers and kept in bamboo trays. From each batch of seed cocoons, a definite quantity of cocoons are taken randomly and they should be subjected to assessment. Flimsy, melted, uzifly infected and deformed cocoons are sorted out and rejected. The batches, which are showing high meltage should be rejected, since they are not economical to process.

SPREADING OF COCOONS

After recording the assessment details, good cocoons are sorted out from flimsy, uzi infested, melted and deformed cocoons. The selected good cocoons should be arranged in bamboo trays in single layer, to avoid pupal mortality during the process.

MOTH EMERGENCE

Seed cocoons, arranged in single layer should be kept in seed cocoon preservation chambers and the rooms are kept dark, as moth emergence is a photo sensitive activity. Silk moths generally emerge on 10th day after cocoon spinning in multivoltine and 11/12 day in bivoltine. The moth emergence is also influenced by temperature and humidity. The pupal period is 1 or 2 days shorter in summer and extended by 1 or 2 days in winter or high altitude places. The moth emergence may prolong for 3 to 4 days; the peak emergence will be on the second and third day. The first day emergence is always called sample emergence. The moths emerged on first day will be more of males and moths emerging on the last days will be more of females in both multivoltine and bivoltine. The moth emergence time also vary in the two sexes. In the case of males in both multivoltine and bivoltine, the emergence start in the night hours and peak emergence is at midnight, whereas females of both these races emerge in the early morning hours. Hence, as a convenient practice the lights are put on at mid night (the time may vary slightly depending upon the race, altitude, day length of the place and season). This varying time of male/ female moth emergence is very useful in grainage operations; as the males and females of each race should be separated immediately after emergence, to prevent selfing; otherwise males and females of the same race will mate and pure race will be produced instead of cross breed or hybrid.

PICKING OF MOTHS

On the moth emergence day, as soon as the light is put on in multivoltine or bivoltine seed cocoon room, the male moths start emerging and they can be identified easily from females. Trained labourers under the supervision of a technical person will pick up the male moths and reject in case of multivoltines (some times they are utilized for reciprocal combinations of multivoltine male moths crossed with bivoltine female moth). The virgin female moths of multivoltine are kept separately in bamboo trays for hybridization, with males of bivoltine to produce cross breed dfls.

For preparation of cross breed silkworm seed, the seed cocoons of multivoltine and bivoltine should be processed synchronously, so that the multivoltine female moths may be crossed with bivoltine male moths.

PAIRING AND ISOLATION

Healthy male (bivoltine) and female (multivoltine) moths are used for pairing. Unhealthy moths and moths with deformed wings are rejected. Coupled moths are transferred to separate trays and not disturbed for 4-5 hours. The coupled moths should be kept in semi-dark rooms for effective mating. After 4-5 hours of copulation, the male moths are separated mechanically and stored in crates in refrigerators at 1°C–10°C and they can be utilized for second time. Male moths should not be used more than two times for seed production.

OVIPOSITION

After separating the male moths, the female moths are allowed for few minutes to excrete the remaining excreta in the abdomen. Then the female moths are picked and placed on egg sheets, covered with black cellules made of plastic. This restricts the movement of the moths during oviposition. After arranging the females on egg sheets, they should be kept in wooden trays and transferred to oviposition room. Oviposition is maintained at 24°C to 25°C and 80% RH. Also, the room should be dark/semi-dark for higher rate of oviposition. Moths are kept in oviposition room for 24 hours. Generally the eggs laid within one day is only considered for supply to farmers.

PRODUCTION OF CROSS BREEDS (CB)

To exploit heterosis in silkworms, two different races are utilized for hybridization programme. In South India Pure Mysore (Multivoltine) is used as female component and bivoltine (NB4D2, NB18, NB7, KA) as male component. There are other multivoltine races also in South India *viz.*, C-nichi, HS6, MY1 etc., which are in limited use. In West Bengal, Nistari is the popular multivoltine race used in cross breed dfl production. Similarly, improved bivoltine races V12, CSR2, CSR4, CSR5 etc. are used as male component in recent times.

PRODUCTION OF RECIPROCAL COMBINATIONS OF CROSS BREED

From the production of above combination of Multivoltine female x Bivoltine male, there is a surplus of bivoltine female and multivoltine male moths. By utilizing bivoltine female and multivoltine male reciprocal cross breed (bivoltine female x multivoltine male) seeds are produced. There is lot of acceptability from sericulturists towards this combination, as it is behaving well in the field. Further, this seed stock can be preserved in Cold Storage similar to bivoltine pure race or bivoltine hybrid seed and released to farmers as per demand. These seeds are produced as buffer stock and generally supplied when there is sudden demand for silkworm seed.

PREPARATION OF BIVOLTINE HYBRID

For preparation of bivoltine hybrid seeds, the grainage requires two different types of bivoltine races *viz.*, oval and dumbbell races (based on the cocoon shape), usually the cocoons of bivoltine races of China and Japan origin are oval and dumbbell shaped respectively, and these races are used as male and female components respectively; *eg.* Oval x dumbbell and dumbbell x oval. The other techniques and procedures of seed production are similar to cross breed silkworm seed production.

MICROSCOPIC EXAMINATION OF MOTHER MOTH

The egg sheets along with moths are taken out after 24 hours of oviposition and subjected to mother moth examination in order to detect pebrine disease, which is transmitted transovarially.

INDIVIDUAL MOTH EXAMINATION

Though individual moth examination is laborious, still it is ideal to check the dreaded disease 'Pebrine'. Individual moth is taken from egg sheet and the abdomen of the moth is crushed by hand in crushing set. A few drop of Potassium carbonate and Potassium hydroxide (KOH) is added to the crushed fluid and smear is prepared. This is observed under a microscope with 600x magnification. This is an old technique.

SAMPLE TESTING

Cross breed layings are produced in large scale and it is not possible to test the moths individually. Hence, 20% of the moths are randomly collected and tested microscopically. Even if any moth shows pebrine spores, the entire lot of seed produced is rejected.

MASS EXAMINATION OF MOTHS

This method is dependable in commercial grainages and requires minimum time. In this method suitable sampling techniques are devised as followed in Japan. The samples consisting of 30 moths are kept in perforated paper covers and stored in hot air oven and dried at 70 °C for 6 hours. After 6 hours, 30 dried moths are put in and 90°C of 0.5 Potassium carbonate solution is added. The material is ground for 2 minutes in a mixie. Then the solution is filtered through a clean cotton cloth to remove scales etc. Later the fluid is centrifuged for 3 minutes at 3000 r.p.m and supernatant is decanted. To the sediment few drops of KOH is added and smears are prepared on a glass slide and examined under microscope for pebrine spores. This is a recent method and very reliable.

EGG PROCESSING

Surface Sterilisation—Washing of eggs

After the microscopic examination of moth smear, the eggs must be surface disinfected in order to avoid any other contamination. The eggs should be dipped for 5-10 minutes in 2% formaldehyde solution and later washed in clean and cold water to remove the traces of formaldehyde. Once the eggs are washed, they should be dried in shade and incubated.

Incubation

Preservation of eggs at the optimum temperature (25°C) and relative humidity (75-80%) makes the eggs to hatch uniformly on the specified date. Proper incubation of eggs facilitates synchronized development of the embryo. The condition of incubation also greatly influences the voltinism of the egg, larval growth, cocoon yield and quality. Hence, great care should be taken to maintain ideal incubation conditions. Once the eggs reach the blue spot stage, the eggs should be kept in black boxes to get uniform hatching.

Acid Treatment

Bivoltine eggs under natural condition hatch after 6 months because of the presence of hormone which inhibits the development of the embryos. This can be neutralized by subjecting the eggs to cold temperature treatment by keeping them in different temperature to dissolve the diapausing hormone or giving them artificial treatment.

Good hatching of bivoltine silkworm seed (pure race and hybrid) depends on proper egg preservation and acid treatment. To obtain higher egg hatching percentage, the eggs are subjected to acid treatment before the release of hibernating substance by the embryo. Ideal duration for acid treatment is between 20-22 hours from the time of oviposition. There are two types of acid treatments *viz.*, hot acid treatment and cold acid treatment. The details of the procedure to be followed for different types of acid treatment are as follows:

- In case of hot acid treatment, the eggs are treated in Hydrochloric acid with specific gravity of 1.075 at temperature 46.1°C., with time duration of 5-6 minutes.

- Artificial hatching can also be obtained by subjecting eggs to cold acid treatment in Hydrochloric acid of 1.1 specific gravity, at room temperature of 20-25°C and time duration of 80-90 minutes. Care must be taken to maintain the required specific gravity of the acid and temperature. Any slight variation in specific gravity of the acid will lead to problem in hatching of eggs. Before dipping the egg sheets in the hydrochloric acid, they are dipped in 2% formaline for two minutes to make the eggs adhere firmly to the egg sheets and for surface disinfection of eggs.

- After the acid treatment, the eggs must be washed in running cold water. Before the eggs are removed from the running water, it must be ensured that there are not traces of acid left over on the eggs. Then the eggs may be dried in shade and sent for incubation, where the eggs should be kept in thin layer, and not accumulated in heaps. The acid treated eggs will hatch in 8-10 days similar to multivoltine silkworm seed.

HIBERNATION

There are two types of eggs in silkworm *i.e.*, non-hibernating eggs or multivoltine eggs and hibernating eggs or bivoltine eggs. Non-hibernating eggs will hatch in 8-10 days from the date of egg laying and hibernating eggs do not hatch in natural conditions unless they are activated either by artificial stimulation *i.e.*

treatment of eggs in hydrochloric acid or by cold storage. Both non-hibernated and acid treated eggs can be stored for a short period upto 20 days to postpone the hatching till the desired date. These eggs can be preserved at 5°C with RH 80%. Whereas hibernated eggs can be preserved for long duration. To activate the hibernating eggs, they are required to be kept in temperature of 15°C, '0°C, 5°C and 2.5°C with RH of 80%.

Hibernated/bivoltine eggs can be made to hatch between 120-300 days (4-10 months) by subjecting them to different temperature schedules in the Cold Storage for a specific period. There are two stages of Cold Storage *i.e.*, preparatory phase or aestivation period at temperature regime of 25°C–15°C and refrigeration phase called hibernation from 10°C–2.5°C. The period of aestivation and duration of cold storage are interlinked. In order to get optimum hatching percentage, hibernated eggs are required to be preserved at appropriate temperature and humidity. The details of the hibernation schedules which are usually followed in the cold storages are furnished here below:

Hibernation Schedule

In tropical conditions 4 months, 6 months and 10 months schedules are followed for preservation of bivoltine eggs in cold storages, depending upon the requirement of these eggs.

FOUR MONTHS SCHEDULE

Under this schedule, bivoltine eggs are required to be preserved in cold storage as detailed below so as to get optimum hatching after 4 months.

Deg. C	No. of day(s)
25	10
20	02
15	02
10	03
5	50
2.5	50
12 to 15	01
Incubation	12

SIX MONTHS SCHEDULE

In this schedule bivoltine eggs are required to be preserved in cold storage as detailed below so as to get optimum hatching after 6 months.

Deg. C	No. of day(s)
25	20
20	15
15	10
10	10
5	50
2.5	60
12 to 15	01
Incubation	12

After keeping the seed at 2.5°C for 60 days, they can be taken out and they should be preserved in intermediate temperature for 2-3 days in each grade and gradually brought up. Incubation is preservation of eggs at 25°C from the day they are released from cold storage at intermediate temperature till hatching i.e. 15 to 25°C.

TEN MONTHS SCHEDULE

Bivoltine hibernating seed can be preserved for 10 months in cold storage as detailed below:

Deg. C	No. of day(s)
25	40
20	50
15	25
10	25
5	60
2.5	50
12 to 15	4 to 5
2.5	30
Incubation	12

Then the dfls should be taken out and kept in intermediate temperature as prescribed for 6 months schedule.

STORAGE AND TRANSPORTATION OF SILKWORM SEED

Silkworm seeds are highly affected by high temperature low and high humidity and fluctuation of temperature and humidity and hence silkworm seeds are ideally preserved at optimum temperature and humidity (*eg.* 22-25°C and 76-80 % R.H). The eggs are spread uniformly and heaping of eggs should be avoided. The multivoltine and cross breed eggs should not be stored for more than 5 to 6 days from the date of oviposition, since the eggs will hatch in 9 to 11 days depending upon the environmental temperature. Whereas, bivoltine eggs can be stored for several months in the cold storage.

The eggs are transported in the cool hours of early morning or late evening in well aerated bamboo or plastic baskets specially fabricated for egg transportation. The eggs should not be exposed to direct sunlight and the eggs should not be kept on hot objects and eggs should not be carried on air tight containers.

ECONOMICS OF SILKWORM SEED PRODUCTION

Silkworm seed production is a highly profitable small scale industry. The silkworm seed act enacted by the different states makes it mandatory to obtain license from Sericulture Department to produce silkworm seed. As per the act, without valid license no person can produce silkworm seed. Presently silkworm seed are produced by Central and State Government agencies. The Central Silk Board (CSB) has set up several grainages (Silkworm Seed Production Centres) in different parts of the country to provide extension and development support. Similarly under the State Government agency the department of sericulture have also established several grainages in different area which are potential for sericulture development. The Government agencies produce and supply only about 25-30% of the silkworm seed requirement of the farmers; whereas the Licensed Seed Producers (LSP) produce and supply the remaining quantity. Though silkworm seed production is highly profitable, yet it depends on several factors. Good quality silkworm seeds are always in demand and hence any LSP who produce good quality seed make considerable profit. Commercial grainage operation requires permanent assets like grainage building and equipments and working capital for raw materials, chemicals and other day-to-day miscellaneous expenditure. A permanent building for grainage may be expensive and hence most of the private seed producers operate from rented buildings. However, Government Departments have proper grainage buildings in order to produce quality seed.

NORMS AND REQUISITES

To produce one cross breed dfl four multivoltine and two bivoltine seed cocoons are required, as some of the unsuitable cocoons have to be rejected and some degree of cocoon selection will facilitate quality seed production. Mating of moths should be ensured to 30-35% and recovery of dfl should be at 20-22% of net available cocoons after rejection of unsuitable and unhealthy cocoons. At each level the unfit, unhealthy ones have to be rejected. Hence, every grainage operation should be perfect and systematic in order to make a good profit, otherwise any default may lead to poor quality seed leading to customer rejection and ultimate loss. The economics of a grainage with a production capacity of one lakh dfl per month (12 lakhs dlf per year) is given here.

(i) ACTIVITY

Production capacity	:	12,00,000 dfl/year
Multivoltine seed cocoons required	:	48,00,000/year
Bivoltine seed cocoons required	:	24,00,000/year
Rejection @ 10%	:	
Rejection of multivoltine seed cocoons	:	4,80,000
Rejection of bivoltine seed cocoons	:	2,40,000
Net available cocoons (Multivoltine + Bivoltine)	:	65,00,000

Mating of moths at 30-35% of total cocoons.
Recovery of dfls (seed) at 20 to 22%
Recovery of dfls @ 20% of 65,00,000 is 13,00,000
Rejection at seed level due to disease or poor quality or other grainage loss : 1,00,000 dfl
Net seed (dfl) production is 12,00,000/year.

(ii) INCOME

The grainage has two sources of income from the sale of silkworm seed and the sale of cocoon shell. After the moths emerge from the cocoon, the cocoon shell comes as a bye-product, which is a valuable commodity for spun silk industry.

The silkworm seed cost vary widely during different seasons. Like any other commodity the rate often depend upon demand and supply. During the year 2002, the silkworm seed were sold at Rs. 200/- to Rs. 300/- in different months. Usually silkworm seed are in high demand during February-March, June-July and September-November every year. The high demand during February-March and June-July is mainly due to low production of silkworm seed due to seed cocoon scarcity; whereas high demand during September-November is mainly because of intensive silkworm rearing by the farmers in large scale as a result of mulberry leaf production facilitated by monsoon and irrigation. Sales of 12,00,000 dfls @ Rs. 250/100 dfl will generate Rs. 30,00,000/- and sale of 72,00,000 cocoon shell will generate Rs. 1,80,000/-. The cocoon shells are called pierced cocoons and they are sold by weight. One kg. of pierced cocoon may cost Rs. 100/- to Rs. 120/- depending upon quality and maintenance. Approximately 4000 pierced cocoons will weigh one kilogram; yielding 1800 kg of pierced cocoon, which @ Rs. 100/- per kg. will generate Rs. 1,80,000/-.

The other details of expenditure are given in Table 75; which indicates remunerative returns and adequate profit to the entrepreneur.

Similar to the seed (dfl) cost, the seed cocoon cost also vary depending upon season and demand and supply and hence the economics are arrived at based on averages and hence the actual profit is likely to be varying, depending upon managerial, operational and technical skills of the entrepreneur.

Table 74: List of grainage equipments

Sl. No.	Name of the equipments
1	Bamboo trays for cocoon sorting and preservation.
2	Wooden trays for preservation of pupae till moth emergence.
3	Wooden stands with multi-tier to arrange the trays.
4	Ant wells with water to keep the wooden stands to prevent ant attack on pupae/ cocoons.
5	Hollow steel stands to facilitate moth picking.
6	Wax coated paper to cover the pupae.
7	Wet and dry bulb thermometer.
8	Humidifiers to increase the humidity when it goes below 75%.
9	Knap sack sprayer to spray formaline for disinfection.
10	Egg sheets for oviposition.
11	Black cellules to provide darkness and isolation of moths.
12	Zinc trays for preservation of moths.
13	Refrigerators for moth preservation.
14	Mortar and pestle for moth crushing.
15	Domestic mixer for moth crushing.
16	Centrifuge for tissue/cell dis-integration.
17	Hot air ovens to dry the moths.
18	Incubators to maintain 25 ºC and 80% RH for eggs incubation.
19	Egg sheet cabinet for egg preservation.
20	Washing equipments like trays and basins.
21	Binocular microscope for moth smear examination.
22	Foot mat and wash basin for personal disinfection.
23	Facial disinfection masks to provide protection during formaline application.
24	Cocoon cutting knife.
25	Wooden table, stool for moth examination.
26	Acid treatment bath.

SEED PRODUCTION TECHNIQUES IN WILD SILKMOTHS (Non-Mulberry Silkworm)

The basic principles and procedures of seed production followed in mulberry silkworm are applicable to wild silk moths also. However, the size of cocoon, pupa and moth are considerably larger and hence the space required for various grainage operations is more in the case of muga, eri tropical and temperate tasar silkworms. Another major difference is the behaviour of wild silk moths; unlike the domesticated *B.mori*, the wild silk moths do not emerge synchronously or uniformly and hence the grainage operations extend for more days; for instance moth emergence from pupa prolongs to more than seven days in wild silk moths, whereas moth emergence is completed within 2-3 days in *B. mori*.

Table 75: Economics of commercial silkworm seed production (Cross Breed)

Sl. No.	Activity	Quantity (Kg.)	Estimated expenditure (Rs.)
1	Seed cocoon cost:		
	(i) @ Rs.120/kg for multivoltine	4800	5,76,000.00
	(ii) @ Rs.200/kg for bivoltine	3350	6,70,000.00
2	Staff salary @ Rs.4000/month for 6 staff + bonus + provident fund		3,00,000.00
3	Labour wages @ Rs.75/day for 12 labourers + bonus + provident fund		3,52,500.00
4	Grainage rent + electricity + water charges @ Rs.10,000/month		1,20,000.00
5	Consumables like chemicals and miscellaneous @ Rs.1500/month		25,000.00
6	Depreciation on equipments @ Rs.50,000/year		50,000.00
7	Interest on capital @ Rs.50,000/year		50,000.00
8	Miscellaneous (cocoon transportation etc.)		50,000.00
	Total Expenditure		21,93,500.00
			or say 22,00,000.00
9	Gross income from sales :		
	@ Rs.250/100 dfls (12,00,000 dfls)		30,00,000.00
	@ Rs.100/kg pierced cocoons (1800 Kg.)		1,80,000.00
	Gross Income		**31,80,000.00**
	Total Expenditure		22,00,000.00
	Net Income		**9,80,000.00**

Note :   The price/rate quoted refer to the year 2002-03 and are subject to market changes. The profit mainly depends on seed cocoon cost and sale price of dfls. The quantity of dfl production depends on the efficiency of staff.

Similarly, oviposition (egg, laying) is completed within 24 hours in *B.mori*, whereas oviposition extends to three days and even more in wild silk moths. Yet another major difference is observed in the oviposition substratum. Though now-a-days eri silkworm eggs are prepared in paper sheets, generally egg laying plastic box or egg laying polythene or nylon bag are used in the case of muga and tasar (tropical and temperate). In fact, in the traditional method of oviposition, straw bands called "Khorikha" is used for egg production in eri and muga silkworm and the wings of the egg laying female moth used to be tied to the 'Khorikha', so that the moths do not fly away. In the case of tasar silk moths, the wings of egg laying female moths are cut at the rudiment, in order to accommodate the moth within the oviposition container. Except these major changes or differences in the egg production practices of wild silk moths, the practices and techniques of egg production in wild silkmoths is similar to the methods described in the case of *B. mori*.

B. SILKWORM REARING (*BOMBYX MORI* L.)

Silkworm rearing requires technical skill to prevent cocoon crop loss, due to diseases and pests. Crop loss affects the economics of sericulturists.

Therefore, in this chapter, the important preventive steps are discussed, which will form a strong foundation for successful silkworm rearing. The silkworm life cycle ranges from 25-30 days depending upon the race, season etc. and in this short life cycle curative measures are not successful and hence only preventive measures are adopted in silkworm rearing to ensure successful cocoon crop.

Various technical aspects are involved in the silkworm rearing and cocoon production. Among them, disinfection of rearing houses, appliances and surrounding areas are considered primary ones. This practice eliminates disease causing pathogens *viz.,* bacteria, viruses, fungi and protozoa. As there is no curative methods for silkworm disease, adoption of disinfection methods brings hygiene during silkworm rearing which is essential for successful cocoon crop.

Rearing appliances required for rearing 200 dfls of mulberry silkworm

The rearing appliances required for rearing 200 dfls of mulberry silkworms are given below. Generally, improved bivoltine hybrids require more space whereas cross breeds like PM x NB4D2 require comparatively less area for rearing. On an average, for rearing 200 dfls about 1000-1100 sq.ft bed area is required.

The following are general materials which are regularly used in silkworm rearing practices.

		Usage	Quantity required
1.	Rocking or power sprayer	Disinfection	1
2.	Face mask and gum shoe	Disinfection	1 set
3.	Electric room heater or charcoal stoves	Incubation/Rearing	2
4.	Blower (Kerosene)	Disinfection of Mountages/trays	2
5.	Thermometer (Wet x Dry type)	Temp./Humidity recording	2

Requirements associated with silkworm rearing (for 200 dfls)

		Quantity
1.	Chawki rearing trays (wooden/plastic, 3'x4'x2.5')	20
2.	Chawki rearing bottom stand	2
3.	Rearing stands for chawki rearing (to accommodate 10 trays in a row, otherwise 20 trays in a stand)	—
4.	Feeding stands	4
5.	Ant wells (cement)	48
6.	Leaf chopping board	1
7.	Leaf chopping knife	1 or 2
8.	Leaf chamber	2
9.	Leaf mat (washable) rexin/plastic	6 mtr

10.	Bed cleaning nets (3'x4') mesh size ¼ inch x ¼ inch.		100
11.	Litter basket/drum (bamboo/plastic)		2
12.	Late age rearing trays (bamboo-round type)	} CB rearing	80
	(For leaf feeding method)	} Bivoltine rearing	100
13.	Late age rearing trays (plastic-rectangular-3'x4'x2.5")	CB rearing	4
		Bivoltine rearing	100
14.	Late age rearing (Iron/wooden) stands to	} CB rearing	4
	accommodate 20 trays in a stand.	} Bivoltine rearing	5
15.	Shoot rearing rack – size 5'x30'x3 tires (for shoot feeding method)		2
16.	Mountages (bamboo chandrike)		80
	(collapsible plastic mountage)	CB rearing	220
		Bivoltine rearing	300
17.	Leaf basin (plastic)		2
18.	Bucket (plastic) .... 15 litre capacity with lid		2
	.... 8 litre capacity with lid		2
19.	Plastic mug		2
20.	Foam pads		40
21.	Leaf basket (bamboo/plastic)		6
22.	Gunny cloth for leaf chamber (leaf preservation)		40 mtrs.

Consumables:

•	Formaline	10 kg
•	Bleaching powder	5 kg
•	Lime (slacked lime powder)	20 kg
•	Bed disinfectant (Vijetha / RKO)	8 kg
•	Old news paper	8 kg
•	Paraffin paper	18 mtr
•	Kerosene	5 litres

Man-power requirements for rearing of 200 dfls.

		Individual leaf feeding method	Shoot rearing method
1.	During chawki rearing (including pre-brushing disinfection, leaf harvest and rearing)	12 man-days	12 man-days
2.	During late age rearing (leaf harvesting IV & V instar and up to spinning)	100 man-days	38 man-days
3.	Cocoon harvesting, marketing	10 man-days	10 man-days

General disinfectants used and method of application in sericulture

At present about four disinfectants are commonly used for disinfection during various activities of sericulture.

Common disinfectants and their usage in sericulture

Name of the disinfectant	Percentage	Quantity to be used	Places of application	Methods of application
Formaline	2%	@ 2 ltr/sq.m. area	Air tight rearing houses, rearing appliances	For 100 layings: 5.5 litres of formaline (36%) + 94.5 litres of water + 50 grams of detergent powder. Spray by rocking/ power sprayer.
Bleaching Powder	2%	@ 2 ltr/sq.m. area	Open type rearing house, rearing appliances	For 100 litres of bleaching powder solution: 2 kgs of bleaching powder + 100 litres of water, 300 g of slaked lime powder. Filter; the solution and spray through rocking/power sprayer.
Slacked Lime	Commercial grade	@ 45 g/sq.m area	Around rearing house, to regulate rearing bed humidity especially during moulting.	Dusting through a muslin cloth bag.
Chlorine Dioxide (Sanitech)	2.5% + 0.5% slaked lime	@ 2 ltr/sq.m. area	Recommended for both closed and open type of rearing houses; rearing appliances, followed by sun-drying of rearing appliances after 24 hours of spray.	For 100 litres of disinfectant: A. 250 g of activator crystals + 2.5 litres of sanitech, stir for some time, allow 5 minutes, solution colour will change to light yellow. B. 500 g of slaked lime + 97.5 litres of water, mix both A and B solutions to get 100 litres of sanitech disinfectant, spray by rocking/ power sprayer.

Transportation of silkworm eggs

The silkworm eggs should be carried in well aerated box or loosely packed cloth and air tight condition of the eggs container should be avoided. The silkworm eggs should be transported only during cool hours of the day.

Silkworm egg incubation

In order to get synchronized hatching of eggs to promote uniform growth of silkworm throughout the rearing period, the silkworm eggs should be properly incubated during the egg stage. Non-provision of suitable environment condition to the eggs affects the uniform growth of embryo, which result in brushing of silkworm two to three days. Such silkworm batches settle for moulting at different times and make the silkworm rearing very cumbersome.

Now a days, various equipments have been developed for incubation of silkworm eggs. However, these equipments (incubators) find a little way to reach sericulturists due to their high cost and continuous electricity dependency.

Therefore, many alternative simple methods *viz.*, earthen flowerpot method, buried earthen pot method; double walled chamber method and low cost incubation chamber are developed. A farmer can choose any one of the methods for successful incubation of silkworm eggs. Among the above methods, low cost incubation chamber provides the optimum temperature (24-25°C) and humidity (80-85%) throughout the incubation period.

i. *Earthen flowerpot method*

About 15" diameter pot is filled with a layer of 2" to 3" thick clean sand and made wet. Rows of holes are made (1/2" dia) above 2-3" of sand level. Silkworm laying sheets are vertically hung from the top in such a way that the sheets are not touching the wet sand. Then it is covered with a clean wet cloth.

ii. *Buried earthen pot method*

In this method, a round wide mouth earthen pot is buried in clean sand upto the neck and the sand is made wet. The silkworm laying sheets are tied to a rod-support and hung inside the pot and the mouth is covered with a wet cloth. The wetness of the cloth/sand should be maintained every day. To avoid ant attack, bleaching powder may be sprayed around the pot. In this method, temperature can be reduced to 2-4°C, and the method is useful upto head-pigmentation stage and thereafter 'Black Boxing' of eggs should be done.

iii. *Double walled chamber method*

The chamber is made with two walls constructed by burnt bricks, sand and cement. The gap between them is 3" which is filled with loose and clean sand. The outer wall size is 6'x4'x3' and the inner wall is 4'x2'x3'. This chamber will accommodate about 5000 dfls.

Egg sheets are tied to a metal rod support with enough space between sheets and are hung in the chamber. The sand has to be soaked periodically with clean water to maintain optimum environment required for incubation. This method may also be used for loose eggs incubation.

iv. *Low cost incubation chamber*

The main chamber consists of two sub-chambers. The outer chamber of 18" height and an opening is made at one end, which is 21" diameter and another opening at bottom which is 15" diameter. The outer chamber wall is provided with 5 mm diameter holes in longitudinal pattern. The inner chamber is made up of soil which is 18" height. This inner chamber is provided with rows of holes of 5 mm size, sand bed of 1" depth is kept at the bottom, moistured by water. The egg sheets are arranged vertically with thin bamboo strips and cover the mouth of both chambers with a fine gunny cloth and keep the eggs till they reach head pigmentation stage.

This method has several advantages because it can be made easily with locally available cheap materials. Fabrication is simple and also ensures optimum condition for the growth of embryo for uniform hatching.

Loose egg incubation

Now-a-days loose eggs are produced and they are becoming more popular among the sericulture farmers. The loose eggs should be incubated in specially designed incubation frames. The incubation frames are made up of 2 cm thick, light, wooden strips with two supporting frames (outer and inner). The outer frame is 36x24 cm. Its bottom is lined with thick black cloth. The inner frame of size 32x20 cm will appropriately fit inside the outer frame. This structure can hold one box of dfls (equivalent to 50 dfls). Depending upon the brushing capacity, the silkworm rearer can have sufficient number of frames and the eggs should be spread uniformly in the incubation frame.

Black Boxing

The aim of 'Black Boxing' is to ensure 90 to 95% of hatching of eggs simultaneously in order to have a uniform batch of rearing throughout rearing period. Almost all the sericulturists now-a-days follow this method realizing its importance. This method is described below.

The embryo reaches its 'Pin head' stage about 48 hours before hatching and attain 'Blue egg' stage about 24 hours prior to hatching. That is why the sericulturists are emphasized with the knowledge of identification of 'Pin head' and 'Blue egg' stage in order to follow the technique successfully. During this period, the egg sheets/loose eggs (still in incubation frames) should be kept into convenient wooden or card board and wrapped up in either black sheet or black cloth. Thereafter, such structure should be left undisturbed for 48 hours or 24 hours. Such arrangement enable the embryos to attain full and uniform growth without hatching. On the date of anticipated hatching of eggs, as shown by the presence of a few hatched larvae, the egg sheets or loose eggs (as the case may be) should be exposed suddenly to bright light between 8 am to 9 am. This technique of photo stimulus will ensure 90 to 95% hatching of eggs in about one or two hours time.

Silkworm Rearing Techniques

1. Brushing of young larvae

Brushing is transferring of newly hatched larvae from egg sheets into rearing trays. Expose the black boxed egg sheets to light on the day of hatching in the morning hours. The newly hatched larvae after one hour of hatching will be ready to feed mulberry leaf. They are to be fed with finely chopped tender mulberry leaves. Sprinkle thin layer of cut mulberry leaves of size 0.5cm x 0.5cm on the hatched larvae. After 10 to 20 minutes, remove the egg sheets and adhered larvae to the trays. With the help of soft bird's feather brush all the larvae to the center of the tray and make a square bed using foam rubber pads around the bed and cover with paraffin papers for maintenance of optimum relative humidity.

2. Young age silkworm rearing

(a) Chawki or Young Silkworm Rearing

Rearing of young age silkworm is called chawki rearing. Here, worms are reared upto second moult and distributed to the rearers for late age rearing. Early instars show best growth under ideal conditions of temperature around 27°C and relative humidity around 80-90% .

Temperature and humidity requirement during young age silkworm rearing

Parameters	I Stage	II Stage	III Stage
Temperature	27°C	27°C	26°C
Humidity	85-90%	85-90%	80%
Leaf size	0.5-2.0 sq.(cm)	2.0-4.0 sq.(cm)	4.0-6.0 sq.(cm)

Feeding of early instar can be regulated by a quantum of three feeds/day. Tender mulberry leaf quantity of 2.5-3.0 kg for I instar; 6-7 kg for II instar; and 20-35 kgs for III instar is required for normal vigorous growth of silkworms. All the early larval instars need to be given only succulent, tender and medium leaves with higher protein and moisture contents.

Bed cleaning in tray is done by various methods using of paddy husk, straw and bed cleaning net. During I instar, bed cleaning is to be done once before pre-moulting and during II instar twice, once after moult and another time just before next moult. During III instar, thrice, after moult, before next moult and once in the middle.

(b) Late age rearing

Late age rearing after third moult does not require high temperature and humidity compared to young silkworm. During late age, the quantity of mulberry leaf required is more. The larval development will be minimum during V instar after 4th moult.

Temperature and humidity required for optimum growth of silkworm

Stage	Temperature	Humidity	Leaf size
4th instar	24-25°C	70-75%	Entire leaf
5th instar	23-24°C	23-24%	Entire leaf/branches

Since the late age worms thrive better under cooler temperature and lower humidity, they are reared without paraffin paper cover. Leaves of mature types containing less moisture, but sufficiently fresh should be fed to the worms.

(c) **Mounting**

This is the last stage of silkworm rearing operation. After the feeding stage the larvae mature and become ready to spin the cocoon. The silkworm after 6-7 days of 4th moult becomes yellowish and translucent, stops feeding and starts searching for a base to crawl and starts spinning. Transferring such matured silkworm on to the mountage or cocoon frame is called "Mounting". While mounting care should be taken to avoid over crowding. Proper spacing should be given to avoid double cocoon spinning and urinated cocoons. The number of worms in each mountage depends upon the size of the mountage and type of silkworm races. The mountages also should be properly disinfected before mounting the silkworm to avoid contamination. There are different types of mountages in use.

- Plastic mountage
- Bamboo made chandrike
- Straw mountage
- Bottle brush
- Revolving mountage

Harvesting of cocoons

Harvesting of cocoons is done on the fifth or sixth day of spinning. Before cocoons are harvested from the mountage, few cocoons may be collected from the mountages and the cocoons should be cut open to find if the larvae have transformed into pupae and if pupa formation is completed, then the cocoons are ready for harvest. Seed cocoons can be harvested on eighth day after spinning depending upon atmospheric temperature. Cocoons meant for silk reeling can be harvested on sixth day from spinning. The spinning worms in the mountages should be kept in well-ventilated room/space with sufficient airflow to ensure good cocoon quality.

Sericulture in India: An Overview

Parameters/ characters	MULBERRY SERICULTURE	NON-MULBERRY SERICULTURE			
		Muga Culture	Tropical Tasar Culture	Temperate Tasar Culture	Eri culture
1. Silkworms	*Bombyx mori*	*Antheraea assama*	*Antheraea mylitta*	*A. pernyi* (China), *A. proylei* *A. frithii, A.roylei* (India) *A. yamamai* (Japan)	*Samia cynthia ricini* (domesticated), *S. cynthia* (wild)
2. Origin of silkworms	China and/or India	Assam (India)	India (Chotanagpur)	China	South east Asia
3. Geographical distribution/Main belt	Karnataka, Andhra Pradesh, West Bengal, Tamil Nadu Jammu & Kashmir in plain to high altitude	Assam, Meghalaya, Manipur and borders of West Bengal adjacent to Assam, upto 6000' ASL	Jharkhand, Bihar, Chattisgarh, M.P., Orissa, W.B., Maharashtra, A.P., U.P. and Karnataka upto 2000' ASL	Manipur, H.P., J. & K. Uttaranchal, Assam, Meghalaya, Arunachal Pradesh, Nagaland etc. of N.E. Region.	Assam, Manipur, Bihar, Jharkhand, West Bengal, Andhra Pradesh and Orissa upto 5000' ASL. Ecoraces: 5, Purelines: 6, Crosses: 7
4. Total No. of known silkworm races/breeds in India	355 (M.V.: 63, B.V.: 292)	4 colour polymorphs	26 ecoraces and 4 colour polymorphs	*A.proylei*: 5 breeds, *A.pernyi*: 2 breeds colour polymorphs: 4	

	Mulberry				
5. Commercial/popular silkworm races	B.V.: CSR2, CSR4, CSR5, NB4D2, KA, NB18, NB7, SH6, KSO1, CSR hybrids, Dun 6 x Dun 22, Dun 12 x Dun 19. M.V: PM, Nistari	———	Daba, Sukinda, Laria, Sarihan, Raily, etc.	PRP12, PRP5, B6	Borduar, Titabar, ES1, ES2
6. Voltinism in silkworms	Uni./Bi./Multivoltine	Mutivoltine	Bivoltine/trivoltine	Univoltine-weak bivoltine	Multivoltine
7. Primary/Main food plant	*Morus alba, M. indica,* and *M. nigra*	*Machilus bombycina, Litsaea polyantha, L. citrata* and *L.salicifolia*	*Terminalia arjuna, T. tomentosa, Shorea robusta, Ziziphus jujuba* and *Lagerstroemia parviflora*	*Q. grifithii, Q. incana* syn. *Q.leucotrichophora, Q. semecarpifolia, Q. serrata* auct. non Thunb.syn. *Q. acutissima* Carr.	*Ricinus communis, Heteropanax fragrans, Evodia fraxinifolia, Manihot utilitissima*
8. Soil type	Clayey-sandy loam, red loam, black cotton, Alluvial	Clayey loam, red loam, sandy loam, black cotton and lateritic red loam	Red, lateritic, black cotton, sandy red/loam.	Clayey loam, black cotton, sandy red	Clayey loam, black cotton
9. Soil pH	6.0-8.0	4.0 - 6.8	5.0 - 7.0	5.6 - 7.0	5.6 - 7.0
10. Annual rainfall (cm.)	60-250	180-210	100-200	100-190	150-200
11. Planting material	Cuttings, seeds, saplings	Seeds, juvenile cuttings, air layerings	Seeds, air layerings, juvenile cuttings	Seeds, air layerings	Seeds, cuttings
12. Seed collecting season	April-July	April-May	April-May	Oct-Nov.	Most part of the year.
13. Planting season	June-July	July-Aug.	June-Aug.	June-Aug.	March-April or Sept.-Oct.
14. Spacing between plants and rows(ft.)	2'x2' (Irrigated) **3'x3' (rainfed)** 4'x2' (rainfed) **(3'+5')x2'(BV culture)**	6'x6', 9'x9', 12'x12'	**4'x4'** 6'x6' 8'x8'	**4'x4'** 4'x6' 6'x6'	**3'x3'** 6'x6' 9'x9',12'x12'
15. Number of the plants/ha. (recommended: **Bold**)	27800(2'x2') 12500(4'x2') **11950 (3'x3')**	2988, 961, 748 respectively	6724, 2988, 1681 respectively	6724, 4483, 2988 respectively	11950 (*R..communis*), 2988 (*H.fragans*), 961, 748 respectively
16. Maximum leaf yield/ ha. (Kg.)(After fertilization) in ruling varieties	40000-75000 (irrigated) 12000 (rainfed)	26489	18000	13000	20000

	300:120:120 (irrigated) 350:140:140 (BV culture) 100:50:50(rainfed)	100:50:50	150:50:50:50	150:50:50	90:40:20
Recommended N.P.K. dose	300:120:120 (irrigated) 350:140:140 (BV culture) 100:50:50(rainfed)	100:50:50	150:50:50:50	150:50:50	90:40:20
17. Available plant accessions/ varieties in India	908 (indigenous: 647), Morus spp.: 15	L. polyantha: 10, M. bombycina: 8, C. glanduliferum: 3	Pentaptera: 130 (plus trees), Seedling orchards: App.40	Q.serrata: 10, Q. grifithii: 3, Q. semecarpifolia: 7	R. communis: 41, H.utilitissimia: 12
18. High yielding/ popular plant variety	V1,S54, S36, S1635,DD, S799,MR2,S1,K2,S146, (tropical), Tr-10, BC259, Gosherami, Chakmajra, China white (temperate)	Som: S6, S3,S5, S4 Soalu: F5, F6, M1, F1	Ds4, N6, Ds2,O1, O2, S1, S2, N3, PBG19	Lanceolate, oblanceolate in Q. serrata, thick non-spiny in Q. semecarpifolia	Aruna, local red (non powdery), red petioled, Hojai.
19. Plant diseases	Leaf spot (Cercospora moricola), Powdery mildew (Phyllactinia corylea), leaf rust(Cerotelium fici), Twig blight (Acidium mori), Twig blight (Fusarium pallidoroseum), Bacterial blight(Pseud omonas mori), Twig blight (Bacterium moricolum), Mosaic (Virus)	Grey blight(Pestalotiopsis dessiminata), Leaf spot (Cercospora sp.) Red rust (Cephaleurus sp.), Leaf curl and wilt(viruses)	Powdery mildew (Phyllactinia terminali), Leaf spot(Pestalotiopsis palmarum), Rust, Leaf curl, damping off of seedlings, stem canker and wood rot	Powdery mildew (Phyllactinia corylea), Rust (Cronartium quercum), Sooty mould (Chaetophoma quercifolia), Leaf blister(Taphrina cacrulescence)	Rust (Melampsora ricini), Blight (Alternaria ricini), Leaf spot (Cercospora ricinella), seedling blight (Phytophthora colocasiae).
20. Plant Pests	Mealy bug (Maconellicoccus hirsutus), Jassid(Empoasca flavescens), Scale insect (Saissetia nigra), Thrips (Pseudodendrothrips mori), Bihar hairy caterpillar (Spilosoma obliqua Walker), Cut worm (Spodoptera litura), Morgonia hairy caterpillar (Eupterote mollifera), tussock caterpillar (Euproctis fraterna), Wasp moth(Amata passalis), Wing less grssshopper (Neorthacris acuticeps nilgirensis), Stem girdler (Sthenias grisator) and white ants, mites, molluscs, termites, wood borers and leaf eating beetles.	Sucking pests (Thrips, aphids, jassids), Stem borer (Zeuzera multistrigata), Leaf minor or semi loopers, leaf galls (Paropsylla besooni and Apantelis sp.), Leaf roller, Hairy caterpillars etc.	May-June beetle (Anomala blancardii), Stem borer (Sphenoptera konbierensis), Gall fly (Trioza fletcheri minor and notatus, Trabala visnou, Phylloplecta hirsuta), Stem canker and white ants (termites)	Sap sucking (Cervaphis quercus,Tuberculatus indicus, Lachnus tropicalis), Defoliating (Lymantria sp. Apoderus mealy wing (Trialeurodes Phallara assimilis, Desichera mendosa, Odonestis pheroba), meristematic(Batocera lanceolata, Cyrlepsistomus sp.) and gall insects.	Semiloopers Achoea janata) Capsule borer (Dichochrocis punctiferalis), Hairy cater pillar (Euproctes lunata), ricini), Jassids(Empoasca flavescens).

21. Capacity of rearing.	100dfls/0.8, 0.6-0.7, 0.8-1.0 and 1.2-1.4 MT foliage for UV, MV, CB and BV respectively.	1000 dfls/3-4 MT foliage	450 dfls/13 MT foliage	1000 dfls/4-5 MT. foliage	100-125 dfls/MT. foliage.
22. Consumption of foliage by each larva (gm)	5.5-8.0 gm	120-150	200-300	65-75	10-15
23. Foliage quality required	Early instars: tender leaves. Adult larvae: mature leaves	I-III instars: tender leaves. IV-V instars: Semimature-mature leaves	Early instars: tender leaves. IV-V instars: Semimature-mature leaves	Early instars: tender and succulent leaves. IV-V instars : Semimature leaves.	
24. Nature of culture	Domesticated (in door)	Semidomesticated (Rearing out door, spinning indoor)	Wild (Outdoor rearing)	Wild/domesticated (outdoor as well as indoor rearing.)	Domesticated (in door)
25. Type of employment	Full time or part time	Full or part time	Part time	Part time	Full time
26. Income(Rs.)/ha/yr.	Above 100000/-	45000/-	12000-15000/-	6000-10000/-	35000/-
27. No. of silkworm crops annually	1, 2, 6 according to ecological conditions	5-6	2-3	1-2	4-6
28. No. of seed crops annually	No demarcation	3	1-2	1	No demarcation
29. No. of commercial crops annually	No demarcation	2	1-2	1	No demarcation
30. Colour of young larvae	Black or dark brown like a hairy caterpillar	Black with brown head and distinct yellow lines in second first, light yellow in second stages	Dull brownish yellow with Black in black head		Greenish yellow changing to pure yellow by 3-4 days & bear black segments.
31. Whether larvae eat a portion of egg shell after hatching.	Not reported	Yes	Yes	Yes	No
32. Colour of larvae	Greenish-pinkish white	Usually dark. green yellow, blue and orange (rarely)	Usually dark, green, yellow, blue, almond (rarely)	Usually dark green. yellow, blue, almond (rarely)	Usually creamy. green, blue or pale ye''ow, unspotted or single, double zebra/semi zebra spotted.
33. Whether cuticle part is eaten after moulting.	Not eaten	eaten	eaten	eaten	Not eaten

34. Length of matured larvae (cm)	6.5-7.2(MV-BV)	9.0	10.0-13.0	6.0-8.5	7.0-8.5
35. Weight of matured larva (gm)	4.0-5.0(MV-BV)	15.0	20-42	13.0-20.0	6.5-8.0
36. Larval duration (days)	23-26 according to race at 22-24°C. Male mature earlier and female later.	Jarua: 50-55 Jethua: 27-33 Aherua : 22-28 Bhadiya:25-31 Kotia :26-32	I: 30-35 II:40-45 III:60-75	I:30-35 II:40-45 III:60-70	Min.:17-20 Max.:45-50 according to season.
37. Completion of moult	More or less uniform	Not uniform	Not uniform	Not uniform	More or less uniform
38. Effect of weather	Highly sensitive to weather fluctuation causing diseases. UV: Weakest MV: Resistant	Reared outdoor before cocooning so directly effected by weather	Directly effected by weather as reared completely outdoor untill cocooning.	Directly effected by weather as reared outdoor completely	Very little direct effect as reared indoors. Resistant to high temp. and R.H.
39. Temp. required for rearing					
Max. (°C)	30°C	34	40-43	28	35
Min. (°C)	20°C	7	9-10	15	15
Optimum (°C)	24-27	24-25	20-30	25-26	24-26
RH required for rearing (%)	65-90 80-90 :Young larvae 65-75: late age worms	75-80	Early stages: 75-80 Late stages:60-70	Early stages:80-90 Late stages:70-80	75-85
40. Chromosome number (2n)	56	30	62	98(A.pernyi) 60 (A. roylei) 79(A. proylei)	28 (P.ricini) 26(P.cynthia)
41. Silkworm diseases	Pebrine (Nosema bombysis), Flacherie and Septicemia (Streptococci, Staphylococci, Bacillus prodigiosus/ Bacillus pyocyaneaus) Sotto (Bacillus sotto syn. B. thuringiensis var. sotto), Court (B. prodigiosus syn. Serratia marcescens)	Pebrine (Nosema sp. transovarian), Grasserie (Viral), Flacherie (Viral and bacterial), Muscardine (fungal) diseases.	Pebrine (Nosema sp.), Polyhedrosis (viral: CPV), Bacteriosis (Sealing of anus, Chaintype excreta and rectal protrusion: Bacillus diplococcus/ B.monococcus) diseases.E. coli).	Polyhedrosis (Viral) and bacteriosis: Chaintype excreta (Bacillus sp.), anal sealing (Proteus mirabilis & Pseudomonas sp.), Rectal protrusion (Staphylococcus and E. coli).	Pebrine (Nosema sp.), flacherie (bacterial) and muscardine (Botrytis bassiana, P. ricini is resistant while P. cynthia susceptible to grasserie.

	Grasserie (Borrelina Virus-NPV), So called flacherie (CPV-Smithia virus), infectious flacherie (Morator virus), Gattine (Virus followed by *Strep tococcus bombycia*), White Muscardine (*Beuveria bassiana* fungi followed by *Serratia marcescens* bacterium), Green muscardine(*Metarrhizium anisspliae*), Green muscardine of Japan (*Spicaria prasina*), Yellow muscardine(*Isaria farinosa*), Aspergillus diseases(*A. flavus, A. oryzae, A. ochraceus, Sterigmatocystis fulva, S. japonica, S. sp.*)				
42. Parasites of silkworms	Uzifly (*Tricholyga bombycis, T. sorbillans* and *Sturmia sericariae*)	Uzifly (*Exorista sorbillans*), Brachonid fly (*Apanteles glomeratus*).	Uzifly (*Blepharipa zebina*), Ichneumon flies (*Xantbopimpla punctatus*).	Uzifly (*B. sugan*), Ichneumon flies (*Xanthopimpla punctator*), Brachonid fly (*Apanteles ruficrus*)	Occasionally attacked by uzifly & flypests (*T. bombycis*).
43. Invertebrate predators/enemies of silkworms	*Anthranus verbasis, Pediculoides ventricosus* (mites), ants, *Hexmermis microamphidis* (Nematodes).	*Vespa orientalis* (Hornet), *Polistes hebraeus*(Yellow wasp), *Solenopsis* sp. (fire ant), *Camponotus* sp. (Carpenter ant), *Oecophylla smargdina* (red ant), *Canthecona furcellata* (Pentatomid), *Reduvius cincticrus*(Reduviid), *Herodula Westwoodi* (Mantis) and spiders.	*Canthecona furcellata, Sycanus collaris, Xanthopimpla* predator, *Polistes hebraeus, Oecophylla smargdina, Herodula westwoodi* and ants.	*Polistes schach, P. olivaceous* (Paper wasps), *Vespa cincta, V. tropica* (Hornet wasps), *Canthecona furcellata* (Sheild bug), *Peristhesancus zetterstedti*(Assasin bugs), *Mantis religiosa*.	Mites and ants.
44. Vertebrate enemies	Lizards, rats, squirrels and birds	Rats, snakes, birds, bats	Rats, snakes, birds, bats	Rats, snakes, birds, bats	Rats, lizards, squirrels and birds.

45. Cocooning ratio	85-95	65-75	65-75	65-75	85-95
46. Effective rate of rearing (ERR)	70-80%	70-80%	50-70%	50-60%	75-80%
47. Cocoon/dfl(No.) (Dfl: cocoon ratio)	200-300 (depending upon race)	70-80	50-65	20-30	200-250
48. Yield/100 dfl(kg/No.)	25 kg (MV hybrid) 25-40 kg (M-B hybrid) 30-50 kg (BV hybrid), CSR hybrids: 70-80 Kg 30-40 kg(UV) 20-28 (MV)	7000-8000	5000-6500	2000-3000	70-90 kg
49. Period of cocooning (days)	3-4, low temp. delays cocooning.	3-4	3-4	3-5	3-4, more in low temp.
50. Cocooning place	A crevice, Chandriki or cocoonage indoor	Jali or cocoonage made from twigs and leaves indoor	On host plants outdoor	On host plants outdoor or their branches indoor.	Chandrika or dry leaves indoor
51. Cocoon quality	Compact, soft leathery and closed. Reelable and flossy.	Compact, solid leathery and closed. Reelable	Compact, solid leathery, hard and closed. Reelable.	Compact, soft leathery and closed. Reelable.	Soft, leathery flossy, open mouthed, unreelable.
52. Cocoon colour	Dull-ivory white, light, deep or golden yellow, green, rose pink etc.	Light-golden, brown, fawn or glossy white.	Grey, yellow or whitish-blackish grey.	Light brown-creamish white.	Creamy white or brick red.
53. Cocoon shape	Oval, spherical, conical, taping, waisted or strangulated.	Oblong	Oblong-oval, pendent.	Oval	Elongated, One end tapering and other end flat and round.
54. Cocoon length (cm)	2.5-3.5	5.2	2.5-5.7	6.5-6.8	4.0-5.0
55. Cocoon breadth (cm)	1.5-2.0	2.4	2.3-3.8	2.5-2.8	2.5-2.8
56. Cocoon weight (gm)	1.5-1.8(pure) 1.85-2.13 (CSR hybrids)	4.1-6.3	6.5-16.0	5.5-6.5	2.8-3.2
57. Presence of oxalate of lime in cocoon	Nil	Present	Present	Nil	Nil
58. Peduncle length in cocoon(cm)	Absent	Rudimentary	3.2-7.2	Absent	Absent

59. Cooking efficiency of cocoons (%)	90-100	100	90-100	90-100	90-100
60. Shell weight of cocoons (g).	0.3-0.4 in Japanese pure races 0.35-0.55 in new Japanese hybrids, 0.466-0.51 in CSR hybrids 0.2-0.3 in UV and MV hybrids 0.15-0.25 in MV 0.29-0.42 in BV and UV	0.28-0.57 according to race/crop	0.87-3.2 according to race	0.5-0.65 according to race	0.38-0.42
61. Shell ratio of cocoons (%)	20-25 in males 18-25 in females (Japanese), 12-14 in Indian MV, 19-23 in Indian BV or UV, 23-24 in CSR hybrids, 14 in MV hybrids, 15-20 in new CB hybrids	5.49-9.5	12-22 according to race	9-11	12-13
62. Renditta (Kg. of cocoons required for yielding 1.0 kg. silk)	UV : 10.0 MV :18.0 BV: 8.0, MV x MV hybrids: 13-15 MV-BV hybrid: 10-11, BV hybrids: 6.5 CSR hybrids: 5-5.6	22-25	8-25 according to race	15-16 (3 in pierced cocoons)	20-25
63. Diapausing stage	Eggs	Pupa (very rarely)	Pupa	Pupa	Eggs in S. cynthia (S.cynthia ricini : Non-hibernating)
64. Pupal weight (g.)	1.65-1.95	3.8-5.7	9.0-12 (10.3) according to race	5.0-6.0 (5.46 avg.)	2.2-2.8 (2.6)
65. Diapause period (days)	UV : 280-300 BV : 90-180	14-55	BV : 180-200 TV : 145-150	260-280	15-18(summer) 35-40(winter)
66. Colour of pupa	Brown, deep brown, black etc.	Brown to copper brown	Dark brown	Brown-dark brown	Brown
67. Size of pupa (cm.)	Male-2.5 Female-2.8 (MV-Smallest)	3.2x1.8	4.5x2.3 according to race	3.5x1.8	2.8x1.5

68. Optimum requirements:

i) For cocoon preservation:					
(a) Temperature (°C)	23-25	26-28	28-30	5-20	22-24
(b) RH%	75-80	70-80	45-50	60-75	70-80
(ii) For moth emergence:					
(a) Temperature (°C)	23-25	26-28	28-30	18-26	20-22
(b) RH (%)	80-85	80-85	75-85	65-80	75-80
iii) For Coupling					
a) Temperature (°C)	24-25	25-28	24-26 (20-22) (intermittantly)	18-25	20-22
b) RH (%)	75-80	75-80	75-85	70-75	70-80
c) duration(hrs.) (max. within parentheses)	3-4 (24)	4-6 (24)	4 (12)	4 (24)	3-6 (24)
iv) For egg laying					
a) Temperature (°C)	24-25	25-28	24-26	24-26	20-22
b) RH (%)	75-80	75-80	75-80	75-80	75-80
c) Device	Egg cards or paste coated craft paper and moth funnel	Khorika (thin bundle of straw)	Sweet boxes or earthen saucers	Bamboo monias	Khorika (bundle of straw)
v) Oviposition duration (days) (Optimum within parenthesis)	1-2(1)	5-6(3)	6-7(3)	3-4(3)	3-4(2)
vi) For Egg hatching					
a) Temperature (°C)	25 for non-hibernating-BV, 10-15-20-25 gradually for hibernating MV	25-27	25-35 (30)	26-28	22-24
b) RH (%)	80-85	80-85	70-80	70-80	75-80
	7 days	7 days	7-8 days	7-8 days	7-8 days

c) incubation period (days) before hatching.	2-3 days at each temp.				
vii) For rearing					
(a) Temperature (°C)	20-30 / 27-28(I-II stage) / 25-26 (III stage) / 22-23 (IV-V stage)	24-25	20-30	25-26	24-26
(b) RH(%)	85 (I-II stage) / 80 (III stage) / 75 (IV stage) / 70 (V stage)	75-80	75-85 (Young) / 60-70(Late age)	80-90 (Young) / 70-80 (Late)	75-85
69. Cocoon:Dfl	3:1	3.5:1	4.5:1	4:1	3:1
70. Dfl:Dfl	1:70	1:15-20	1:10	1:1.5	1:80
71. Disinfection of eggs.	Eggs laid on egg cards are soaked in 2% formalin for 6 min. and washed in cold water thoroughly. Eggs laid on the paste-coated craft paper detached by soaking egg sheets in water for 60 min. and loosened eggs washed in salt solution of 1.06-1.10 specific gravity for rejecting floating eggs before disinfection with 2% formalin as above.	Soaking in 2% formalin for 5 min. and use of high chlorine bleaching powder during loose egg preparation and washing with water.	Washing in soap soln. and soaking in 5-10% formalin soln. for 5 min. followed by thorough washing with water.	Washing in soap soln. and surface sterilization in 2% formalin for 2-3 min. and washing with water.	Washing with soap soln. and soaking in 5% formalin for 5 min. followed by thorough washing with water.
72. Preservation of eggs. a) at temp. °C	5-2.5(Hibernation) / 12-15(Aestivation)	—	8-10 (within 48 hrs. or 120 hrs. after oviposition)		5-10

b) Duration (days)	(within 15 hrs. of deposition.) 90-120 (Hiber.) 5 (Aestiv.) Eggs may be preserved from one spring to following spring or autumn to following spring by following specific schedules.	—	7	20-25	—
73. Fecundity	400-600 according to race	100-250	160-240 according to race	80-120	300-500
74. Hatching(%)	90-95, if treated with HCL (15%) for 5 min. washed and dried.	70-90	50-90	60-70	85-90
75. Period for hatching after oviposition(days)	10-12 (delayed in winter season)	7-8(summer) 14-15(Winter)	9-10(Summer) 10-12 (Winter)	9-10	9-10(Summer) 14-15(Winter)
76. Hatching duration (days)	1-2	3-4	3-4	3-4	2-3
77. Worms suitable for rearing	1 day hatching	2 days hatching	3 days hatching	2 days hatching	2 days hatching.
78. Egg Characters:					
a) Weight of 100 eggs (gm)	UV-0.05 gm. MV-0.025 gm (0.25-0.5 mg/egg)	0.9 gm (9 mg/egg)	0.9-1.2 gm (9-12 mg/egg depending upon race.)	0.8 gm (8 mg/egg)	0.2-0.6 gm (2-6 mg/egg) depending upon race).
b) No. of eggs in one gm.	UV-2000 MV-4000	110-120	100-110	125-130	160-170
c) Layings required for a gm of seed.	UV-4 MV-8	0.5-1	0.6-0.4	1.0-1.50	0.5-0.4
d) Shape of eggs	Round, dorsoventrally flattened	Round, streak less, dorsoventrally flattened	Disk, streak and edge present. Round, dorsoventrally flattened.	Round, streak absent dorsoventrally flattened	Round

	1	2	3	4	5
e) Size of eggs					
Length	1.0-1.3 mm European UV:Big Japanese UV:medium Indian UV:Small Indian MV: Smallest	2.8 mm	3.0 mm	2.8 mm	1.5 mm
Breadth	0.9-1.2 mm	2.5 mm	2.5 mm	2.2 mm	1.0 mm
f) Colour of eggs.	Hibernating type European : Lead grey Japanese: Brown Chinese: Green Non-hibernating type Indian-White	Streakless, brownish	Dark brown turning to white, light yellow or creamy on washing.	Bluish green or brownish	Candid white.
g) Hatching season	UV-spring BV-Spring and late summer MV-All the year round.	Jarua-Winter Chotua-Early spring Jethua-Spring Aherua-Early summer Bhodia-Late summer Kotia-Autumn	I-Late summer(June-July) II-Rainy(Aug.-Sept.) III-Winter(Oct.-Dec.)	Spring summer and autumn (16 hrs. photo-periodic treatment for 21 days resorted for summer & autumn crops.)	All the year round 4-6 broods a year.
h) State of hatching	UV:Uniform with proper hibernation and incubation. MV: Uniform at constant room temperature	Not uniform	Not uniform	Uniform at constant room temperature	
79. Moth characters:					
a) Emergence time	Usually morning	17-20 hrs. peak, continues upto 22.00 hrs.	19-21 hrs.	14-20 hrs., Peak 16-18 hrs.	Morning upto 12 a.m.
b) Body colour	White. In some grey marking in the wing. Black apical spot on the forewing. Markings prominent in males.	Females-yellowish light brown. Males-Copper brown to dark brown.	Females-grey and yellow rarely brown Males-brown, rarely yellow and grey.	Females-brownish, grey wings. Males-greenish, grey wings. Male wings have pink with white border while female wings have pink with faint white/ grey border	Brown & black with pink border, abdomen woolly white, crescent like marks on all wings.

c) Size	Wing expansion in male 2.3 cm and in female 3.1 cm in MV races. UV races have bigger wings.	Females - 3.5cm, with 15 cm wing expansion. Males-3.0 cm with 13.0 cm wing expansion.	Females-4.5 cm, antennae 1.5 cm. Males-4.0 cm ,wing expansion in females is about 18 cm. and 16 cm in males	Females 3.5 and Males 4.0 cm long. Wing expansion in males 14.0 cm and in females 16.0 cm.	Wing expansion in Male-11.7 cm. Female-13.7 cm.
d) Preservation of males	4-5°C upto 7 days	4-5 °C up to 7 days	4-5°C upto 7 days	4-5°C upto 7 days	4-5°C upto 7 days.
e) Copulation	Pairs almost immediately after emergence in the morning	30-40 minutes after emergence, prefers darkness	After 2-3 hrs. of emergence. Peak period 12-20 hr. midnight.	3-4 hrs. after emergence	Active by night, pairs in morning and unpairs towards evening.
f) Duration of coupling (hrs.)	3-12 (3 enough)	3-24 (3 enough)	10-12 (4 enough)	4-24 (4 enough)	3-24 (3-6 enough)
g) Nature of egg deposition.	Deposited densely in a single layer in circular order and neat manner	Prefers complete darkness. Eggs laid in batches of 8-10 irregularly.	Prefers darkness. Eggs laid in batches of 5-10 irregularly.	Prefers complete darkness. Eggs laid irregularly in batches of 5-10.	Eggs laid in clusters irregularly.
80. Discrimination of sex					
a) larva	Genital markings in the form of Herolds bud in male and Ishiwatas point in females are present. Larvae with sex linked characters also found. In CSR-8 females have yellow and males have white cocoons	Like A. mylitta D.	Genital markings appear as milky white spots on ventral surface of VIII and IX abdominal segments in the form of Herolds glands in male and Ishiwatas fore and hind glands in female larvae.	Like A. mylitta D.	Genital markings in male and female larvae found like mulberry silkworms.
b) Pupa	Presence of an X mark in male and a dot mark in female at the posterior ventral end.	Like A. mylitta D.	Presence of a dot in male and oval marking in females at the IX abdominal segment.	Like A. mylitta D.	Like mulberry pupae X and a dot marks are found in male and female pupae. Females bigger and heavier.

c) Moth	Males have hairy antennae and are more active with slender and tapering abdomen while females are passive with large abdomen.	Like A. *mylitta* D. The postmedian line is bordered by a single white lining on either side in males and by a white inner and a pink and narrow bipectinate outer line in the famels. antennae.	Males are smaller with a narrow abdomen and broad antennae while females are bigger with a distended abdomen and narrow bipectinate antennae.	Like A. *mylitta* D.	Males have curved and broad antennae, females have slender. Male active, females lethargic. Male have tapering wing and female broad wing with large abdomen.
d) Egg	A special sex linked strain has black eggs in female and white eggs in male.	—	—	—	—
81. Silk filament					
a) Colour	White or yellow or light green according to presence of carotenoids or its components and flavanones in the sericin enveloping fibroin.	Golden yellow light brown.	Copperish-deep brown-yellow-light.	Yellow-grey.	White or brick red in *P. ricini* and light brown in *P. cynthia*
b) Section	Circular or oval	Circular	Flat like a ribbon	Flat	Flat circular
c) Length(m)	UV-800 to 1500 BV-800-1000 MV-300 to 600 M-B.hybrids-600 to 700 hybrids - 1000 to 1300	204-500	650-1300 according to race.	650-750	Unreelable.
d) Texture	Fine and smooth	Fine and coarse	Coarse and stiff	Fine and coarse	Fine and smooth.
e) Size (Denier) (Weight of 9000 meter yarn in gm)	UV-2 to 2.75 BV-2.0 to 3.0 MV-1.5 to 2.5 CSR hybrids-2.94 to 3.44	4.0-9.6	8.0-12.0	4.0-5.5	2.2-2.5
f) Elongation(%)	8-18	24-36	19-22	28-32	20-22
g) Tensile strength or tenacity (g/d)	UV-3-4 MV-3-3.5	3.5-4.0	1.8	2.8-3.0	3.0-3.5
h) Impurities	Absent	Present	Present	Present .	Absent

i) Lustre	Highly lustrous	Absent	Absent	Absent	Absent
j) Dyeability	Rich affinity with dyes.	Little or no affinity with dyes.	Little or no affinity with dyes	Little or no affinity with dyes.	Little familiarity with dye stuffs.
k) Reelability(%)	67-85	40-50	50-65	50-60	Unreelable.
82. Leaf : Yarn (kg.) ratio	UV-120:1 PM-409.5:1 CB-240.5 : 1 BV 225:1	200-225:1	300-600 : 1 depending upon ecorace.	600:1	120:1
83. Income per ha/year (Rs.) Income/100 dfl/crop.	50000-250000 (UV-BV in Kashmir to Karnataka) 9000-12000/- (Karnataka) 2500-5000/-(Kashmir)	35000-45000 (Assam)	12000-15000	8000-10000 (Manipur) (Bihar & M.P.)	25000-35000 (Assam)

References

Agarwal, H. O. (1995). History of sericulture. In: Proc. Training course in sericulture-II. Himachal Pradesh University. Shimla, pp. 174-192.

Agarwal, V.S. (1986). Economic plants of India. Kailash Prakashan, Kolkata.

Ahsan, M.M. (2000). Sericulture status and technologies for cold and temperate conditions in India. In Souvenir: National Conference on Strategies for Sericulture Research and Development, Central Sericultural Research and Training Institute, Mysore. pp. 41-46.

Anonymous 1949-1979 (Revised 1985 todate). The Wealth of India. Raw materials. I-XI Vols. CSIR, New Delhi.

Anonymous (1976). Wealth of India, CSIR, New Delhi.

Anonymous (1987). Oil seed production: Agricultural situation in India. Directorate of Economics and Statistics. Min. of Ag., New Delhi, 12: 926-928.

Anonymous (1995-96). Annual Report, C.S.G.R.C. (erstwhile S.M.G.S.), Hosur

Anonymous (1995). Package and practices of vegetable crops. U.A.S., Dharwad, p. 276.

Annymous (1997). State of Forest Report, For. Sur. of Ind., Dehradun.

Anonymous (1999a). Background information note on collaboration in evaluation of germplasm between CSB units. In: Seminar on Breeders-Scientist Interaction: Issues related to germplasm maintenance, protection and utilisation, CSGRC, Hosur. Feb.10, 1999.

Anonymous (1999b). Background information note on guidelines for uniform maintenance of germplasm. Ibid.

Anonymous (1999c). Background information note on measures to promote utilisation of germplasm. Ibid.

Anonymous (1999d) Background information note on "Registration" of host plant and silkworm germplasm. Ibid.

Anonymous (2000). CSGRC. Central Sericultural Germplasm Resources Centre, Hosur.

Anonymous (2001). Vision 2010. CSGRC, Perspective Plan. CSGRC, Hosur.

Ansari, A.A. (1997). Medicinal plants of Madhulia forest of Gorakhpur. Jour. Non-timber For. Prod. 4(34): 138-150.

Arora, G.S. and Gupta, J.J. (1979a). Taxonomic studies on some of the Indian Non-mulberry silkmoths (Lepidoptera: Saturiniidae). Mem. Zool. Sur. of Ind. 16(1): 25-28.

Arora, G.S. and Gupta, J.J. (1979b). Non-mulberry silkmoths, Memo. Zool. Sur. of Ind. 17(1): 25-28.

Ashok Kumar, K., Chinya, P.K., Chowdhary, N.B. and Datta, R.K. (2000). Silkworm powder—a potential ingredient in fermenting media for production of lysine. Sem. on Seri. Tech.: An appraisal, p. 52.

Bahadur, K.N. and Gaur, R.C. (1980). A note on the *Terminalia tomentosa* complex. *Ind. Jour. For.* **3**: 367-369.

Balasubramaniam,V. (2000). Sericulture Industry in India—Past, Present and Future. In Souvenir: National Conference on Strategy for Sericulture Research and Development, CSRTI, Mysore, pp. 1-4.

Bari, M.A. (1990). Self-supporting mulberry dike—fish pond eco-system. *Ind. Silk*, **28**(10): 22-23.

Baruah, A.M., Reddy, D.N.R. and Mallesh, B.C. (1999). Potentiality of eri silkworm waste on oyster mushroom culture. Proc. Natl. Semn. On Trop. Seri., 28-30 Dec. p. 57.

Basavaraja, H.K., Nirmal Kumar, S., Suresh Kumar, N. Mal Reddy, N. and Kshama Giridhar, R.K. (1995). New productive bivoltine hybrids. *Ind. Silk*, **34**(2): 5-9.

Basavaraja, H.K., Suresh Kumar, N., Mal Reddy, N., Kalpana, G.V., Nirmal Kumar, S., Jayaswal, K.P., Joge, P.G., Palit, A.K. and Datta, R.K. (2001). New system of maintenance and multiplication of CSR breeds. *Ind. Silk*, **40**(2): 5-100.

Beck, S.D. (1965). Resistance of plants to insects. *Ann. Rev. Ento.* **10**: 207-232.

Benchamin, K.V. (2000a). Muga and eri culture: Subsistence to substantial farming. *Ind. Silk*, **38**(9&10): 51-54.

Benchamin, K.V. (2000b). An analysis of production, productivity and profitability in sericulture. In Souvenir: National Conference on Strategy for Sericulture Research and Development, CSRTI, Mysore, pp. 55-59.

Beroza, M. and Green, N. (1963). Materials tested as insect attractants. Agriculture Hand Book No. 239, U.S.D.A., Washingtom, D.C.

Beroza, M. (1970). Current usage and some recent developments with insect repellants and attractants in the U.S.D.A. In: Chemicals controlling insect behaviour. Ed. M. Beroza. Academic Press. pp. 145-163.

Bhatia, K., Lal, J., Gulati, A.S. and Ayyar, K.S. (1987). Tannins from *Terminalia* bark—A strengthening material for soil. *Ind. For.* **109**(8): 521-523.

Bongale, U.D. (1990). Guava—A new associate of mulberry in the non-traditional areas of Karnataka. *Ind. Silk*, **28**(10): 20-21, 23.

Bongale, U.D. (2000). Productivity alleviation in mulberry cultivation—An appraisal. Lead paper in : Natl. Conf. on Strat. for Seri. Res. & Dev., CSRTI, Mysore, pp. 34-37.

Bose, A.N. (1992). Protein from insect. Ananda Bazar Patrika, Kolkata, 21stJune, p. 16.

Bose, P.C. and Majumder, S.K. (1990). Biochemical composition of pupae waste and its utilization. *Ind. Silk*, **29**: 45-46.

Bose, P.C., Majumder, S.K. and Sengupta, K. (1989). Role of silkworm *Bombyx mori* L. nutrition and their occurrence in haemolymph, silk gland and silk cocoons-a review. *Ind. Jour. Seri.* **28**: 17-31.

Brandis, D. (1906). Indian Trees. Periodical Experts Book Agency, Delhi and International Book Distributors, Dehradun.

Calkin. V.P. (1947). *J. Amer. Chem. Soc.* **169**: 384. In: Lecithin and its importance. *J. Sci. Ind. Res.*, 1995, 14A: 290.

Central Silk Board (1986). Silk In India : Statistical Biennial. CSB, Bangalore, India.

Central Silk Board (1999). Compendium of statistics of silk industry. CSB, Bangalore, India.

Central Silk Board (2001). Latest sericultural statistics in India, CSB, Bangalore, India.

Chinnaswamy, K.P., Doreswamy, C., Devaiah, M.C. and Venkatesh, M. (1993). Pharmaceutical value of mulberry. *Ind. Silk*, **31**(11): 47-48.

Chinnaswamy, K.P and Hariprasad, K.B. (1998). Utilization of mulberry crop residue as fuel wood. *Ind. Silk*, **37**: 17-18.

Chopra, R.N., Badhwar, R.L. and Ghose, S. (1965). Poisonous plants of India. ICAR, New Delhi.

Choudhury, S.N., Ahmed, R., Bhattacharya, P.R., Dutta , S. Das., A.M. and Rajkhowa, A. (1998). Performance of muga silkworm (*Antheraea assama* Ww.) on different Som (*Persea bombycina* King ex Hook. (F) Kost) Plant collections from Assam, India. Proc. III Inter.Conf.on Wild Silkmoth, Bhubaneswar, pp. 66-69.

Chowdhury, S.N. (1984). Mulberry Silk Industry. Dibrugarh, Assam.

Dandin, S.B. and Jolly, M.S. (1986). Mulberry descriptor. Central Sericultural Research & Training Institute, Mysore.

Dandin, S.B. and Ramesh, S.R. (1987). A kulpa vruksha called mulberry. *Ind. Silk*, **26**: 49-53.

Dandin, S.B. (1997). Mulberry improvement programmes—An overview. *Ind. Silk*, **36**(1): 5-9.

Dandin, S.B. (1998). Sericulture in China and India: A comparison. *Ind. Silk*, **37**(4): 5-9.

Dandin, S.B. (2000). Developing national strategy and action plan for seribiodiversity conservation and sustainable use. Base paper in : National Workshop on Management of Sericultural Germplasm for Posterity. CSGRC, Hosur, pp. 1-7.

Dandin, S.B., Mukherjee, P. and Sinha, R.K. (2000). Research strategy of silkworm and mulberry germplasm. *Ind. Silk*, **38**(9 & 10): 65-69.

Das, P.K., Sahu, A.K. and Babulal (2000). Management of muga silkworm and host plant germplasm. Base paper in: National Workshop on Management of Sericultural Germplasm for Posterity, CSGRC, Hosur.

Dash, B. (1993). Ayurvedic Cures for Common Diseases. Hind Pocket Books Pvt.Ltd., Delhi.

Datta, R.K., Basavaraja, H.K., Suresh Kumar, N. and Nirmal Kumar, S. (1997). Evolution of robust hybrids of bivoltine silkworm (*Bombyx mori.* L) for tropics. XVII Congress of the International Sericulture Commission, 22-26 April 1997.

Datta, R.K. (2000a). Sericulture in India and future research strategies. *Ind. Silk*, **38**(9&10): 21-25.

Datta, R.K. (2000b). A break through in tropical bivoltine sericulture. In Souvenir: National Conference on Strategies for Sericulture Research and Development. Central Sericultural Research and Training Institute, Mysore, pp. 21-25.

Datta, R.K. (2000c). Silkworm breeding in India: Present status and new challenges. Lead Paper in: National Conference on Strategies for Sericultural Research & Development. Central Sericultural Research and Training Institute, Mysore, pp. 12-20.

Datta, R.K., Basavaraja, H.K., Mal Reddy, N., Nirmal Kumar, S., Ahsan, M.M., Suresh Kumar, N. and Ramesh Babu, N. (2000a). Evolution of new productive hybrids CSR2 x CSR4 and CSR2 x CSR5. *Sericologia*, **40**(1): 151-167..

Datta, R.K., Basavaraja, H.K., Mal Reddy, N., Nirmal Kumar, S., Ahsan, M.M., Suresh Kumar, N. and Ramesh Babu, M. (2000b). Evolution of new productive bivoltine hybrids, CSR 3 x CSR 6. *Sericologia*, **40**(3): 407-416..

Debraj, Y., Sarmah, M.C., Datta, R.N., Singh, L.S., Das, P.K. and Benchmin, K.V. (2001). Field trial of elite crosses of eri silkworm, *Philosamia ricini* Hutt. *Ind. Silk*, **40**(2): 15-16.

Delthier, V.G., Brown, L.B. and Smith, C.N. (1960). The designation of chemicals in terms of the responses they elicit from insects. *J. Econ. Ent.* **53**: 134-136.

Delthier, V.G. (1982). Mechanism of host plant recognition. *Ent. Exp. App.* **31**: 49-56.

Devaiah, M.C., Govindan, R.and Thippeswamy,C. (1981). Economics of rearing eri silkworm, *Samia cynthia ricini* Boisd. on Castor, *Ricinus communis* L. In: Proc.Seri. Symp.& Sem., T.N.A.U., Coimbatore, pp. 117-120.

Dookia, B.R. (1986). Biological studies of the eri silkworm, *Philosamia ricini* Hutt. on four castor cultivars in semi-arid climate of Rajasthan. Natl.Sem.on Propsects & Problems of Seri. in India, March 27-30.

Dwivedi, N.K. (1988). Importance of white mulberry as a fodder tree. *Ind. Silk*, **27**(6): 18-20.

Dhar, A.K. (1997). Silk carpets of Kashmir. *Ind. Silk*, **36**: 33-35.

Dhingra, R.K., Roy, A.K. and Bardaiyar, V.N. (2002). Coriander as a mixed crop in mulberry nursery. *Ind. Silk*, **41**(4): 11.

Emanuel, C.J.S.K., Kapoor, M.L. and Sharma, V.K. (1990). Achievements in tree improvement and their significance in gene conservation. *Jour. Econo. Bot. & Phytochem.* **1**(2-4): 48-54.

FAO (1976). Manuals on Sericulture. FAO Agriculture Service Bulletin. FAO. Rome I-IV (Reprinted 1987 by CSB Bangalore).

Feinsinger, P. (1983). Co-evolution and pollination. In: Co-evolution. Eds. D.J. Futuyma and M. Slatkin. Sinauer Associates Inc. Sunderland, Massachusettes, pp. 282-327.

Fotedar, R.K., Dhar, A., Bindroo, B.B. and Khan, M.A. (2003). Differentiation of two elite sub-tropical mulberry genotypes. *Ind. Silk*, **42**(2): 11-13.

Fraysmouth, W.A and Pilgrim (1918). Extract from the Indian Tanstuffs and their tannage-Bulletin No. 1.

Futuyma, D.J. (1983). Evolutionary interactions among herbivorous insects and plants. In: Co-evolution. Eds. D.J. Futuyma and M. Slatkin. Sinauer Associates. Sunderland, Massachusettes, pp. 207-231.

Gamble (1922). A Manual of Indian Timbers, London, p. 671.

Gangwar, S.K. and Thangavelu, K. (1991). Earning before harvest of mulberry leaf by intercropping. *Ind. Silk*, **29**(11): 39-41.

Gargi, Shukla, P., Kumar, D., Kumar, R. and Pandey, R.K. (1997). Intercropping for profitable sericulture in Purvanchal. *Ind. Silk*, **35**(11): 31-32.

Geiger, Charles (2000). Silk industry and sericulture, 2000. *Ind. Silk*, **38**(9 & 10): 17-18.

Hamamura, Y. (1970). The substances that control the feeding behaviour and growth of the silkworms, *Bombyx mori* L. In: Control of Insect Behaviour by Natural Products. Eds. D.L. Wood, R.M. Silverstein and M. Nakajima, Academic Press. N.Y., pp. 55-80.

Hanson, F.E. and Delthier, V.G. (1973). Role of gestation and olfaction in food plant discrimination in the tobacco hornworm, *Maduca sexta. J. Insect Physiol.* **19**: 1019-1034.

Hanson, F.E. (1976). Comparative studies on induction of food choice preferences in Lipidopterous larvae. In: The Host Plant in Relation to Insect Behaviour and Reproduction. Ed. T. Jermy, Plenum Press, pp. 71-77.

Hazarika, P.K. and Hazarika, L.K. (1996). Effect of castor varieties on performance of eri silkworms. *Ind. J. Entomol.* **58**(4): 284-290.

Hongulu, J., Shangguan, Z. and Zhujiang, D. (1997). *Acta Sericologia Sinica*, **22**: 4.

Hsiao, T.H. and Fraenkel, G. (1968). The influence of nutrient chemicals on the feeding behaviour of the Colorado potato beetle, *Leptinotarsa decemlineata* (Coleoptera : Chrysomelidae). *Ann. Ent. Soc. Amer.* **61**: 44-54.

Hyde, Nina (1986). The Queen of Textiles. *Ind. Silk*, **25**(6): 2-18.

Ibohal Singh, N., Iotombi Singh, N., James Keisha, T., Rajendra Singh, Y. and Chaoba Singh, K. (1998). Conservation and utilisation of Indian oak fed *Antheraea fauna*. In: Proc.: The 3rd Intern. Conf. Wild Silkmoths, CSB, Bangalore, pp. 330-331.

Ibohal Singh, N. (2000). Management of temperate tasar silkworm germplasm. Base paper in: National Workshop on Management of Sericultural Germplasm for Posterity. CSGRC, Hosur.

Iizuka, T., Koike, S. and Mizutrani, J. (1976). Isolation and identification of caﬁeic acid and 3-hydroxyanthranilic acid form faeces of silkworm larvae reared on artificial diet. *Jour. Seri. Sci.* Japan, **45**: 321-327.

Ito, T. (1960). Effect of sugars on feeding of larvae of the silkworms, *Bombyx mori. J. Insect Physiol.* **5**: 95-107.

Ito, T. (1961). Effect of dietary Ascorbic acid on the silkworm, *Bombyx mori. Nature.* **192**: 951-952.

Iyengar, M.N.S. (1998). Anti-bacterial peptides from silkworms to fight leukemia cells. *Ind. Silk,* **37**: 10.

Iyengar, M.N.S. (2000). Research brief-Antifungal fraction from silkworm haemolymph. *Ind. Silk,* **39**: 12.

Jain, P.P., Suri, R.K., Mathur, K.C. and Goel, C.K. (1990). Scope of utilisation of oil seed from the Garhwal region. *J. Econ. Bot. & Phytochem.* **1**(2-4): 34-37.

Jain S.K. (1979). Medicinal Plants. National Book Trust, New Delhi.

Jermy, T. (1966). Feeding inhibitors and food preferences in chewing phytophagus insects. *Ent. Exp. App.* **9**: 1-12.

Jermy, T. Hanson, F.E. and Delthiers, V.G. (1968). Induction of specific food preferences in Lepidopterous larvae. *Ent. Exp. App.* **11**: 211-230.

Jayaramaiah, M. and Sannappa, B. (1998). Influence of castor genotypes on rearing performance of different eri silkworm breeds. In: Proc. III Int. Conf. on Wild Silkmoth, Bhubaneswar, pp. 29-30.

Jayasawal, K.P., Basavaraja, H.K. and Sinha, R.K. (2000). Management of mulberry (*B. mori* L.) germplasm. In: National Workshop on Management of Sericultural Germplasm for Posterity. CSGRC, Hosur, pp. 1-10.

Jolly, M.S., Sen, S.K. and Ahsan, M.M. (1974). Tasar Culture. Ambika Publishers, Mumbai.

Jolly, M.S., Sen, S.K., Sonwalkar, T.N. and Prasad, G.K. (1979). Non-Mulberry Silks, FAO. Agricultural Services Bulletin. 29. FAO, Rome Vol. 4, 124.

Jolly, M.S. and Dandin, S.B. (1986). Collection, conservation and evaluation of mulberry (*Morus* sp.) germplasm. CSRTI, Mysore.

Juyal, A.C., Singh, B.D., Mohan, R., Ramakant and Maurya, Ghanshyam, S. (2003). A suitable genotype for North-Western India. *Ind. Silk,* **41**(9): 9-10.

Kanjilal, U., Ranjilal, P.C., De, R.N. and Das, A. (1940). Flora of Assam, **4**: 304-325, Govt. of Assam.

Kanan, V.A. (1989). The elegant, lustrous and unique soft silk jari sarees. *Ind. Silk,* **28**(2): 35-36.

Kanan, V.A. (1986). Adulteration in silk sarees. *Ind. Silk,* **25**(6): 19.

Kang Sun Ryu, Heu Sam Lee and Iksoo Kim. (2002). Effects and mechanisms of silkworm powder as a blood glucose-lowering agent. *Int. J. Indust. Entomo.* **4**(2): 93-100.

Kaur, R., Mir, M.R., Khan, M.A. and Nazir, S. (2002). Intercropping of mulberry with saffron. *Ind. Silk,* **41**(2): 5-6.

Kato, M. and Yamada, H. (1966). Silkworm requires 3,4-dihydroxybenzene structure of chlorogenic acid as a growth factor. *Life Sci.* **5**: 712-722.

Katti, M.R., Kaur, R. and Gowri, S. (1996). Pupa skin—A useful waste. *Ind. Silk,* **35**: 5-8.

Kawakami, K. (2001). Illustrated working process of new bivoltine silkworm rearing technology. Japan International Cooperation Agency, CSRTI, Mysore, India.

Kennedy, J.S. (1978). The concept of "olfactory arrestment" and "attraction" *Physiol. Ent.* **3**: 91-98.

Khanna, P., Pacholi, R.K. and Singh, V.P. (1998). Fuelwood and leaf fooder yield tables of *Anogeissus latifolia* and *Terminalia tomentosa. Ind. For.* **124**(3): 198-205.

Knorr, D. (1982). Functional properties of chitin and chitosan. *J. Food Sci.* **47**: 593-595.

Kochhar, S.L. (1981). Economic botany in the tropics. Macmillan India Ltd., Delhi.

Kogan, M. (1977). The role of chemical factors in insect/plant relationships. Proc. XV Intl. Cong. Ent., Washington, pp. 211-224.

Koshy, T.D. (1998). Indian wild silks export: Problems and prospects. *Ind. Silk,* **37**(6 & 7): 75-78.

Koul, Gupta, A., Singh, D. and Gupta, S.P. (1996). Intercropping of vegetables in mulberry. *Jour. of Seri.* **4**(20): 48-51.

Krishnaswamy, S. (!994). A practical guide to mulberry silk cocoon production in tropics. Sriramula Sericulture Consultant, Bangalore.

Kurian, J. (1995). Plants That Heal. Oriental Watchman Publishing House, Pune, India.

Kumar, D. and Das. R.C. (2000). Changing trends in the world silk production scenario. *Ind. Silk,* **39**(1): 21-24.

Kumar, N., Suresh and Yanamoto, T. (1996). Sericulture industry in Japan. *Ind. Silk,* **34**(12): 16-18.

Lakshmi Raju, D. and Nataraju, M.S. (1998). Silk and milk: Participation of farm youths. *Ind. Silk,* **36**(12): 15-17.

Li Long, Zhang Jain, Fan Xun and Qian Youqing (2002). China mulberry sericulture in 1990s. *Ind. Silk,* **40**(10): 26-29.

Ma, W.C. (1972). Dynamics of feeding responses in *Pieris brassica* L. as a function of chemosensory input: a behavioural ultrastructural and electrophysiological study. Meded. Landb. Hogesch. Wageningen, 72-11.

Machida, Y. (1992). Application of chitin and chitosan to drugs. *Pharm. Tech. Jap.* **8**: 15-20.

Machii, H. (1990). On gamma-aminobutyric acid contained in mulberry leaves. *J. Seri. Sci. Jap.* **59**: 381-382.

Maheshwari, J.K. and Singh, H. (1991). Ethno-botanical notes from Banda district, Uttar Pradesh. *J. Econ. Bot. & Phytochem.* **2**(1-4): 16-20.

Majhi, S.K. and Thangavelu, K. (1991). Quality improvement in non-Mulberry silks. *Ind. Silk,* **30**(1): 33-35.

Majumder, S.K. (1992). Industrial entomology—Integrated production and utilization of bye-products through biotechnology of silkworms. Natl. Conf. on Mulb. Seri. Res., CSR & TI, Mysore.

Majumder, S.K., Dutta, R.N. and Kar, R. (1994). The silkworm chrysalis may be a food source for human nutrition. *Sericologia,* **34**: 739-742.

Majumder, S.K., Dutta, R.N., Kar, R. and Pavankumar, T. (1996). Pupa waste, an alternative resource for isolation of lecithin. *Sericologia,* **36**: 555-557.

Majumder, S.K. (1997). Scope for new commercial products from sericulture. *Ind. Silk,* **35**(12): 13-18.

Mandel, P.K. (1991). Sericulture waste utilization in China. *Ind. Silk,* **30**: 32.

Mane, J., Vage, M., Kulkarni, S.V. and Patil, G.M. (2000). An yield analysis of vegetable interrcropping in mulberry. *Advances in Plant Sciences Research,* **12**: 54-56.

Manna, S.S. (1999a). Identification of silk. *Ind. Silk,* **37**(11): 27-28.

Manna, S.S. (1999b). Upkeeping of silk fabrics. *Ind. Silk,* **38**(2): 24.

Masilamani, P. and Vadivelu, K.K. (1997). Effect of growth regulator and nutrient on viability and vigour of preconditioned seeds of Anjan (*Hardwickia binata* Roxb.) *Ind. Jour. For.* **20**(3): 223-226.

Mathur, S.K. and Singh, T. (1989). If it is silk, it needn't be soft always. *Ind. Silk,* **28**(2): 42.

Mathur, S.K. and Vishwakarma, S.R. (1997). Chinese oak tasar farm technology and its relevance to Indian conditions. *Ind. Silk,* **36**(5): 11-15

Mishra, S.N. (2002). Sericulture industry in Indonesia, *Ind. Silk,* **41**(7): 24-25.

Mishra, P.N., Sharma, K.K., Kimothi, R.C. and Thangavelu, K. (2000). Maintenance of genetic resources of *Quercus*. Base paper in: National Workshop on Management of Sericultural Germplasm for Posterity, CSGRC, Hosur., pp. 1-15.

Mori, M. (1982). n-Hexacosanol and n-Octa cosanol: feeding stimulants for larvae of the silkworms *Bombyx mori . J. Insect Physiol.* **28**: 969-973.

Mukherjee, P. (1999). Sericulture in the Republic of Ukraine. *Ind. Silk,* **38**(4): 16-20.

Mukherjee, P. (2000). Management of mulberry germplasm in India. Base Paper in: National Workshop on Management of Sericultural Germplasm for Posterity. pp. 1-10.

Nadiger, G.S. (1998). Non-mulberry silks-A synonym to eco-friendly textile. *Ind. Silk,* **37**(6&7): 71-73.

Nair, K.K.N. and Jayakumar, R. (1998). Ethnobotany of Muthuva tribe in the context of biodiversity rehabilitation at Chinnar Wildlife sancutary, Western Ghats of India. *Jour. Non timber. For. Prod.* **5**(3/4): 159-172.

Najma, Ganapathy, P.M., Sasidharan, N., Bhat, K.M. and Ganaharan, R. (1981). A handbook of Kerala timber, *KFRI Res. Rep.* **9**: 201-207.

Nayar, J.K. and Fraenkel, G. (1962). The chemical basis of host plant selection in the silkworms, *Bombyx mori* L. *J. Insect. Physiol.* **8**: 505-525.

Negi, S.S. and Naithani, H.B. (1995). Oaks of India, Nepal and Bhutan. International Book Distributors, Dehradun.

Nomura, T. (1988). Phenolic compounds of the mulberry tree and related plants. In: Progress in the Chemistry of Organic Natural Products. Eds. W. Herg, H., Grisebach, G.W., Kirbyt and Ch. Tamm. Springer-Verlag, New York, **53**: 88-201.

Omori, M., Yano, T., Okamoto, J., Tsushida, T., Murai, T. and Higuchi, M. (1987). Effect of anaerobically treated tea (Gabaron tea) on blood pressure of spontaneously hypersensitive rats. *Nippon Nogeikagaku Kaishi,* **61**: 1449-1451.

Oommen, P. Joy. (2000). The Road Ahead. In Souvenir : Natl. Conf. on Strat. for Seri. Res. & Develop., 16-18th Nov., 2000, CSRTI, Mysore, pp. 15-16.

Pandey, B.P. (1980). Economic Botany. S. Chand & Co. Ltd., New Delhi.

Pandey, B.P. (1992). A text book of Botany: Angiosperms. S. Chand & Co. Ltd., Delhi.

Parker, R.N. (1925). Hybrid *Terminalia* (Arjun + Asan) and general remarks on tree hybrid. *Ind. For.* **51**(12): 599-603.

Parkinson, C.E. (1936). Indian *Terminalias* of the section *Pentaptera. Ind. For. Res.* **1**(1): 1-27.

Patil, G.M., Kulkarni, K.A., Patel, R.K. and Badiger, K.S. (1998). Performance of eri silkworm, *Samia cynthia ricini* Boisd. on different castor genotypes. Proc. III Int. Conf. on Wild Silkmoths, Bhubaneswar, 11-14 Nov., 1998, pp. 193-195.

Patil, S.G.V. (2002). Evaluation of promising genotype S1635—under irrigated conditions. *Ind. Silk,* **41**(2): 7-9.

Pavan Kumar, T., Shankar, Sharmila and Matsuo, M. (2000a). Guidelines for maintenance and multiplication of CSR breeds. National Silkworm Seed Project—Japan International Cooperation Agency. Deepak Enterprises, Bangalore.

Pavan Kumar, T. (2000 b). Role of National Silkworm Seed Project in silkworm seed preparation. In Souvenir: National Conference on Strategies for Sericulture Research and Development, Central Sericultural Research & Training Institute, Mysore, pp. 33-39.

Pavan Kumar, T., Khatri, R.K. and Bharadwaj, N.J. (2000c). Impact of mega cellules on fecundity and disease free layings recovery (unpublished).

Pearson, R.S. and Brown, H.P. (1932). Commercial timbers of India, **2**: 975, Kolkata.

Phillip, T. (1988). The story of the conquests of silk through the ages. *Ind. Silk,* **27**(1): 27-29.

Phillip, T. (1989). Mulberry fruit—an ideal food. *Ind. Silk,* **28**(8): 13-14.

Prasad, D.A., Reddy, K.J. and Subba Reddy, C.J. (1974). Chemical composition and nutritive value of Anjan. (*Hardwickia binata* Roxb.) leaves. *Ind. J. Animal Sci.* **44**(3): 150-152.

Prasad, D.N. (1989). Multilocational trials of mulberry. In: Genetic Resources of Mulberry and Utilisation, Eds.: Sengupta, K. and Dandin, S.B., CSR & TI, Mysore, pp. 203-210.

Prasad, D.N., Sinha, A.K. and Bania, H.R. (1992). *Terminalia arjuna* can be grown in various soil pH. *Ind. Silk,* **30**(12): 37-38.

Prasad, R. and Parvel, Z. (1987). Observation on the seed viability of Sal (*Shorea robusta*). *Ind. For.* **113**: 89-94.

Priya Ranjan, Srivastav, P.K. and Thangavelu, K. (1994). Strategies for conservation of *Terminalia*—the principal source of valuable MFPs. *Jour. Non-timber For. Prod.* **1**(1/2): 65-73.

Radhakrishna, P.G., Sekharappa, B.M. and Maribashetty, V.G. (2000). Silk and milk : An economic package for rural upliftment. *Ind. Silk,* **39**(5): 11-12,18.

Rahman, M.A., Jaikishan and Kukreti, D.P. (1956). The stem bending properties of Indian timbers. *Ind. For.* **82**(9): 469-478.

Rajaram, Sengupta, A.K., Das, R., Devnath, M. and Samson, M.V. (1993). Collection, identification and evaluation of *Machilus bombycina* King (Laurales: Lauraceae) germplasm, The muga food plant. I. *Sericologia,* **33**(1): 109-124.

Raja Ram, Kumar, S., Roy, G.C., Sinha, A.K. and Sinha, B.R.R.P. (1998). Effect of the different morphotypes of *Quercus semecarpifolia* on the rearing of *Antheraea proylei* J. In : Proc. 3rd. Intl. Conf. on Wild Silkmoths. pp. 97-98.

Rajaram and Samson, M.V. (1999). A natural hybrid of *Litsaea salicifolia* and *L polyantha. Ind. Silk,* **37**(11): 24-26.

Rajiv, S. and Kumar, V. (1996). Sericulture bye-products of China. *Ind. Silk,* **35**: 39.

Ramadevi, O.K. and Khali, D.P. (1996). Blending of silviculture with moriculture. *Ind. Silk,* **34**(12): 13-15.

Ramakanth and Anantharaman, K.V. (1997). Cocoon pelade for better health. *Ind. Silk,* **35**.

Ramakrishna, Naika, Doreswamy, C. and Venkatesh Gowda, T.N. (2000). Medicinal values of non-mulberry host plants. *Ind. Silk,* **39**(3): 15-17.

Ramani, S. and Singh, S.P. (2000). Registration, import/export including quarantine of mulberry (*Bombyx mori* L.) germplasm. Base paper in: National Workshop on Management of Sericultural Germplasm for Posterity, CSGRC, Hosur, pp. 1-23.

Randhawa, M.S. (1983). Flowering Trees. National Book Trust, New Delhi.

Rao, J.V.K., Singh, R.N. and Singh, C.M. (1998). The tribal development and wild silks. *Ind. Silk*, **37**(6 & 7): 79-83.

Rao, Manibhushan, K. (1991). Text Book of Horticulture. Macmillan India Ltd., Delhi.

Ray, Indrajit (1990). Defy diabetes with mulberry. *Ind. Silk*, **28**(10): 34-35.

Revenasiddaiah, H.M. and Yashoda Bai, S. (1999). Extraction of paste chlorophyll from faeces of different larvae of bivoltine silkworm, *Bombax mori* L. Proc. Natl. Sem. on Trop. Seri., 28-30 Dec., p. 56.

Ronald Curriy (1997). Global Silk Industry: Today and Tomorrow. *Ind. Silk*, **35**(12): 5-7.

Ronald Currie (2000). Global Silk Marketing. *Ind. Silk*, **38**(9 & 10): 13-15.

Sagar, M. Hamlyn, P. and Wales, D. (1991). Manufacture of wound dressings from micro-fungal fibres. *Eur. Pat. Appl.* **460**: 774.

Sagreiya, K.P. (1967). Forests and Forestry. National Book Trust, New Delhi.

Sahu, A.K. and Das, P.K. (1999). Background information note on germplasm of *Antheraea assama* Ww and *Philosamia ricini* Hutt. Base paper in: Breeders-Scientist Interaction: Issues Related to Germplasm Maintenance, Protection and Utilization, CSGRC, Hosur, pp. 1-4.

Saikia, S. and Goswami, B.C. (1997). Mejankari: The unexplored Assam silk. *Ind. Silk*, **35**(12): 33-55.

Samson, M.V. (2000). Cocoon production and silkworm protection. Lead paper in: Natl. Conf. on Strat. for Seri. Res. and Devel., CSR & TI, Mysore, pp. 38-48.

Sanjappa, M. (1989). Geographical distribution and exploration of the genus *Morus* L. (Moraceae). In: Genetic Resources of Mulberry and Utilisation. Eds. Sengupta, K. and Dandin, S.B., CSR & TI, Mysore, pp. 4-7.

Saratchandra, B. and Joshi, K.L. (1988). A note on the rearing of eri silkworm on different varieties/cultivars of tapioca (*Manihot utilitissima* Phol.). *Sericologia*, **28**(3): 425-427.

Saratchandra, B. (2000a). Development of mulberry silk industry in the eastern region of India. *Ind. Silk*, **58**(9 & 10): 27-32.

Saratchandra, B. (2000b). Role of extension in popularizing improved sericulture technologies in different regions of India. In: Natl. Conf. on Seri. Res. & Devel., CSR & TI, Mysore, pp. 53-54.

Sarkar, A. and Prasad, S. (1991). Cultivation of mulberry in water logged areas. *Ind. Silk*, **29**(11): 9-12.

Sarkar, A., Jalaja, S. Kumar and Datta, R.K. (1999). Potentiality of Victory-1 under irrigated conditions of South India. *Ind. Silk*, **38**(1): 5-8.

Sarkar, A. (2000). Improvement in mulberry: Current status and future strategies. Lead paper in: National Conference on Strategies for Sericultural Research and Development, 16-18 Nov., 2000, CSR & TI, Mysore, pp. 1-11.

Sarkar, D.C. (1988). Ericulture in India. Central Silk Board, Bangalore.

Sarker, D.D. (1998). The Silkworm : Biology, Genetics and Breeding. Vikas Pub. House Pvt. Ltd., New Delhi.

Sarmah, M.C. (2000). Management of eri silkworm and host plant germplasm. In: National Workshop on Management of Sericulture Germplasm for Posterity. CSGRC, Hosur.

Saxena, K.N. and Schoonhoven, L.M. (1982). Induction of orientational feeding preferences in *Mandusa sexta* larvae for different food sources. *Ent. Exp.* **32**: 173-180.

Schoonhoven, L.M. and Meerman, J. (1978). Metabolic cost of changes in diet and neutralization of allelochemics. *Ent. Exp.* **24**: 689-693.

Schoonhoven, L.M. (1981). Chemical mediators between plants and phytophagous insects. In: Semiochemicals: Their Role in Pest Control. Eds. Nordlund, D.A., Jones, R.L. and Lewis, W.J. John Wiley & Sons, N.Y. pp. 31-50.

Sengupta, A.K., Yadav, G.S., Raja Ram, Das, R., Devnath, M. and Basumatary, B.K. (1991). Genetic resources of muga host plant and its propagation. Base paper in: Workshop on Muga Culture, 19th Sept. 1991, CSB, Guwahati, pp. 1-20.

Sengupta, A.K., Yadav, G.S., Raja Ram, Das, R., Devnath, M. and Basumatary, B.K. (1993). Genetic diversity in muga host plant. *Ind. Silk,* **29**(10): 28-33.

Sengupta, A.K., Das, S.K., Rao, P.R.T., Ghosh, B. and Saratchandra, B. (1997). Silkworm breeds and their hybrids. *Ind. Silk,* **35**(11): 18-20.

Sengupta, K. (1986). Tasar silk industry in India. Central Tasar Research & Training Institute, Ranchi, pp. 1-28.

Sengupta, K. & Dandin, S.B. (1989). Genetic Resources of Mulberry and Utilisation, CSR & TI, Mysore.

Sengupta, K. (1991). Sericulture in Pakistan. *Ind. Silk,* **29**(11): 31-32.

Seth, M.K. (1995a): Food plants of tasar silkworms. In Proc.: Training Course in Sericulture IV. Eds. Agarwal, H.O. and Seth, M.K., H.P.Univ., Shimla, pp. 787-206.

Seth, M.K. (1995b). Food plants of oak tasar silkworms. In Proc.: Training Course in Sericulture IV. Eds. Agarwal, H.O. and Seth, M.K., H.P.Univ., Shimla, pp. 852-859.

Seth, M.K. (1995c). Primary and secondary food plants of eri silkworms. In Proc.: Traing. Course in Seri. IV. Eds. Agarwal, H.O. and Seth, M.K., H.P. Univ., Shimla, pp. 886-893.

Seth, M.K. (1995d). Food plants of muga silkworms. In Proc.: Training Course in Sericulture IV. Eds. Agarwal, H.O. and Seth, M.K., H.P. Univ., Shimla, pp. 894-901.

Shamim Baksh, Quadri, S.N., Khan, M.A., Trag, A.R. and Kaur, R. (1995). Introduction of autumn rearing of silkworm (*Bombyx mori* L.) in temperate region. I. Technical feasibility. In Proc.: Training Course in Sericulture II. Eds. Agarwal, H.O. and Seth, M.K., H.P. Univ., Shimla, pp. 396-406.

Sharma, K.R. and Sinha, R.K. (2000). Korean Sericulture—Adding new dimensions. *Ind. Silk,* **39**(7): 24-26.

Shimzu, Y., Yazwa, M. and Takeda, N. (1992). Aromatic amino acids in the leaves of *Morus alba* and their possible medicinal value. *Sericologia,* **32**: 633-636.

Shukla. P. and Mishra, S.P. (1979). An Introduction to Taxonomy of Angiosperms. Vikas Pub. House Pvt. Ltd., New Delhi.

Shukla, P., Mishra, R.K. and Chowdhury, P.C. (1989). Intercropping: A new vista in mulberry cultivation. *Ind. Silk,* **31**: 39-41.

Siddiqui, A.A. and Das, P.K. (1999). Augmentation of muga food plants-a success story. *Ind. Silk,* **38**(2): 19-21.

Siddiqui, A.A., Chauhan, T.P.S., Singh, B.D., Lochan, R. and Saraswat, S.B. (2002). New bivoltine breeds. *Ind. Silk,* **40**(10): 17-19.

Silkmans Companion (1992). Central Silk Board, Bangalore.

Singh, B.D., Singh, P.K. and Kant, R. (2003). Lentil and Moong—intercropping with multerry. *Ind. Silk*, **41**(11): 10-12.

Singh, B.M.K. and Srivastava, A.K. (1997). Ecoraces of *Antheraea mylitta* D. and exploitation strategy through hybridization. CTR & TI, Ranchi. Current Technology Seminar in Non-Mulberry Sericulture. Base Paper **6**: 1-39.

Singh, B.M.K. and Sinha, A.K. (2000). Role of sericulture on sustenance of forest management. *Ind. Silk*, **39**(6): 14-18.

Singh, C., Vishwanathan, M.K., Kumar, N. and Agarwal, M.C. (1998). Growth survival and mean annual increment of *Quercus leucotrichophora* (Ban Oak) on degraded lands of Doon Valley. *Ind. For.* **124**(9): 732-738.

Singh, J.S., Rawat, Y.S. and Chaturvedi, O.P. (1984). Replacement of Oak forests with Pine in Himalaya affects the nitrogen cycle. *Nature*, **311**: 54-56.

Singh, K., Srivastava, A., Prakash, D., Das, P.K., Siddiqui, A.A. and Raghuvanshi, S.S. (2000). Ranking of foliar constituents in foliar morphotypes of muga food plants, *Machilus bombycina* King. *Sericologia*, **40**(2): 279-283.

Singh, K.C. and Das, P.K. (1991). Understanding the Oak *Quercus acutissima* Carr. in India. *Ind. For.* **117**(12): 1070-1072.

Singh, P. (1990). Ethnobotanical studies on some forest trees of Jabalpur Forest Division (Madhya Pradesh). *J. Econo. Bot. & Phytochem.* **1**(2-4): 43-47.

Singh, R., Kalpana, G.V. and Yamamoto, T. (2002). Modern trends in Japanese sericulture research. *Ind. Silk*, **40**(12): 17-20.

Singh, R.Y. (1982). Fodder trees of India. Oxford & IBH Pub. Co. New Delhi, p. 663.

Singh, U., Wadhwani, A.M. and Johri, B.M. (1983). Dictionary of Economic Plants in India. ICAR, New Delhi.

Singhal, B.K., Dhar, A., Sharma, A., Qadri, S.M.H. and Ahsan, M.M. (2001). Sericultural bye-products for various valuable commercial products as emerging bioscience industry. *Sericologia*, **41**(3): 369-391.

Singhvi, N.R. and Sinha, A.K. (1991). Mulberry: The Kalpavriksha of the twenty first century. *Ind. Silk*, **30**(2): 24-26.

Sinha, A., Sarkar, A., and Das, B.C. (1987). Technology for intercropping in mulberry. *Indian Farming*, **36**(11): 11-12.

Sinha, B.R.R. Pd. (1998). Ecoraces of Indian wild silk moths. *Ind. Silk*, **37**(647): 38.

Sinha, B.R.R.Pd. (1999). Background information note on germplasm maintenance of Indian tasar silkworm. Base paper in: Breeders—scientists interaction: issues related to germplasm maintenance, protection and utilization. SMGS, Hosur.

Sinha, R.K., Kumareshan, P., Mahadevamurthy, T.S. and Thangavelu, K. (2002). Management and utilization of silkworm germplasm. Base paper in: Works. on Seri. Germp. Manage. and Utiliz. CSGRC, Hosur, p. 24.

Sinha, S.C. (1996). Medicinal plants of Manipur. Mass & Sinha, Imphal.

Slansky, F. (1976). Phagism relationship among butterflies. *J.N.Y. Ent. Soc.* **84**: 91-105.

Smith, M.A. and Cornell, H.V. (1979). Hopkins host selection in *Nasonia vitripennis* and its implications for sympatric speciation. *Anim. Behav.* **27**: 365-370.

Somashekhar, T.H. (2000). Quality improvement—A dire necessity. Lead paper in: National Conference on Strategies for Sericulture Research & Development, CSR & TI, Mysore, pp. 61-61.

Sonwalkar, T.N. (1998). Utilization of bye-products in silk industry. *Ind. Silk*, **37**: 24-27.

Sreekumar, S., Sudhakaran Nair, Appasamy, P., Vijayaraghavan, K. and Thiagarajan, V. (1994). Now mulberry on your dining table. *Ind. Silk*, **32**(10): 45.

Srivastava, A., Geetika, Prakash, D., Singh, R., Siddiqui, A.A., Das, P.K., and Raghuvanshi, S.S (1998). Nutritional status of muga food plants—*Litsaea polyantha*. *Sericologia* **38**(4): 693-695.

Srivastav, P.K. (1991). *Terminalia*: The wonder tree of India. *Ind. Silk*, **29**(10): 15-19.

Srivastav, P.K. and Goyal, A.K. (1991). Fruit diversity in 24 genotypes of *Terminalia. J. Cytol. Genet.* **26**: 141-149.

Srivastav, P.K. Siddiqui, A.A. and Goyal, A.K. (1992). Genetic diversity in half-sib seedlings of *Terminalia arjuna.* Bedd. *Sericologia,* **32**(3): 469-475.

Srivastav, P.K. and Thangavelu, K. (1995). Genetic resources and utilisation in *Terminalia arjuna* and *T. tomentosa.* In: Proc. Training course in Seri. UGC Workshop, H.P. Univ. Shimla, Eds: Agarwal, H.O. and Seth, M.K., pp. 318-328.

Srivastav, P.K., Ranjan, P. and Sinha, S.S. (1996). Conservation of *Terminalia* genetic resources: The prinicipal source of non-wood forest products in India. *For. Genet. Resources,* **24**: 54-58. FAO, Rome.

Srivastav, P.K. Beck, S., Siddiqui, A.A., Brahmachari, B.N. and Thangavelu, K. (1997a). Genetic divergence in leaf characters of *Terminalia arjuna* (Roxb.) W & A. and *T. tomentosa* W & A. *J. Non-timber For. Prod.* **4**(3/4): 103-111.

Srivastav, P.K., Srivastav, D.P. and Thangavelu, K. (1997b). *Terminalia* (section: *Pentaptera*) descriptor. *Sericologia,* **27**(2): 289-299.

Srivastav, P.K. and Thangavelu, K. (1997) Genetic diversity and conservation in *Terminalia arjuna* and *T. tomentosa. J. Non-timber For. Prod.* **46**(1/2): 9-16.

Srivastav, P.K. and Singh, N.I. (1997). Genetic resources and conservation of biodiversity in *Quercus serrata.* Int. Symp. Sustainable utilization of biodiversity, SAI Instt. Environ. Dehradun, p. 77.

Srivastav, P.K., Thangavelu, K., Khare, R. and Singh, R.N. (1998a). Treatise on *Terminalias.* Indian Publishers Distributors, Delhi.

Srivastav, P.K., Sinha, U.S.P. and Thangavelu, K. (1998b). Foliar characters and constituents in hybrid genotypes of *Terminalia* (*Pentaptera*). *Ind. Jour. Seri.* **37**(1): 76-78.

Srivastav, P.K. and Thangavelu, K. (2000). Genetic resources and utilization in *Terminalia arjuna* and *T. tomentosa.* In: Sericulture in India. Eds. Agarwal, H.O. and Seth, M.K. Bishen Singh Mahendra Pal Singh, Dehradun, pp. 793-811.

Srivastav, P.K., Singh, N.I. and Thangavelu, K. (2000a). Genetic resources of *Quercus*: Their maintenance, evaluation and utilization. *Proc. Ind. Aca. Seri.* (in press).

Srivastav, P.K., Thangavelu, K. and Singh, N.I. (2000b). Non-mulberry food plants: Present status and prospects for improvement in new millenium. Lead paper in: Natl. Seminar on Seri. Res. & Devel. Strategies for Improvement in New Millenium, CSR & TI, Mysore. Nov. 16-18, 2000, pp. 21-33.

Srivastav, P.K., Ibohal Singh, N. and Sinha, B.R.R.P. (2002a). Effect of different morphotypes of *Quercus serrata* on the rearing of *Antheraea proylei* J. XIX Cong. Intl. Seri. Comm. Bangkok (in press).

Srivastav, P.K., Sinha, U.S.P., Beck, S. and Thangavelu, K. (2002b). Foliar characters and constituents in *Pentaptera* genotypes of *Terminalia. Ind. Jour. Seri.* (in press).

Srivastav, P.K. and Singh, N. Ibohal (2002). Management of genetic resources of oak tasar food plants in India. Base paper in: Works. on Germp. Manage. and Utiliz., Feb. 6-7, 2002, CSGRC, Hosur, pp. 1-18.

Staedler, E. and Hanson, F.E. (1978). Food discrimination and induction of preference for artificial diets in the tobacco hornworm *Mandusa sexta*. *Physiol. Ent.* **3**: 121-133.

Sugun, R. and Kumar, V. (1996). *Ind. Silk*, **34**: 19.

Surendranath, B., Samson, M.V. and Jayaprakash, N. (2000). New mulberry silkworm race authorization system in India. *Ind. Silk*, **39**(7): 13-16.

Thangavelu, K., Chakraborty, A.K., Bhagowati, A.K. and Isa, M. (1988). Handbook of Muga Culture. Central Silk Board, Bangalore.

Thangavelu, K. (1997). Silkworm breeding in India—An overview. *Ind. Silk*, **36**(2): 5-13.

Thangavelu, K., Mukherjee, P., Sinha, R.K., Mahadevamurthy, T.S., Mukheree, S., Sahni, N.K., Kumaresan, P., Rajarajan, P.A., Mohan, B., Sekhar, S. (1997). Catalogue on Silkworm (*Bombyx mori* L.) Vol.-1, Silkworm and Mulberry Germplasm Station, Hosur, India, p. 138.

Thangavelu, K., Srivastav, P.K. and Srivastava, A.K. (2000). Management of tropical tasar silkworm and host plant germplasm. Base paper in: Natl. Works. on Manage. of Seri. Germp. for Post. CSGRC, Hosur.

Thangavelu, K. and Rai, S. (2000). Prospects of tasar culture in new millennium. In Souvenir: Natl. Conf. on Strat. for Seri. Res. and Develop. CSR & TI, Mysore, pp. 65-71.

Thangavelu, K. (2000). Present status and future research strategy in non-mulberry silkworm. Lead paper in: National Conference on Strategies for Sericultural Research and Development, Central Sericultural Research & Training Institute, Mysore, pp. 49-59.

Thangavelu, K., Sinha, R.K., Mahadevamurthy, T.S., Radhakrishnan, S., Kumaresan, P., Mohan, B., Rayaradder, F.R., and Sekar, S. (2000). Catalogue on Silkworm (*Bombyx mori* L.) Germplasm. Vol-2. Central Sericultural Germplasm Resources Centre, Hosur, India, p. 138.

Thangavelu, K., Sinha, R.K., and Mohan, B. (2003). Silkworm germplasm and their potential use. Concept papers—Mulberry silkworm breeders summit, APSSRDI, Hindupur (A.P.), pp. 14-23.

Tikader, A. (1991). Turmeric as an intercrop with mulberry. *Ind. Silk*, **30**(5): 29-30.

Tikader, A., Rao, A.A. and Mukherjee, P. (1999). *Ex situ* conservation of oldest mulberry tree. *Ind. Silk*, **38**(2): 17-18.

Tikader, A., Saraswat, R.P., Amarnath, A. and Thangavelu, K. (2002). Management and utilization of mulberry germplasm. Base paper in Works. on Seri. Germp. Manage. and Utiliz. CSGRC, Hosur, pp. 1-23.

Troup, R.S. (1913). Indian Forest Utilization (2nd Ed.), Kolkata.

Troup, R.S. (1921). The silviculture of Indian trees, Oxford.

Tsushida, T., Murai, T., Omori, M. and Okamoto, J. (1987). Production of a new type tea containing a high level of gamma-aminobutyric acid. *Nippon Nogeikagaku Kaishi*, **61**: 817-822.

Uma, H.P., Gouramma, V., Bajpai, A.K. and Sinha, A.K. (1993). Resham ke hastashilpa: Labhkari aya ke shroth. *Ind. Silk*, **32**: 53-56.

Ullal, S.R. and Narasimhanna, M.N. (1994). Handbook of Practical Sericulture, Central Silk Board, Bangalore.

Vathsala, T.V. (1997). Creativity in cocoon crafts. *Ind. Silk*, **36**: 17-22.

Yadav, G.S., Isa, M. and Thangavelu, K. (1985). Studies on the taxonomy and floral biology of Som, *Machilus bombycina* King Laurales: Lauraceae. *Sericologia*, **25**(1): 63-70.

Yokoyama, T. (1962). Synthesized science of sericulture. Central Silk Board, Bangalore, India, pp. 242-253.

Venkatesh, M., Chinnaswamy, K.P., Raju, M. and Devaiah, M.C. (1993). Mulberry for environmental food security. *Ind. Silk*, **31**(12): 34.

Yamaguchi, A. (2000). Future direction of bivoltine silkworm breeding in India. Souvenir in: National Conference on Strategies for Sericulture Research and Development. 16-18, Nov. 2000, Central Sericultural Research & Training Institute, Mysore, pp. 17-19.

Youzhe, Chen (1996). Unify the understanding for restoring prestige of China Silk. *Ind. Silk*, **34**(12): 23-26.

www.ingramcontent.com/pod-product-compliance
Lightning Source LLC
Chambersburg PA
CBHW080456200326
41458CB00012B/3989